T0142066

Advances in Intelligent Systems and Computing

Volume 348

Series editor

Janusz Kacprzyk, Polish Academy of Sciences, Warsaw, Poland
e-mail: kacprzyk@ibspan.waw.pl

About this Series

The series "Advances in Intelligent Systems and Computing" contains publications on theory, applications, and design methods of Intelligent Systems and Intelligent Computing. Virtually all disciplines such as engineering, natural sciences, computer and information science, ICT, economics, business, e-commerce, environment, healthcare, life science are covered. The list of topics spans all the areas of modern intelligent systems and computing.

The publications within "Advances in Intelligent Systems and Computing" are primarily textbooks and proceedings of important conferences, symposia and congresses. They cover significant recent developments in the field, both of a foundational and applicable character. An important characteristic feature of the series is the short publication time and world-wide distribution. This permits a rapid and broad dissemination of research results.

Advisory Board

Chairman

Nikhil R. Pal, Indian Statistical Institute, Kolkata, India
e-mail: nikhil@isical.ac.in

Members

Rafael Bello, Universidad Central "Marta Abreu" de Las Villas, Santa Clara, Cuba
e-mail: rbellop@uclv.edu.cu

Emilio S. Corchado, University of Salamanca, Salamanca, Spain
e-mail: escorchado@usal.es

Hani Hagras, University of Essex, Colchester, UK
e-mail: hani@essex.ac.uk

László T. Kóczy, Széchenyi István University, Győr, Hungary
e-mail: koczy@sze.hu

Vladik Kreinovich, University of Texas at El Paso, El Paso, USA
e-mail: vladik@utep.edu

Chin-Teng Lin, National Chiao Tung University, Hsinchu, Taiwan
e-mail: ctlin@mail.nctu.edu.tw

Jie Lu, University of Technology, Sydney, Australia
e-mail: Jie.Lu@uts.edu.au

Patricia Melin, Tijuana Institute of Technology, Tijuana, Mexico
e-mail: epmelin@hafsamx.org

Nadia Nedjah, State University of Rio de Janeiro, Rio de Janeiro, Brazil
e-mail: nadia@eng.uerj.br

Ngoc Thanh Nguyen, Wroclaw University of Technology, Wroclaw, Poland
e-mail: Ngoc-Thanh.Nguyen@pwr.edu.pl

Jun Wang, The Chinese University of Hong Kong, Shatin, Hong Kong
e-mail: jwang@mae.cuhk.edu.hk

More information about this series at http://www.springer.com/series/11156

Radek Silhavy · Roman Senkerik
Zuzana Kominkova Oplatkova · Zdenka Prokopova
Petr Silhavy

Editors

Intelligent Systems in Cybernetics and Automation Theory

Proceedings of the 4th Computer Science
On-line Conference 2015 (CSOC2015),
Vol 2: Intelligent Systems in Cybernetics
and Automation Theory

 Springer

Editors
Radek Silhavy
Faculty of Applied Informatics
Tomas Bata University in Zlín
Zlín
Czech Republic

Roman Senkerik
Faculty of Applied Informatics
Tomas Bata University in Zlín
Zlín
Czech Republic

Zuzana Kominkova Oplatkova
Faculty of Applied Informatics
Tomas Bata University in Zlín
Zlín
Czech Republic

Zdenka Prokopova
Faculty of Applied Informatics
Tomas Bata University in Zlín
Zlín
Czech Republic

Petr Silhavy
Faculty of Applied Informatics
Tomas Bata University in Zlín
Zlín
Czech Republic

ISSN 2194-5357   ISSN 2194-5365   (electronic)
Advances in Intelligent Systems and Computing
ISBN 978-3-319-18502-6   ISBN 978-3-319-18503-3   (eBook)
DOI 10.1007/978-3-319-18503-3

Library of Congress Control Number: 2015938157

Springer Cham Heidelberg New York Dordrecht London
© Springer International Publishing Switzerland 2015
This work is subject to copyright. All rights are reserved by the Publisher, whether the whole or part of the material is concerned, specifically the rights of translation, reprinting, reuse of illustrations, recitation, broadcasting, reproduction on microfilms or in any other physical way, and transmission or information storage and retrieval, electronic adaptation, computer software, or by similar or dissimilar methodology now known or hereafter developed.
The use of general descriptive names, registered names, trademarks, service marks, etc. in this publication does not imply, even in the absence of a specific statement, that such names are exempt from the relevant protective laws and regulations and therefore free for general use.
The publisher, the authors and the editors are safe to assume that the advice and information in this book are believed to be true and accurate at the date of publication. Neither the publisher nor the authors or the editors give a warranty, express or implied, with respect to the material contained herein or for any errors or omissions that may have been made.

Printed on acid-free paper

Springer International Publishing AG Switzerland is part of Springer Science+Business Media
(www.springer.com)

Preface

This book constitutes the refereed proceedings of the Intelligent Systems in Cybernetics and Automation Control Theory Section of the 4th Computer Science On-line Conference 2015 (CSOC 2015), held in April 2015.

The volume Intelligent Systems in Cybernetics and Automation Control Theory brings 30 of the accepted papers. Each of them presents new approaches and methods to real-world problems and exploratory research that describes novel approaches in the field of cybernetics and automation control theory.

Particular emphasis is laid on modern trends in selected fields of interest. New algorithms or methods in a variety of fields are also presented.

CSOC 2015 has received (all sections) 230 submissions, 102 of them were accepted for publication. More than 53 % of all accepted submissions were received from Europe, 27 % from Asia, 10 % from America and 10 % from Africa. Researches from 26 countries participated in CSOC 2015 conference.

CSOC 2015 conference intends to provide an international forum for the discussion of the latest high-quality research results in all areas related to Computer Science. The addressed topics are the theoretical aspects and applications of Computer Science, Artificial Intelligences, Cybernetics, Automation Control Theory and Software Engineering.

Computer Science On-line Conference is held on-line and broad usage of modern communication technology improves the traditional concept of scientific conferences. It brings equal opportunity to participate to all researchers around the world.

The editors believe that readers will find the proceedings interesting and useful for their own research work.

March 2015

Radek Silhavy
Roman Senkerik
Zuzana Kominkova Oplatkova
Zdenka Prokopova
Petr Silhavy
(Editors)

Organization

Program Committee

Program Committee Chairs

Zdenka Prokopova, Ph.D., Associate Professor, Tomas Bata University in Zlin, Faculty of Applied Informatics, email: prokopova@fai.utb.cz

Zuzana Kominkova Oplatkova, Ph.D., Associate Professor, Tomas Bata University in Zlin, Faculty of Applied Informatics, email: kominkovaoplatkova@fai.utb.cz

Roman Senkerik, Ph.D., Associate Professor, Tomas Bata University in Zlin, Faculty of Applied Informatics, email: senkerik@fai.utb.cz

Petr Silhavy, Ph.D., Senior Lecturer, Tomas Bata University in Zlin, Faculty of Applied Informatics, email: psilhavy@fai.utb.cz

Radek Silhavy, Ph.D., Senior Lecturer, Tomas Bata University in Zlin, Faculty of Applied Informatics, email: rsilhavy@fai.utb.cz

Roman Prokop, Ph.D., Professor, Tomas Bata University in Zlin, Faculty of Applied Informatics, email: prokop@fai.utb.cz

Program Committee Members

Boguslaw Cyganek, Ph.D., DSc, Department of Computer Science, University of Science and Technology, Krakow, Poland.

Krzysztof Okarma, Ph.D., DSc, Faculty of Electrical Engineering, West Pomeranian University of Technology, Szczecin, Poland.

Monika Bakosova, Ph.D., Associate Professor, Institute of Information Engineering, Automation and Mathematics, Slovak University of Technology, Bratislava, Slovak Republic.

Pavel Vaclavek, Ph.D., Associate Professor, Faculty of Electrical Engineering and Communication, Brno University of Technology, Brno, Czech Republic.

Miroslaw Ochodek, Ph.D., Faculty of Computing, Poznan University of Technology, Poznan, Poland.

Olga Brovkina, Ph.D., Global Change Research Centre Academy of Science of the Czech Republic, Brno, Czech Republic & Mendel University of Brno, Czech Republic.

Elarbi Badidi, Ph.D., College of Information Technology, United Arab Emirates University, Al Ain, United Arab Emirates.

Luis Alberto Morales Rosales, Head of the Master Program in Computer Science, Superior Technological Institute of Misantla, Mexico.

Mariana Lobato Baes,M.Sc., Research-Professor, Superior Technological of Libres, Mexico.

Abdessattar Chaâri, Professor, Laboratory of Sciences and Techniques of Automatic control & Computer engineering, University of Sfax, Tunisian Republic.

Gopal Sakarkar, Shri. Ramdeobaba College of Engineering and Management, Republic of India.

V. V. Krishna Maddinala, Assistant Professor, GD Rungta College of Engineering & Technology, Republic of India.

Anand N Khobragade, Scientist, Maharashtra Remote Sensing Applications Centre, Republic of India.

Abdallah Handoura, Assistant Prof, Computer and Communication Laboratory, Telecom Bretagne - France

Technical Program Committee Members

Ivo Bukovsky
Miroslaw Ochodek
Bronislav Chramcov
Eric Afful Dazie
Michal Bliznak
Donald Davendra
Radim Farana
Zuzana Kominkova
   Oplatkova
Martin Kotyrba
Erik Kral

David Malanik
Michal Pluhacek
Zdenka Prokopova
Martin Sysel
Roman Senkerik
Petr Silhavy
Radek Silhavy
Jiri Vojtesek
Eva Volna
Janez Brest
Ales Zamuda

Roman Prokop
Boguslaw Cyganek
Krzysztof Okarma
Monika Bakosova

Pavel Vaclavek
Olga Brovkina
Elarbi Badidi

Organizing Committee Chair

Radek Silhavy, Ph.D., Tomas Bata University in Zlin, Faculty of Applied Informatics,
email: rsilhavy@fai.utb.cz

Conference Organizer (Production)

OpenPublish.eu s.r.o. Web: http://www.openpublish.eu
Email: csoc@openpublish.eu

Conference Website, Call for Papers

http://www.openpublish.eu

Contents

Extraction of Referential Heading-Entries in Recognized Table of Contents Pages

Phuc Tri Nguyen and Dang Tuan Nguyen

Faculty of Computer Science,
University of Information Technology, VNU-HCM,
Ho Chi Minh City, Vietnam
{phucnt,dangnt}@uit.edu.vn

Abstract. This paper presents our research focusing on extracting referential heading-entries in recognized table of contents (TOC) pages. This task encounters two issues: the complexity of layouts (e.g., a referential heading-entry can have one or many lines, with "decorate" texts, etc.), and some text data errors caused by OCR processing in training data. Our approach uses several layout-based and content-based features to classify textual lines of TOC pages in datasets. Also, we propose synthesis rules to combine related and classified lines into identify referential heading-entries. The experiments are conducted on ICDAR Book Structure Extraction Datasets 2009, 2011, and 2013. The results of experiments show that proposed approach is more efficient than previous methods of referential heading-entries extraction.

Keywords: table of content recognition, document structure extraction, referential heading-entries extraction.

1 Introduction

In this paper, we propose an approach for extracting referential heading-entries from recognized table of contents (TOC) pages of OCRed books. However, there are two challenges of this task:

- TOC pages not only have referential heading-entries but also non-referential heading-entries such as header, footer, and "decorate" texts. Identification and removal of non-referential heading-entries are the first challenge.
- A referential heading-entry may be either single line or multi-lines. Therefore, the text data errors caused by optical character recognition (OCR) processing can greatly affect the performance of extracting referential heading-entries.

There are some existing methods for resolving these problems, but they are focused primarily on identification of some specific TOC formats only. Since the TOC formats of books are diverse, the performance of these methods is unpredictable on general and large datasets. In this research, we do not focus on working on some specific TOC formats, but we propose a labeling and synthesis approach for extracting referential heading-entries in recognized TOC pages. Our proposed approach consists of two steps:

© Springer International Publishing Switzerland 2015
R. Silhavy et al. (eds.), *Intelligent Systems in Cybernetics and Automation Theory*,
Advances in Intelligent Systems and Computing 348, DOI: 10.1007/978-3-319-18503-3_1

– Labeling step: we use a classifier model to label each line of TOC pages. Specifically, Support Vector Machine (SVM) classifier with Radial Basis Function (RBF) kernel is used in our experiments.
– Synthesis step: extract referential heading-entries in recognized TOC pages by applying some synthesis rules based on label information from the labeling step.

The main aim of this research is to focus on enhancing the extraction of referential heading-entries in recognized TOC pages. Then this task will be evaluated by "Titles" indices of ICDAR Book Structure Extraction 2013 [1].

The remaining of this paper is organized as follows: Section 2 reviews related works on referential heading-entries in recognized TOC pages. Section 3 describes our proposed method in details. Experiments and results are given in Section 4. Finally, Section 5 concludes the paper.

2   Related Works

There are various works available in the extraction referential heading-entries. However, these approaches primarily focus on exploiting specific TOC format.

Caihua Liu et al. [2] use rule-based approach for extracting referential heading-entries. Firstly, they use some rule to recognize the beginning and the ending of referential heading-entries based on the characteristics of referential heading-entry such as referential heading-entry usually begins with key words Chapter, Part, etc.; begin or end with a number or Roman numeral. After that, line font size is applied to deal with the error cases referential heading-entries have multi-line.

Lukas Gander et al. [3] use some characteristics about the coordinates of lines, blocks, and strings, the distances between consecutive physical layout elements, the line indents fuzzy logic to handle the variations in the style of books.

Guillaume Lazzara et al. [4] use the information about text location in TOC pages to cluster referential heading-entries. They use some features about line alignments, lines are end with a number or not, textual line have some special character or not, etc.

Bodin Dresevic et al. [5] propose a supervised method to extract referential heading-entries based on pattern occurrences detected in training set.

3   Proposed Approach

Since TOC of books from different domain and genres have format differently, we propose a labeling method for determining the components of heading-referential entries. After that, some synthesis rules are used to group the labeled lines become referential heading-entries. The referential heading-entries extraction module is shown in Fig 1.

Input   Output

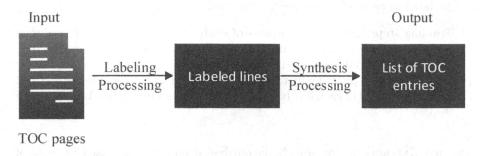

TOC pages

Fig. 1. Referential heading-entries extraction module

The input of referential heading-entries extraction module is a list of lines in TOC pages and output is a list of referential heading-entries.

3.1 Labeling Process

In this section we present the labeling step. Fig 2 is an example of label processing for each line in a TOC page.

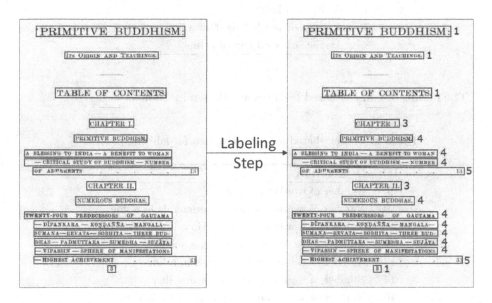

Fig. 2. Example of label processing for each line in a TOC page of the book "Primitive Buddhism: its origin and teachings" (1896) in ICDAR Book Structure Extraction 2013 dataset [1]

The labeling process consists of 2 steps: training step and testing step.

- Training step: Extract the features of each line in TOC pages in training set and label them. Labeled lines are used to build a classifier model. The ground truth of labeled lines is created manually.
- Testing step: Extract the features of each line in TOC pages in testing set. Predict the label of each line based on the classifier model built from training step.

Feature Extraction. We use the layout-based and content-based features of line for feature extraction module. The details of layout-based features are shown in Table 1.

Table 1. Layout-based features designed for labeling process

ID	Description
1	Upper-left corner coordinates of a line
2	Upper-right corner coordinates of a line
3	Lower-left corner coordinates of a line
4	Lower-right corner coordinates of a line
5	Height of a line
6	Width of a line
7	Average of words height of a line
8	Average of word gap of a line

The details of content-based features are shown in Table 2.

Table 2. Content-based features designed for labeling process

ID	Description
9	Number of words in a line
10	Number of words are Roman numerals in a line
11	Number of words are numeral in a line
12	Number of words are upper case in a line
13	Number of words are lower case in a line
14	Number of words are title case in a line
15	Number of words are special character in a line
16	Letter case of the first word in a line
17	Letter case of the last word in a line
18	Line have "Chapter", "Section" terms or not

To take advantage of information between a line and other adjacent lines, we extract a feature vector consisting of 18 described features of current line, 18 described features of the previous line and 18 described features of the next line.

Label. We observe that TOC pages consist of the following components:

- Referential heading-entries: are heading texts refer to the chapter or section of a book. A referential heading-entry may have a single line or multi lines.
- Non-referential heading-entries: are components which do not refer to any part in a book such as header, footer or decorate text.

Based on the information described above, we propose 6 labels for representing each line in TOC pages. Details of the label are shown in Table 3.

Table 3. Label of lines in TOC pages

Label	Description
1	A non-referential heading-entry
2	A referential heading-entry
3	Beginning of a referential heading-entry
4	Body of a referential heading-entry
5	Ending of a referential heading-entry
6	Adding text of a referential heading-entry

3.2 Synthesis Process

The synthesis process is accomplished all of lines in TOC pages based on the information label is assigned from labeling process. Details of a referential heading-entry is combined as the following Table 4.

Table 4. A referential heading-entry is combined from labeled lines

Labels	Description
$\{2\}$	A line is a referential heading-entry
$\{3,5\}$	A referential heading-entry consists of lines which are beginning and ending of a referential heading-entry
$\{3,\{4,4\ldots\},5\}$	A referential heading-entry consists of lines which are beginning, one or multi body and ending of a referential heading-entry
$\{3,5,\{6,6\ldots\}\}$	A referential heading-entry consists of lines which are beginning, ending and one or multi adding text of a referential heading-entry
$\{3,\{4,4\ldots\},5,\{6,6\ldots\}\}$	A referential heading-entry consists of lines which are beginning, one or multi body, ending, and one or multi adding text of a referential heading-entry

The following Fig 3 shows two referential heading-entries are extracted from synthesis process.

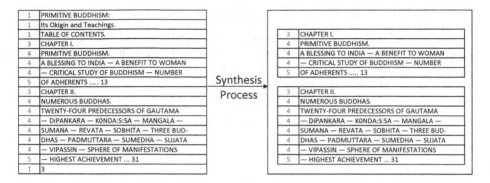

Fig. 3. The example shows the result of the synthesis process

4 Experiment

4.1 Datasets

We use the datasets published by ICDAR Book Structure Extraction Competition 2009 [6] (including 527 books), 2011 [7] (including 513 books) and 2013 [1] (including 967 books). For each dataset, ICDAR Book Structure Extraction Competition [8] also provided a ground truth which contains labeled information about TOC pages and referential heading-entries of books.

4.2 Evaluation Metrics

We use the evaluation method which was stipulated by ICDAR Book Structure Extraction Competition [8]. Since a referential heading-entry consists of 3 components: title, page number, and level, the metrics of evaluation are Precision, Recall and F-measure which are measured on 5 following indices:

- Titles: evaluate the accuracy of extracted referential heading-entries based on ground truth.
- Levels: evaluate the accuracy of extracted TOC structures based on ground truth.
- Links: evaluate the accuracy of extracted links based on ground truth.
- Complete entries: evaluate full matching titles, links and levels.
- Entries disregarding depth: evaluate only matching referential heading-entries and links.

4.3   Experiment Setup

To build the classifier model in labeling process, we use the LIBSVM library [9] with RBF kernel and parameters optimized by the 5-fold cross-validation strategy.

We setup 3 runs:

- Run 1: we use the dataset from 2009 for training set, and the dataset from 2013 for testing set.
- Run 2: we use the dataset from 2011 for training set, and the dataset from 2013 for testing set.
- Run 3: we combine two datasets from 2009 and 2011 for training set, and the dataset 2013 for testing set.

4.4   Results

Table 5 shows the testing results of our Run 1 on the dataset from 2013 [1], with the dataset from 2009 [6] used as training set.

Table 5. Results of Run 1 on ICDAR Book Structure Extraction dataset 2013 [1]

Aspects	Precision	Recall	F-Measure
Titles	55.32%	57.26%	56.00%
Levels	38.69%	40.03%	39.16%
Links	43.96%	45.54%	44.57%
Complete entries	30.87%	31.94%	31.28%
Entries disregarding depth	43.96%	45.54%	44.57%

Table 6 shows the testing results of our Run 2 on the dataset from 2013 [1], with the dataset from 2011 [7] used as training set.

Table 6. Results of Run 2 on ICDAR Book Structure Extraction dataset 2013 [1]

Aspects	Precision	Recall	F-Measure
Titles	53.46%	55.20%	53.97%
Levels	36.69%	37.86%	37.02%
Links	41.99%	43.40%	42.48%
Complete entries	29.00%	29.99%	29.35%
Entries disregarding depth	41.99%	43.40%	42.48%

Table 7 shows the testing results of our Run 3 on the dataset from 2013 [1], with the datasets from 2009 [6] and the dataset from 2011 [7] used as training set.

Table 7. Results of Run 3 on ICDAR Book Structure Extraction dataset 2013 [1]

Aspects	Precision	Recall	F-Measure
Titles	63.57%	64.46%	63.91%
Levels	44.85%	45.38%	45.04%
Links	51.02%	51.73%	51.29%
Complete entries	36.04%	36.44%	36.19%
Entries disregarding depth	51.02%	51.73%	51.29%

We compare our method with the best system of Dresevic et al. [7] at ICDAR Book Structure Extraction 2013 [3]. We consider system of Dresevic at al [7] as "baseline".

Table 8 shows the comparison of our method of referential heading-entries extraction with baseline on the ICDAR Book Structure Extraction Dataset 2013 [1].

Table 8. The comparison of our 3 Runs with the baseline on ICDAR Book Structure Extraction dataset 2013 [1]

Aspects	Run	Precision	Recall	F-Measure
Titles	Run 1	-3.06%	-5.80%	-3.59%
	Run 2	-4.92%	-7.86%	-5.62%
	Run 3	**+5.19%**	**+1.40%**	**+4.32%**
Levels	Run 1	-7.73%	-9.98%	-8.20%
	Run 2	-9.73%	-12.15%	-10.34%
	Run 3	-1.57%	-4.63%	-2.32%
Links	Run 1	-9.52%	-11.95%	-9.97%
	Run 2	-11.49%	-14.09%	-12.06%
	Run 3	-2.46%	-5.76%	-3.25%
Complete entries	Run 1	-11.90%	-13.98%	-12.33%
	Run 2	-13.77%	-15.93%	-14.26%
	Run 3	-6.73%	-9.48%	-7.42%
Entries disregarding depth	Run 1	-9.52%	-11.95%	-9.97%
	Run 2	-11.49%	-14.09%	-12.06%
	Run 3	-2.46%	-5.76%	-3.25%

In the table 8, our Run 3 attains the best accuracy on "Titles" indices.

5 Conclusion

This paper proposed a method based on labeling and synthesis approach for extracting referential heading-entries in recognized TOC pages. Our approach consists two steps. Firstly, label each textual line of TOC pages with the SVM classifier model. Secondly, identify referential heading-entries by applying synthesis rules based on classified and labeled lines of TOC pages. Our method reached the higher score on "Titles" indices in comparison with baseline and other systems at ICDAR Book Structure Extraction 2013 [1].

In the future work, we will focus on improving 2 indices "Links" and "Levels" of referential heading-entries extraction.

References

1. Doucet, A., Kazai, G., Colutto, S., Mühlberger, G.: Overview of the ICDAR 2013 Competition on Book Structure Extraction. In: Proceedings of the Twelfth International Conference on Document Analysis and Recognition (ICDAR 2013), Washington DC, USA, p. 6 (2013)
2. Liu, C., Chen, J., Zhang, X., Liu, J., Huang, Y.: TOC Structure Extraction from OCR-ed Books. In: Geva, S., Kamps, J., Schenkel, R. (eds.) INEX 2011. LNCS, vol. 7424, pp. 98–108. Springer, Heidelberg (2012)
3. Gander, L., Lezuo, C., Unterweger, R.: Rule based document understanding of historical books using a hybrid fuzzy classification system. In: Proceedings of the 2011 Workshop on Historical Document Imaging and Processing, HIP 2011, pp. 91–97. ACM, New York (2011)
4. Lazzara, G., Levillain, R., Géraud, T., Jacquelet, Y., Marquegnies, J., Crépin-Leblond, A.: The scribo module of the olena platform: A free software framework for document image analysis. In: Proceedings of the Eleventh International Conference on Document Analysis and Recognition (ICDAR 2011), pp. 252–258 (2011)
5. Dresevic, B., Uzelac, A., Radakovic, B., Todic, N.: Book layout analysis: Toc structure extraction engine. In: Geva, S., Kamps, J., Trotman, A. (eds.) INEX 2008. LNCS, vol. 5631, pp. 164–171. Springer, Heidelberg (2009)
6. Doucet, A., Kazai, G., Dresevic, B., Uzelac, A., Radakovic, B., Todic, N.: ICDAR 2009 Book Structure Extraction Competition. In: Proceedings of the Tenth International Conference on Document Analysis and Recognition (ICDAR 2009), Barcelona, Spain, pp. 1408–1412 (2009)
7. Doucet, A., Kazai, G., Meunier, J.L.: ICDAR 2011 Book Structure Extraction Competition. In: Proceedings of the Eleventh International Conference on Document Analysis and Recognition (ICDAR 2011), Beijing, China, pp. 1501–1505 (2011)
8. Doucet, A., Kazai, G., Dresevic, B., Uzelac, A., Radakovic, B., Todic, N.: Setting up a competition framework for the evaluation of structure extraction from ocr-ed books. International Journal of Document Analysis and Recognition (IJDAR), Special Issue on Performance Evaluation of Document Analysis and Recognition Algorithms 14, 45–52 (2011)
9. Chang, C.C., Lin, C.J.: LIBSVM: A library for support vector machines. ACM Transactions on Intelligent Systems and Technology 2, 27:1–27:27 (2011), http://www.csie.ntu.edu.tw/~cjlin/libsvm

Correlation Coefficient Analysis of Centrality Metrics for Complex Network Graphs

Natarajan Meghanathan

Jackson State University, Jackson, MS, USA
natarajan.meghanathan@jsums.edu

Abstract. The high-level contribution of this paper is a correlation coefficient analysis of the well-known centrality metrics (degree centrality, eigenvector centrality, betweenness centrality, closeness centrality, farness centrality and eccentricity) for network analysis studies on real-world network graphs representing diverse domains (ranging from 34 nodes to 332 nodes). We observe the two degree-based centrality metrics (degree and eigenvector centrality) to be highly correlated across all the networks studied. There is predominantly a moderate level of correlation between any two of the shortest paths-based centrality metrics (betweenness, closeness, farness and eccentricity) and such a correlation is consistently observed across all the networks. Though we observe a poor correlation between a degree-based centrality metric and a shortest-path based centrality metric for regular random networks, as the variation in the degree distribution of the vertices increases (i.e., as the network gets increasingly scale-free), the correlation coefficient between the two classes of centrality metrics increases.

Keywords: Centrality, Complex Networks, Correlation Coefficient, Degree, Shortest Paths.

1   Introduction and Related Work

Network Science is an actively researched area focusing on the analysis and visualization of large complex real-world network graphs. Several metrics (a.k.a. measures) are used in the analysis of complex network graphs, and centrality is one of the key metrics. The centrality of a node is a measure of the topological importance of the node with respect to the other nodes in the network. Centrality-based ranking of nodes is purely a link-statistic based approach and does not depend on any offline information (such as reputation of the particular entity representing the node, etc) [1]. The widely-used centrality metrics are: Degree, Betweenness, Eigenvector and Closeness, all of which are studied in this paper, along with Farness centrality and Eccentricity.

We model a large real-world network as a graph of vertices and edges (vertices represent the nodes - entities like players, countries, airports, books, etc constituting the real-world network, and edges represent the interactions between these entities in the form of competitions played against, airline connections, books related to a particular domain, etc). The currently available application software (e.g., [2-3]) can be used to visualize the similarities/dissimilarities between any two centrality metrics by

© Springer International Publishing Switzerland 2015
R. Silhavy et al. (eds.), *Intelligent Systems in Cybernetics and Automation Theory*,
Advances in Intelligent Systems and Computing 348, DOI: 10.1007/978-3-319-18503-3_2

varying the color and size of a vertex proportional to the centrality values (with respect to the two metrics being compared) for that vertex. However, it is difficult to deduce a numerical value for the correlation between the centrality metrics from the visualization alone. To the best of our knowledge, the only literature study available so far on correlation analysis of centrality metrics on real-world network graphs is the recently published paper, [4]: here, the authors determine the correlation coefficients between a suite of centrality metrics on several real-world network graphs and simply list the values observed. They do not analyze the topological structure of these real-world network graphs and correlate them with the centrality metric values observed. Moreover, the paper [4] does not attempt at classifying the centrality metrics into one or more classes and arrive at conclusions for correlations across one or more classes of centrality metrics.

Our contribution in this paper is two fold: (i) We classify the major centrality metrics into two categories: degree-based (degree and eigenvector centralities) and shortest path-based (betweenness, closeness, farness and eccentricity centralities). We analyze the correlation between any two of the above six centrality metrics as well as analyze the correlation between the two classes of centrality metrics on real-world network graphs. In this pursuit, we choose six different real-world network graphs, representing diverse domains (like Social club network, Birds' social network, Word adjacency network, Airports network, Games network, Related books network) with the number of nodes ranging from 34 to 332. (ii) We analyze the degree distribution of the vertices in the real-world network graphs as well as measure the variation in the node degrees with respect to the average node degree and the spectral radius. We study the impact of the topological structure of the real-world network graphs on the correlations obtained between any two of the six centrality metrics and the two classes of centrality metrics. We restrict ourselves to undirected network graphs in this paper.

The rest of the paper is organized as follows: Section 2 reviews the various centrality metrics that are available for network analysis studies and categorizes them to the two classes (degree-based or shortest path-based). The section also shows how the centrality metrics are related to each other. Section 3 presents the real-world network graphs analyzed in this research and analyzes the degree distribution of the vertices in these graphs. Section 4 presents in detail the correlation coefficient analysis of the real-world network graphs with respect to the six centrality metrics chosen for comparison. Section 5 concludes the paper. We use the terms 'node' and 'vertex', 'link' and 'edge' interchangeably. They mean the same.

2   Centrality Metrics

In this section, we review the various centrality metrics that have been typically considered for network analysis studies. Centrality metrics have been traditionally used to rank the vertices on the basis of their contribution to the topological structure of the network. Vertices that are located at the core of the network are typically ranked higher than vertices located at the periphery of the network. In this pursuit, we would prefer a centrality metric to take real values (continuous) than integer values (discrete) so that there is not much ambiguity in uniquely ranking the vertices. Several of the centrality metrics use the adjacency matrix of the network graph as the basis; the adjacency matrix for a graph is a binary matrix where there is a 1 in the ith row and j^{th} column if there is an edge from vertex i to vertex j. For undirected graphs, the adjacency matrix is symmetric.

Degree Centrality: The degree centrality (DegC) of a vertex is simply the number of neighbors of that vertex. It is obtained by counting the number of 1s in the row corresponding to the vertex in the adjacency matrix of the graph. The larger the number of neighbors, the higher is the rank of a vertex with respect to degree centrality. The degree centrality measure is likely to lead to tie among one or more vertices and may not be an accurate measure to unambiguously rank the vertices.

Eigenvector Centrality: The Eigenvector centrality (EVC) of a vertex is a measure of the degree of the vertex and the degree of its adjacent vertices. The EVC of the vertices in a network graph is the principal eigenvector of the adjacency matrix of the graph. The principal eigenvector has an entry for each of the n-vertices of the graph. The larger the value of this entry for a vertex, the higher is its ranking with respect to EVC. The principal Eigenvector of the adjacency matrix can be determined using the power-iteration method [5]: The eigenvector X_{i+1} of a network graph at the end of the $(i+1)^{th}$ iteration is given by: $X_{i+1} = \dfrac{AX_i}{\|AX_i\|}$, where $\|AX_i\|$ is the normalized value of the product of the adjacency matrix A of a given graph and the tentative eigenvector X_i at the end of iteration i. The initial value of X_i is the transpose of [1, 1, ..., 1], a column vector of all 1s, where the number of 1s correspond to the number of vertices in the graph. We continue the iterations until the normalized value $\|AX_{i+1}\|$ converges to that of the normalized value $\|AX_i\|$. The value of the column vector X_i at this juncture is the principal eigenvector of the graph; the entries corresponding to the individual rows in X_i represent the EVC of the vertices in the graph. The converged normalized value of the principal eigenvector is referred to as the Spectral radius.

Betweenness Centrality: Betweenness Centrality (BWC) is a measure of how significant a node is in facilitating communication between any two nodes in the network. BWC for a node is the ratio of the number of shortest paths a node is part of for any source-destination node pair in the network, summed over all possible source-destination pairs that do not involve the particular node. If the number of shortest paths between two nodes j and k that go through node i as the intermediate node is denoted as $sp_{jk}(i)$ and the total number of shortest paths between the two nodes j and k is denoted as sp_{jk}, then $BWC(i) = \displaystyle\sum_{j \neq k \neq i} \dfrac{sp_{jk}(i)}{sp_{jk}}$. The number of shortest paths from a node j to all other nodes k in an undirected graph can be determined by running the Breadth First Search (BFS) algorithm [6] on the graph, starting from vertex j (which is also considered to be at level 0 for this BFS run). All the vertices that are directly reachable from vertex j are said to be at level 1; the two hop neighbors of j are at level 2 and so on. The number of shortest paths from the root j (at level 0) to itself is set to be 1. For any other vertex k (at level l, where $l > 0$) on this shortest path BFS tree rooted at j: the number of shortest paths from j to k (sp_{jk}) is the sum of the number of shortest paths from j to each of the neighbors of k (in the original graph) that are at level l-1 on the BFS tree. The number of shortest paths between two nodes j and k that go through node i (i.e., $sp_{jk}(i)$) is simply the maximum of the number of shortest paths from vertex j to i and the number of shortest paths from vertex k to i. This can be determined from the BFS trees rooted at vertices j and k (as described earlier).

Closeness Centrality: The Closeness centrality (ClC) of a vertex is the inverse of the sum of the shortest path distances from the vertex to every other vertex in the graph. The ClC of a vertex is determined by running the BFS algorithm starting from that vertex; summing up the levels of the other vertices in the shortest path tree and finally inverting the sum. The sum of the shortest path distances (and hence its inverse) is likely to be the same for two or more vertices and hence the Closeness centrality may not be accurate measure to unambiguously rank the vertices.

Farness Centrality: The Farness centrality (FarC) of a vertex is a measure of the sum of the shortest path distances of the vertex to every other vertex as well as the variation in the shortest path distances. The Farness centrality of the vertices can be computed by running the power-iteration algorithm on the shortest path distance matrix (the entry in the ith row and j^{th} column is the number of hops on the shortest path from vertex i to vertex j). The smaller the values for a vertex in the principal eigenvector of the shortest path distance matrix, the more closer is the vertex to the rest of the vertices and higher the ranking. Farness centrality of a vertex is the inverse of the value for the vertex in the principal eigenvector of the shortest path distance matrix; the larger is its Farness centrality value, the higher the ranking. Though Farness centrality breaks the tie among vertices that have identical value for the sum of the shortest path distances, as seen in Section 4, we observe this metric to be very highly correlated with the Closeness centrality (correlation coefficient of 0.99) for all the real-world network graphs; hence, we opine that Farness centrality does not contribute any additional information of its own with respect to the location of the vertices and the topological structure of the network, other than that known through Closeness centrality.

Eccentricity: Eccentricity (Ecc) of a vertex is the largest value among the shortest path distances to every other vertex. The smaller the Eccentricity value of a vertex, the higher is its ranking with respect to centrality. Eccentricity is another centrality metric that could take only discrete integer values; the Eccentricity values observed for the vertices in each of the real-world network graphs are not far apart from each other, leading to lots of ambiguity in the ranking of the vertices. While Closeness centrality of a vertex is based on the sum of the shortest path distances; Eccentricity of a vertex is based on the maximum of the shortest path distances.

3 Network Graphs Analyzed and Their Degree Distribution

The network graphs analyzed are briefly described as follows (in the increasing order of the number of vertices): (i) *Zachary's Karate Club* [7]: Social network of friendships (78 edges) between 34 members of a karate club at a US university in the 1970s; (ii) *Dolphins' Social Network* [8]: An undirected social network of frequent associations (159 edges) between 62 dolphins in a community living off Doubtful Sound, New Zealand; (iii) *US Politics Books Network* [9]: Nodes represent a total of 105 books about US politics sold by the online bookseller Amazon.com. A total of 441 edges represent frequent co-purchasing of books by the same buyers, as indicated by the "customers who bought this book also bought these other books" feature on Amazon; (iv) *Word Adjacencies Network* [10]: This is a word co-appearance network representing adjacencies of common adjective and noun in the novel "David Copperfield" by Charles Dickens. A total of 112 nodes represent the most commonly occurring adjectives and nouns in the book. A total of 425 edges connect any pair of words that occur in adjacent position in the text of the book; (v) *American College Football Network* [11]: Network represents the teams that played in the Fall 2000 season of the American Football games and their previous rivalry - nodes (115 nodes) are college teams and there is an edge (613 edges) between two nodes if and only if the corresponding teams have competed against each other earlier; (vi) *US Airports 1997 Network*: A network of 332 airports in the United States (as of year 1997) wherein the vertices are the airports and two airports are connected with an edge (a total of 2126 edges) if there is at least one direct flight between them in both the directions. Data for networks (i) through (v) can be obtained in the form of .gml files from http://www-personal.umich.edu/~mejn/netdata/. Data for network (vi) can be obtained from: http://vlado.fmf.uni-lj.si/pub/networks/pajek/data/gphs.htm.

Figure 1 presents the degree distribution of the vertices in the six network graphs in the form of both the Probability Mass Function (the fraction of the vertices with a particular degree) and the Cumulative Distribution Function (the sum of the fractions of the vertices with degrees less than or equal to a certain value). We also compute the average node degree and the spectral radius degree ratio (ratio of the spectral radius and the average node degree); the spectral radius (bounded below by the average node degree and bounded above by the maximum node degree) is the largest Eigenvalue of the adjacency matrix of the network graph, obtained as a result of computing the Eigenvector Centrality of the network graphs. The spectral radius degree ratio is a measure of the variation in the node degree with respect to the average node degree; the closer the ratio is to 1, the smaller the variations in the node degree and the degrees of the vertices are closer to the average node degree (characteristic of random graph networks). The farther is the ratio from 1, the larger the variations in the node degree (characteristic of scale-free networks). Figure 1 presents the degree distribution of the network graphs in the increasing order of their spectral radius ratio for node degree (1.01 to 3.23). The American College Football network exhibits minimal variations in the degree of its vertices (each team has more or less played against an equal number of other teams). The US Airports network exhibits maximum variations in the degree of its vertices (there are some hub airports from which there are flights to several other airports; whereas there are several airports with only fewer connections to other airports). In between these two extremes of networks, we have the other four network graphs, all of which have a spectral radius ratio around 1.4-1.7, indicating a moderate variation in the node degree (compared to the spectral radius ratios observed for the American College Football network and the US Airports network).

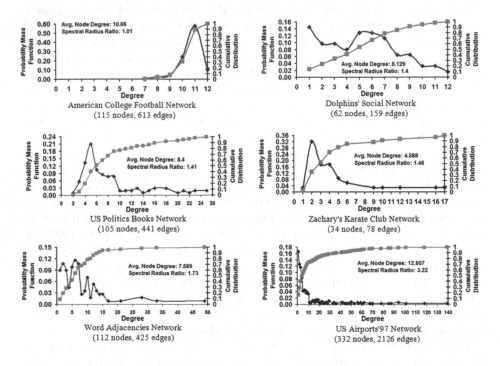

Fig. 1. Node Degree: Probability Mass Function and Cumulative Distribution

4   Correlation Coefficient Analysis of Centrality Metrics

In this section, we discuss the correlation coefficients obtained between any two of the six centrality metrics (Degree, Eigenvector, Betweenness, Closeness, Farness and Eccentricity) on the six real-world network graphs analyzed in Section 3. We developed our own Java code to implement the algorithms to determine each of the above six centralities on complex network graphs. We use the well-known and commonly used Pearson Correlation Coefficient formulation (equation 1) to evaluate the correlation coefficients between the centrality metrics. If $\overline{X}$ and $\overline{Y}$ are the average values of the two centrality measures (say X and Y) observed for the vertices (IDs 1 to n, where n is the number of vertices) in the network, the formula used to compute the Correlation Coefficient between two centrality measures X and Y is:

$$CorrCoeff(X,Y) = \frac{\sum_{ID=1}^{n}(X[ID]-\overline{X})*(Y[ID]-\overline{Y})}{\sqrt{\sum_{ID=1}^{N}(X[ID]-\overline{X})^2}\sqrt{\sum_{ID=1}^{N}(Y[ID]-\overline{Y})^2}} \tag{1}$$

We define three levels of correlation: high, moderate and low. The ranges of the correlation coefficients for these three levels are: high (0.75 and above); moderate (0.40 to 0.74) and low (less than 0.40). Accordingly, in Tables 2 and 3, we highlight the cells in yellow when two centrality metrics exhibit high-levels of correlation and in light blue when two centrality metrics exhibit low-levels of correlation. Cells corresponding to centrality metrics exhibiting moderate levels of correlation are not highlighted. In Table 1, we map the network index used in Tables 2 and 3 with that of the real-world network graphs studied; this mapping is in the increasing order of the number of vertices in the real-world network graphs (in the same order they are listed in Section 3). In Table 2, the order in which the rows are listed is also in the increasing order of the number of vertices in the network graphs. In Table 3, the order in which the rows are listed is in the increasing order of the spectral radius ratio for node degree, a measure of the variation in the degree distribution of the vertices with respect to the average node degree.

Table 1. Network Index for Real-World Network Graphs Listed in Tables 2 and 3

Network Index	Real-World Network Graph	# Nodes	# Edges	Special Radius Degree Ratio
(i)	Zachary's Karate Club Network	34	78	1.46
(ii)	Dolphins' Social Network	62	159	1.40
(iii)	US Politics Books Network	105	441	1.41
(iv)	Word Adjacencies Network	112	425	1.73
(v)	American College Football Network	115	613	1.01
(vi)	US Airports 1997 Network	332	2126	3.22

Table 2. Correlation Coefficients: Centrality Metrics for Real-World Network Graphs (Networks Listed in the Increasing Order of Number of Nodes)

Network Index	Deg EVC	Deg BWC	Deg ClC	Deg FarC	Deg Ecc	EVC BWC	EVC ClC	EVC FarC	EVC Ecc	BWC ClC	BWC FarC	BWC Ecc	ClC FarC	ClC Ecc	FarC Ecc
(i)	0.90	0.92	0.77	0.77	0.46	0.79	0.91	0.90	0.60	0.72	0.72	0.44	0.99	0.78	0.79
(ii)	0.77	0.60	0.71	0.73	0.36	0.33	0.71	0.68	0.05	0.67	0.71	0.62	0.99	0.62	0.67
(iii)	0.93	0.71	0.58	0.59	0.14	0.58	0.53	0.53	0.01	0.78	0.79	0.51	0.99	0.76	0.76
(iv)	0.95	0.92	0.84	0.84	0.54	0.82	0.93	0.92	0.60	0.66	0.66	0.43	0.99	0.71	0.71
(v)	0.87	0.28	0.29	0.29	0.14	0.19	0.28	0.28	0.17	0.82	0.83	0.53	0.99	0.59	0.60
(vi)	0.95	0.70	0.80	0.80	0.41	0.52	0.85	0.84	0.37	0.49	0.51	0.46	0.99	0.53	0.55

Table 3. Correlation Coefficients: Centrality Metrics for Real-World Network Graphs (Networks Listed in the Increasing Order of Spectral Radius Degree Ratio)

Network Index	Deg EVC	Deg BWC	Deg ClC	Deg FarC	Deg Ecc	EVC BWC	EVC ClC	EVC FarC	EVC Ecc	BWC ClC	BWC FarC	BWC Ecc	ClC FarC	ClC Ecc	FarC Ecc
(v)	0.87	0.28	0.29	0.29	0.14	0.19	0.28	0.28	0.17	0.82	0.83	0.53	0.99	0.59	0.60
(ii)	0.77	0.60	0.71	0.73	0.36	0.33	0.71	0.68	0.05	0.67	0.71	0.62	0.99	0.62	0.67
(iii)	0.93	0.71	0.58	0.59	0.14	0.58	0.53	0.53	0.01	0.78	0.79	0.51	0.99	0.76	0.76
(i)	0.90	0.92	0.77	0.77	0.46	0.79	0.91	0.90	0.60	0.72	0.72	0.44	0.99	0.78	0.79
(iv)	0.95	0.92	0.84	0.84	0.54	0.82	0.93	0.92	0.60	0.66	0.66	0.43	0.99	0.71	0.71
(vi)	0.95	0.70	0.80	0.80	0.41	0.52	0.85	0.84	0.37	0.49	0.51	0.46	0.99	0.53	0.55

We observe the Degree Centrality and the Eigenvector Centrality to have the highest correlation coefficient (average of 0.89, across all the networks, with a minimum of 0.77 and a maximum of 0.95); for undirected graphs, the Eigenvector Centrality of a vertex is merely a function of the degree of the vertex and the degrees of its adjacent vertices. Hence, it is natural to expect a high correlation between these two centrality measures for undirected graphs that are the focus of study in this paper.

Among the shortest path-based centrality metrics, we observe a very high correlation between the Closeness and Farness centrality metrics, indicating that there is nothing new that could be learnt from analyzing the Farness centrality of the nodes in the network (in addition to what is inferred from Closeness centrality). The correlation between any other pair of the shortest path-based centrality metrics is moderate for a majority of the networks and is slightly higher in networks that are random/regular and are relatively less scale-free. For networks in which the variation in the degree distribution of the vertices is on the lower side (i.e., spectral radius ratio of 1.4 or less, in the case of the real-world network graphs studied here), all the nodes are more or less equally contribute in facilitating shortest path communication between any two nodes in the network. Hence, the ranking of the vertices according to the Betweenness or Closeness metrics is almost the same. On the other hand, as the network gets increasingly scale-free, we start observing a decrease in the correlation between the shortest path-based metrics. A vertex (through which no shortest path goes through) that is simply closer to a higher degree vertex may exhibit a higher closeness (farness), but a zero betweenness. For the same reason, we observe the Eccentricity of the vertices to be relatively weakly correlated to the Betweenness centrality (and relatively better correlated with the Closeness and Farness centralities).

Overall, we could say that the two degree-based centrality measures (Degree centrality and Eigenvector centrality) consistently exhibit high correlation coefficient across all the networks, and the four shortest path-based centrality measures consistently exhibit a moderate correlation across all the networks (except the very high correlation observed between Closeness and Farness centralities for all networks). The correlations between pairs of centrality metrics from each of the two classes (i.e., a degree-based centrality metric vs. shortest path-based centrality metric) are not consistent across the networks; the type of the network does impact the correlation between these pairs of centrality measures. Networks that are very random and regular (i.e., those that have little variation in the node degrees) like the American College Football network exhibit a very poor correlation between a degree-based centrality measure and a shortest path-based centrality measure. This could be attributed to the presence of certain core central vertices (though not having a high degree) that are on the shortest path to every other vertex in the periphery in a regular random graph, whereas vertices in the periphery of the network are not likely on the shortest path to other vertices as well as not connected to a vertex on the shortest path to several other vertices. The American College Football network exhibits a poor correlation for all of the eight pairs of centrality measures (pairs involving a centrality measure based on degree and a centrality measure based on shortest path). The Eigenvector Centrality and the Betweenness Centrality measures appear to be the least correlated (average correlation coefficient of 0.54, with values of 0.19 and 0.33 observed for the American College Football network and Dolphins' social network respectively - incidentally, these are the two networks for which we observe a lower variation among the node degrees of the vertices). Likewise, the Eccentricity metric is very poorly correlated to both the degree and Eigenvector centralities for networks that are either regular/random or relatively less scale-free.

On the other hand, for networks with moderate variation in node degrees and exhibiting higher levels of scale-free nature (i.e., those that have few nodes with very high degree and the rest of the nodes are of relatively lower degree and connected to one or more of the high-degree vertices), we observe a relatively higher correlation between the degree-based and shortest-path centrality measures. Vertices that have very low degree in such scale-free networks are not on the shortest path to other vertices (hence have a low degree centrality and low betweenness centrality) and vertices that have a high degree are likely to be on the shortest paths between several other pairs of vertices (hence have a high degree centrality and high betweenness centrality). Also, in networks with moderate-high variation in node degrees, though a vertex may not have a high degree, it is likely to be connected to a vertex with moderate-high degree through which it can reach the majority of the rest of the vertices (if not all of the vertices) on shorter paths (with fewer hops), exhibiting good correlation between the Eigenvector centrality and Closeness centrality.

5   Conclusion

We show that the degree-based centrality metrics (degree and Eigenvector centralities) are consistently highly correlated for all the six real-world network graphs considered. Likewise, though the shortest path-based centrality metrics are only moderately correlated for most of the real-world network graphs, we observe such a correlation to be consistent across the network graphs without much variation in the correlation coefficient values. The level of correlation between a degree-based centrality metric and a shortest path-based centrality metric increases with increase in variation of node degree: the two classes of metrics are poorly correlated in regular/random networks and are at the low-end of moderate-level of correlation for real-world networks that are less scale-free. As the real-world networks get more scale-free, the level of correlation between the two classes of centrality metrics is likely to increase. The shortest path-based centrality metrics correlate better for regular/random networks and the level of correlation decreases as the networks get increasingly scale-free. We also observe that correlation analysis involving two centrality metrics whose values are real numbers is more accurate (compared to analysis of metrics in which at least one of the two take discrete values). Especially, with a discrete-value metric like Eccentricity whose range of values could be very limited, the correlation analysis could be less accurate. Also, such discrete-value centrality metrics may not be an appropriate measure to unambiguously rank the vertices. Nevertheless, since the degree centrality metric could take a broader range of discrete values, it is a relatively better measure for ranking the vertices. As part of future work, we plan to conduct correlation coefficient analysis between the various theoretical network graph models and their parameters.

References

1. Newman, M.: Networks: An Introduction. Oxford University Press, Oxford (2010)
2. Gephi, http://gephi.github.io/
3. Pajek, http://vlado.fmf.uni-lj.si/pub/networks/pajek/
4. Li, C., Li, Q., Van Mieghem, P., Stanley, H.E., Wang, H.: Correlation between Centrality Metrics and their Application to the Opinion Model. arXiv:1409.6033v1 (2014)
5. Strang, G.: Linear Algebra and its Applications. Cengage Learning, Boston (2005)
6. Cormen, T.H., Leiserson, C.E., Rivest, R.L., Stein, C.: Introduction to Algorithms. MIT Press, Cambridge (2009)
7. Zachary, W.W.: An Information Flow Model for Conflict and Fission in Small Groups. Journal of Anthropological Research 33, 452–473 (1977)
8. Lusseau, D., Schneider, K., Boisseau, O.J., Haase, P., Slooten, E., Dawson, S.M.: The Bottlenose Dolphin Community of Doubtful Sound Features a Large Proportion of Long Lasting Associations. Behavioral Ecology and Sociobiology 54, 396–405 (2003)
9. Orgnet.com, http://www.orgnet.com/divided.html
10. Newman, M.E.J.: Finding Community Structure in Networks using the Eigenvectors of Matrices. Physics Review. E 74, 36104 (2006)
11. Clauset, A., Newman, M.E.J., Moore, C.: Finding Community Structure in Very Large Networks. Physics Review. E 70, 066111 (2004)

Models Adaptation of Complex Objects Structure Dynamics Control

Boris V. Sokolov[1,3], Vyacheslav A. Zelentsov[1], Olga Brovkina[2,4],
Victor F. Mochalov[1], and Semyon A. Potryasaev[1]

[1] Russian Academy of Science,
Saint Petersburg Institute of Informatics and Automation (SPIIRAS), St. Petersburg, Russia
sokol@iias.spb.su
[2] Global Change Research Centre Academy of Science of the Czech Republic,
Brno, Czech Republic
[3] University ITMO, St. Petersburg, Russia
[4] Mendel University in Brno, Brno, Czech Republic

Abstract. In this paper we present a dynamic multiple criteria model of integrated adaptive planning and scheduling for complex objects (CO). Various types of CO are in use currently, for example: virtual enterprises, supply chains, telecommunication systems, etc. Hereafter, we refer to CO as systems of those types. The adaptation control loops are explicitly integrated within the model of analytical simulation. The mathematical approach is based on a combined application of control theory, operations research, systems analysis, and modeling and simulation theory. In particular, a scheduling problem for CO is considered as a dynamic interpretation. New procedures of dynamic decomposition help us to find the parameter values of the model's adaptation. The example demonstrates a general optimization scheme to be applied to the problem of division of competencies between the coordinating and operating levels of the CO via parametric adaptation of the model's described structure dynamics control processes.

Keywords: complex technical – organizational system, structure dynamic control, planning and scheduling, parametric and structure adaptation of models.

1   Introduction

In practice, the processes of complex objects (CO) operations are non-stationary and nonlinear. It is difficult to formalize various aspects of a CO. CO models have high dimensionality. There are no strict criteria for decision making for CO management and information about many CO parameters are not a priori. Besides, CO operations are always influenced by external and internal, objective and subjective, perturbations. These perturbations initiate the CO structure dynamics and predetermine a sequence of control inputs compensating for the perturbations.

In other words we always experience CO structure dynamics in practice. There are many possible variants of CO structure dynamics control (Ohtilev et al., 2006).

© Springer International Publishing Switzerland 2015
R. Silhavy et al. (eds.), *Intelligent Systems in Cybernetics and Automation Theory*,
Advances in Intelligent Systems and Computing 348, DOI: 10.1007/978-3-319-18503-3_3

The CO peculiarities mentioned above do not let us produce an adequate description of the structure dynamics control processes in both pre-existing and designed CO' based on single-classed models. That is why the concept of integrated modeling (comprehensive simulation) that was proposed by the authors can be useful here. Possible directions of realization were considered in (Ohtilev et al., 2006, Zaychik et al., 2007). In this paper we propose a new approach to the problem of parametric adaptation of models describing control of CO structure dynamics. The existence of various alternative descriptions for CO elements and control subsystems provides an opportunity for adaptive models selection (synthesis) for program control under a changing environment.

Our investigations are based on results of the CO adaptive control theory, which is now being developed by Professor Skurihin V.I. in Ukraine (Skurihin et al., 1989). The analysis of known investigations of the subject (Skurihin et al., 1989, Rastrigin, 1981, Bellman, 1972, Fleming, 1975, Nillson and Darley, 2006; Rabelo et al. 2002), confirms that the traditional tasks of CO control should be supplemented with procedures of structural and parametric adaptation of special control software (SCS) (see fig.1 blocks 3, 7). Here the adaptive control should include the following main phases:

— Parametric and structural adaptation of structure dynamics control (SDC) models and algorithms, to previous and current states of objects-in-service (SO), of control subsystems (CS), and of the environment (see fig.1, blocks 1,2,3)
— Integrated planning and scheduling of CO operations (construction of SDC programs) (blocks 4,5)
— Simulation of CO operations according to the schedules, for different variants of control decisions in real situations, and analysis of planning and scheduling simulation (blocks 6)
— Structural and parametric adaptation of the schedule, control inputs, models, algorithms, and SDC programs to possible (predicted by simulation) states of SO, CS, and of the environment (blocks 7)
— Realization of CO structure dynamics control processes (blocks 8)

To implement the proposed concept of adaptive control let us consider two groups of parameters (Skurihin et al., 1989, Rastrigin, 1980, 1981; Fischer et al. 2004) for CO SDC models and algorithms: parameters that can be evaluated on the basis of real data available in the CO, and parameters that can be evaluated via simulation models for different scenarios of future events.

The adaptation procedures can be organized in two blocks (models) (Ohtilev, 2006, Skurihin et al., 1989): internal, and external adaptation of planning and scheduling models.

When the parametric adaptation of SCS does not provide adequate simulations then structural transformations can be necessary. The two main approaches to structural model adaptation are usually characterized as (Bellman, 1972, Rastrigin, 1980, 1981). The first approach lies in the selection of a model from a given set. The model selected must be the most adequate for the SO and CS. The second approach is the construction of elementary CO SDC models (modules) in compliance with given requirements. This second approach provides a more flexible adjustment of the SO and CS for particular functioning conditions. However, the first is faster and can be effective if the application knowledge base is sufficiently large.

Both approaches need active participation of system analysts and decision-makers who interact with special control software of the simulation system (SIS) and who consider hard-formalizing factors and dependences within the general procedure of the CO SDC program selection.

Let us consider the formal statement of structural and parametric adaptation problems for CO SDC models and after that we are going to investigate the problem of parametric adaptation for models describing CO structure dynamics control. Adaptation of algorithms and control software is out of the scope of this paper.

We have implemented the conceptual model and technology of parametric and structural adaptation of models describing CO SDC processes via an original simulation system (SIS). This simulation system consists of the following elements (Moiseev, 1974, Sowa, 2002; Huang et al. 2005; Kuehnle et al. 2007; Teich 2003; Wu N et al. 1999, 2005): a) simulation models (a hierarchy of models), b) analytical models (a hierarchy of models) for a simplified (aggregated) description of the objects being studied, c) an informational subsystem that is a system of databases (knowledge bases), and d) a control-and-coordination system for interrelation and joint use of the other elements and interaction with the user (decision-maker).

In this paper we want to describe and to investigate the actual algorithm of the model's adaptation via integrated modeling and simulation procedures, which are realized in the SIS.

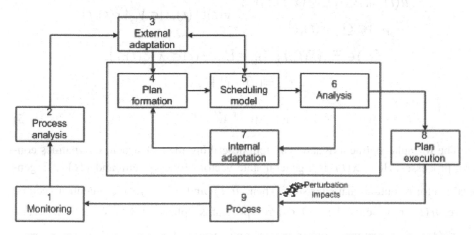

Fig. 1. Conceptual model of parametric and structural adaptation (Skurihin et al., 1989)

2   Problem Statement

We assume that there are several variants of CO SDC models inscribed in the set
$$\overline{\overline{M}} = \{M_1, M_2, ..., M_W\} = \{M_\Theta, \Theta \in I\}, \quad I = \{1, ..., W\},$$
moreover, the vector $\vec{\beta}$ of CO parameters includes the subvector $\vec{\beta}_0$ of fixed CO characteristics and also

subvector $\vec{w} =\parallel \vec{w}^{(1)\mathsf{T}}, \vec{w}^{(2)\mathsf{T}}, \vec{w}^{(3)\mathsf{T}} \parallel^{\mathsf{T}}$ of parameters being adjusted through SS internal or external adapters or defined within a structural adapter. According to (Skurihin et al., 1989), these parameters can be divided into the following groups: $\vec{w}^{(1)}$ is a vector of parameters being adjusted through the internal adapter, $\vec{w}^{(2)}$ is a vector of parameters being adjusted through the external adapter, $\vec{w}^{(3)}$ is a vector of parameters being adjusted within a structural adaptation of CO SDC models.

Now we can present the modified multi-model multi-criteria description of CO SDC problems:

$$\vec{J}_{\Theta}\left(\vec{x}(t),\vec{u}(t),\vec{\beta},\vec{\xi}(t),t\right)\to \underset{\vec{u}(t)\in\Delta_{\Theta}}{\text{extr}} \ , \tag{1}$$

$$\Delta_{\Theta} = \{\vec{u}(t)\mid \vec{x}(t)= $$
$$\vec{\varphi}_{\Theta}\left(T_0,\vec{x}(T_0),\vec{x}(t),\vec{u}(t),\vec{\xi}(t),\vec{\beta}_{\Theta},t\right)\} \ , \tag{2}$$

$$\vec{y}(t) = \vec{\psi}_{\Theta}\left(\vec{x}(t),\vec{u}(t),\vec{\xi}(t),\vec{\beta}_{\Theta},t\right), \tag{3}$$

$$\vec{x}(T_0)\in X_0(\vec{\beta}_{\Theta}), \quad \vec{x}(T_f)\in X_f(\vec{\beta}_{\Theta}), \tag{4}$$

$$\vec{u}(t) =\parallel \vec{u}_{pl}^{\mathsf{T}}(t), \vec{v}^{\mathsf{T}}\left(\vec{x}(t),t\right)\parallel^{\mathsf{T}}; \quad \vec{v}(t)(\vec{x}(t),t)\in V_{\Theta}(\vec{x}(t),t);$$
$$\vec{u}_{pl}(t)\in Q_{\Theta}(\vec{x}(t),t);$$
$$\vec{\xi}(t)\in \Xi_{\Theta}(\vec{x}(t),t); \ \vec{\beta}_{\Theta}\in \mathbf{B}; \ \vec{x}(t)\in X\left(\vec{\xi}(t),t\right);$$
$$\vec{\beta}_{\Theta} =\parallel \vec{\beta}_0^{\mathsf{T}} \ \vec{w}^{\mathsf{T}} \parallel^{\mathsf{T}};$$

$$\vec{w} =\parallel \vec{w}^{(1)\mathsf{T}} \ \vec{w}^{(2)\mathsf{T}} \ \vec{w}^{(3)\mathsf{T}} \parallel^{\mathsf{T}} \ . \tag{5}$$

The formulas define a dynamic system describing the CO structure dynamics control processes. Here $\vec{x}(t)$ is a general state vector of the system, and $\vec{y}(t)$ is a general vector of output characteristics. Then, $\vec{u}(t)$ and $\vec{v}(\vec{x}(t),t)$ are control vectors. Here $\vec{u}(t)$ represents the CO control programs (plans of CO functioning), and $\vec{v}(\vec{x}(t),t)$ is a vector of control inputs compensating for the perturbation influences $\vec{\xi}(t)$). The vector $\vec{\beta}_{\Theta}$ is a general vector of CO parameters. The vector for CO effectiveness measurements is described as (6).

$$\vec{J}_{\Theta}\left(\vec{x}(t),\vec{u}(t),\vec{\xi}(t),\vec{\beta}_{\Theta},t\right)= \parallel \vec{J}_{\Theta}^{(g)\mathsf{T}},\vec{J}_{\Theta}^{(o)\mathsf{T}},\vec{J}_{\Theta}^{(k)\mathsf{T}},\vec{J}_{\Theta}^{(p)\mathsf{T}},\vec{J}_{\Theta}^{(n)\mathsf{T}},\vec{J}_{\Theta}^{(e)\mathsf{T}},\vec{J}_{\Theta}^{(c)\mathsf{T}},\vec{J}_{\Theta}^{(v)} \parallel \tag{6}$$

Its components state control effectiveness for motion, interaction operations, channels, resources, flows, operation parameters, structures, and auxiliary operations

(Ivanov et al., 2006). The indices «g», «o», «k», «p», «n», «e», «c», «v» correspond to the following models: order progress control ($M_{<g,\Theta>}$), operations control ($M_{<o,\Theta>}$), technological chains control ($M_{<k,\Theta>}$), resources control ($M_{<p,\Theta>}$); flows control ($M_{<n,\Theta>}$), operations parameters control ($M_{<e,\Theta>}$), structures control ($M_{<c,\Theta>}$), and auxiliary operations control ($M_{<v,\Theta>}$). In (5) the transition function $\vec{\varphi}_\Theta\left(T_0, \vec{x}(T_0), \vec{x}(t), \vec{u}(t), \vec{\xi}(t), \vec{\beta}_\Theta, t\right)$ and the output function $\vec{\psi}_\Theta\left(\vec{x}(t), \vec{u}(t), \vec{\xi}(t), \vec{\beta}_\Theta, t\right)$ can be defined in analytical or algorithmic form within the proposed simulation system; $Q_\Theta\left(\vec{x}(t), t\right)$, $V_\Theta\left(\vec{x}(t), t\right)$, $\Xi_\Theta\left(\vec{x}(t), t\right)$ correspond to allowable areas for program control, real-time regulation of control inputs, and perturbation inputs; B is the area of allowable parameters; $X\left(\vec{\xi}(t), t\right)$ is the area of allowable states of CO structure dynamics. Expression (4) determines end conditions for the CO state vector $\vec{x}(t)$ at time $t = T_0$ and $t = T_f$ (T_0 is the initial time interval of the CO investigation, and T_f is the final time interval).

The problem of CO structure dynamics control includes tasks of three main classes: Class A problems (problems of structured analysis and CO structure dynamics analysis with or without perturbation influences); Class B problems (estimation (observation), monitoring, and CO structural state identification); Class C problems (control-inputs selection and CO parameters selection, i.e., multi-criteria control problems for CO structures, modes, and parameters, and multi-criteria problems of CO structure-functional synthesis). In the general case, the formal statement of a CO structure dynamics control problem can be written as follows:

We are given: space-time, technical, and technological constraints (2)-(5) determining variants of CO SDC at the operation phase; a vector (1) of CO effectiveness measurements.

We have to determine: $\vec{u}_{pl}(t)$, $\vec{v}\left(\vec{x}(t), t\right)$, and $\vec{\beta}_\Theta$ meeting the constraints (2)-(5) and returning an optimal value of the measurement of general effectiveness $J_\Theta^{(ob)} = J_\Theta^{(ob)}\left(\vec{x}(t), \vec{y}(t), \vec{\beta}_\Theta, \vec{u}(t), \vec{\xi}(t)\right)$.

There are two main groups of methods that can be chosen for $\vec{\beta}_\Theta$ (Skurihin et al., 1989, Rastrigin, 1980, 1981, Bellman, 1972, Zypkin, 1969, Bryson, 1969): identification methods of parametric adaptation, and simulation methods of parametric adaptation. In this paper we propose an integrated procedure that helps us to find vector $\vec{w}^{(1)}$ only. Results that interrelate with the parametric model adaptation are described in the book (Ohtilev et al., 2006).

3   Methods

3.1   Algorithm of Parametric Adaptation of Models Describing Structure Dynamics Control Processes

The input data for the CO planning and scheduling model's adaptation is gathered during CO functioning at the state S_{i-1} and is input to state S_i. Thus we get the formulas (Ohtilev et al., 2006, Moiseev, 1974, Chernousko, Zak, 1985, Singh, Titli, 1978):

$$P_H = P_H\left(\vec{x}(t), \vec{u}_{pl}(t), \vec{v}\left(\vec{x}(t), \vec{\xi}\right), \vec{\xi}\right) \to \max_{\vec{u} \in \Delta}, \tag{7}$$

$$\Delta = Q(\vec{x}(t)) \times V\left(\vec{x}(t), \vec{\xi}, t\right), \quad \vec{u} = \vec{u}_{pk}(t) \times \vec{v}\left(\vec{x}(t), \vec{\xi}\right), \tag{8}$$

where P_H is the possible variant of CO efficiency index such as (1), $\vec{u}_{pk}(t)$ is the main vector of control inputs, in other words it is a control program for the CO dynamics, $\vec{v}$ is a vector of control inputs compensating the perturbation influences over the control program; $Q(\vec{x}(t))$ and $V\left(\vec{x}(t), \vec{\xi}, t\right)$ are the sets' allowable controls $\vec{u}_{pk}(t)$ and $\vec{v}\left(\vec{x}(t), \vec{\xi}\right)$ respectively; $\vec{\xi}(t)$ is a vector of perturbation influences, where $\vec{\xi}(t) \in \Xi(\vec{x}(t), t)$.

The general efficiency measure (7) can be evaluated as a function of $\left(\vec{x}(t), \vec{u}_{pl}(t), \vec{v}\left(\vec{x}(t), \vec{\xi}\right), \vec{\xi}\right)$ via simulation experiments with the CO operation model. Unfortunately, direct analytical expressions cannot be obtained. Therefore, the search for concrete control programs $\left(\vec{u}_{pl}(t)\right)$ is very difficult since the high dimensionality of the vectors $\left(\vec{x}(t), \vec{v}\left(\vec{x}(t), \vec{\xi}\right), \vec{\xi}\right)$ hinder optimization through experiments of directed simulation. This is why we propose the following heuristic decomposition of models (7)-(8). The decomposition is based on the structural peculiarities of these models.

The general efficiency measure of the forecasted states can be evaluated as a function of the enumerated values via simulation experiments with the CO operation model. In this case the following group of tasks substitutes for the initial problem of the CO control problem (7), (8):

$$P_H = P_H\left(\vec{x}(t, \vec{\lambda}'), \vec{u}_{pl}(t, \vec{\lambda}'), \vec{v}\left(\vec{x}(t, \vec{\lambda}'), \vec{\xi}\right), \vec{\xi}\right) \to \max_{\vec{\lambda}' \in \Delta'} \tag{9}$$

$$\Delta' = \{\vec{\lambda}' \mid \vec{u}_{pl}(t, \vec{\lambda}') \times \vec{v}\left(\vec{x}(t, \vec{\lambda}'), \vec{\xi}\right) \in \\ Q\left(\vec{x}(\vec{\lambda}')\right) \times V\left(\vec{x}(\vec{\lambda}'), \vec{\xi}\right)\} \tag{10}$$

$$\sum_{\gamma\in\Gamma'}\lambda'_{\gamma}J_{\gamma}(\vec{x}_{\gamma}) \to \underset{\vec{x}_{\gamma}\in D_{\gamma}\left(T_{f},T_{0},\vec{x}_{\gamma}(T_{0})\right)}{extr}, \tag{11}$$

$$\sum_{\gamma\in\Gamma'}\lambda'_{\gamma}=1,\ \lambda'_{\gamma}\geq 0,\ \vec{x}_{\gamma}=\|\ \vec{x}^{(\gamma)\mathsf{T}}\vec{x}^{(o)\mathsf{T}}\ \|^{\mathsf{T}},\ \gamma\in\Gamma' \tag{12}$$

Here the vectors $\vec{x}_{\gamma}(T_{f})$ returning optimal values to function (16) are being searched for, while the vector $\vec{\lambda}'_{(l)}$ is fixed ($l = 0,1,2,...$ is the number of the current iteration). The set Δ' includes indices of analytical models obtained via the proposed decomposition. The problems of mathematical programming have several important features. The search for components of the vector $\vec{x}^{(l)}_{\gamma}$ can be fulfilled over subsets of the *attainability sets* $D_{\gamma}\left(T_{f},T_{0},\vec{x}_{\gamma}(T_{0})\right)$ rather than over the whole sets of allowable alternatives. The subsets include non-dominated alternatives of the enumerated models. The non-dominated alternatives can be obtained via the orthogonal projection of the goal sets to the *attainability sets* $D_{\gamma}\left(T_{f},T_{0},\vec{x}_{\gamma}(T_{0})\right)$. Each particular model includes the state vector $\vec{x}^{(o)}$ of the operations model $M_{<o>}$ besides its own state vectors $\vec{x}^{(g)}$, $\vec{x}^{(k)}$,..., $\vec{x}^{(c)}$. The structural features of problems (11) and (12) above, let us use decomposition to overcome the problem of high dimensionality.

When the vector $\vec{x}^{(l)}_{\gamma}(T_{f})$ is known, the optimal programs $\vec{u}^{(l)}_{pl}(t,\vec{\lambda}'_{(l)})$ for CO control can be defined within each analytical model M_{γ}, $\gamma'\in\Gamma'$ via numerical methods (for example, via Krylov and Chernousko's method (Chernousko, Zak, 1985, Petrosjan, Zenkevich, 1996, Roy, 1996; Nilsson et al. 2006). The programs $\vec{u}^{(l)}_{pl}(t,\vec{\lambda}'_{(l)})$ are used for evaluation of a new approximation of the vector $\vec{\lambda}'_{(l+1)}$ in the simulation model $M_{<u>}$ describing CO functioning under the perturbation influences.

The problem of $\vec{\lambda}^{*}$ search is similar to the problem of optimal experiments design. Here, elements of the vector $\vec{\lambda}'$ are endogenous variables and the efficiency measure (14) is exogenous. Different methods can be used for optimal design of experiments, for example the method of quickest ascent, or the methods of random search. In conclusion we note that components of the vector $\vec{\lambda}'$ can define the preferable control inputs $\vec{v}\left(\vec{x}(t,\vec{\lambda}'),\vec{\xi}\right)$ for compensation of a mismatch in the planned trajectory of CO dynamics against the predicted (via simulation) trajectory. Finally we propose an example that illustrates one aspect of analytical-simulation procedure (9)-(12).

Structure dynamics control processes in a CO (see formulas (1)-(5)) depend upon internal and external stochastic factors. These factors should be described through more detailed simulation models of CO operation. These models include a deterministic component $\left(\vec{u}_{pl}(t),\vec{v}\left(\vec{x}(t),\vec{\xi}\right)\right)$ describing schedules of activities and a stochastic component $\left(\vec{\xi}\right)$ describing random events (see (5)). The considered activities contributed to managerial work and documentation. These activities could be performed at different nodes of the CO, however the sequence of activities was strictly determined within a previously obtained schedule (plan). The stochastic factors included random background and unexpected operations diverting personnel from the plan. The stochastic factors and complex interrelations of activities necessitated simulation application. Hence, CO optimization was to be performed via a series of simulation experiments. In this situation, analytical models were used to narrow the parameter's variation range.

This example demonstrates a general optimization scheme to be applied to the problem of competencies division between the coordinating and operating levels of the CO. The network included a coordinating level (center) and six operating levels (subsystems). Here the structural adaptation of the CO models implied testing of different plans. The parametric adaptation of the models requires a determination of the optimal ratio of competencies (functions or tasks) expressed in person-hours. Thus, λ'_{γ} (see (11)) expresses the share of node γ' in the total amount of work. The optimization was carried out in order to reduce the plan's duration.

3.2   Algorithm of Structure Adaptation of CO Structure Dynamics Control Processes Models

Generalized formal statements of the structure adaptation CO model can be represented as:

$$AD\left(M_{\Theta}^{(l)}, \overline{P}_{cs}\right) \rightarrow \min, \tag{13}$$

$$t_{st}\left(\vec{w}^{(3)}, M_{\Theta}^{(l)}\right) \leq \overline{t}_{st}, \tag{14}$$

$$M_{\Theta}^{(l)} \in \overline{\overline{M}}, \vec{w}^{(3)} \in W^{(3)}, M_{\Theta}^{(l)} = \overline{\overline{\Phi}}\left(M_{\Theta}^{(l-1)}, \vec{w}^{(3)}, \overline{P}_{cs}\right), l = 1, 2, ..., \tag{15}$$

where $AD\left(M_{\Theta}^{(l)}, \overline{P}_{cs}\right)$ is a function characterizing the adequacy of the model $M_{\Theta}^{(l)}$ for the CO. The latter is described in turn with a set $\overline{P}_{cs}(t) = \left\{\overline{P}_{g}^{(cs)}, \overline{g} = 1, ..., \overline{G}\right\}$ of characteristics; t_{st} is a total time of the CO model's structural adaptation; $\overline{t}_{st}$ is the maximum allowable time of structural adaptation; $\overline{\overline{\Phi}}$ is an operator of iterative construction (selection) of the model $M_{\Theta}^{(l)}$, l is the current iteration number; $W^{(3)}$ is a set of allowable values for the vectors of structure-adaptation parameters.

The analysis of expression (13) shows that the structural adaptation starts and stops according to a criterion characterizing the similarity of a real object and an object described via models (the condition of the model's adequacy is applied) (Sokolov et al., 2012). The adequacy of CO models does not mean a description of all "details" but that the simulation results meet the changes and relations observed in reality.

The listed equations can be interpreted in relation to the forest monitoring models (for the forest structure, biomass estimation, and forest growth models).

The main purpose of quantitative estimation of the model adequacy is to raise the decision-maker's confidence in drawing conclusions in a real situation. Therefore, the utility and correctness of the CO simulation results can be measured via the degree of adequacy of models and objects. Analysis of relations defining a common procedure of the proactive structural adaptation planning models and operational control of the CO demonstrates that its implementation needs a set of specific algorithms of iterative selection of the models (multiple-complexes). These models are denoted as $M_\Theta^{<k,l>}$, where k is the current management cycle (k=1,...,K); l is the current iteration where the design (selection) image is performed (l=1,...,L); Θ is the current model from the model bank or model repository (Θ=1,...,Θ). The model repository is named as the plurality:

$$\overline{\overline{M}}^{<k>} = \{ M_\Theta^{<k,l>}; k = 1,..., K; l = 1,..., L; \Theta = 1,..., \Theta^{<k>} \} \qquad (16)$$

The proposed algorithms for structural adaptation of CO models are based on the evolutionary (genetic) approach. Let us exemplify these algorithms in the structural adaptation of a model describing structure dynamics of one CO output characteristic (of one element of the vector $\vec{y}(t_{\langle k \rangle})$).

The residual of its estimation via the model M_θ, compared with the observed value of the characteristic, can be expressed as:

$$Q_{\langle k \rangle}^{\langle \theta \rangle} = \left[\psi_{\langle \theta, k \rangle}(\vec{x}\left((t_{\langle k \rangle - 1}), \vec{u}(t_{\langle k \rangle}), \vec{\xi}(t_{\langle k \rangle}), \vec{\beta_\theta}, t_{\langle k \rangle}\right) - \tilde{y}(t_{\langle k \rangle}) \right], \qquad (17)$$

To simplify the formula, we assume that the perturbation influences $\vec{\xi}(t)$ are described via stochastic models. Thus, the following quality measure can be introduced for the model M_θ:

$$\overline{Q_{\langle k \rangle}^{\langle \theta \rangle}} = \sum_{k=1}^K g^{(K-k)} \overline{Q_{\langle k \rangle}^{\langle \theta \rangle}}, \qquad (18)$$

where $0 \le g \le 1$ is a "forgetting" coefficient that "depreciates" the information received in the previous steps (control cycles) (Ivanov et al., 2010). If g=0 then $\overline{Q_{\langle K \rangle}^{\langle \theta \rangle}} = Q_{\langle K \rangle}^{\langle \theta \rangle}$, i.e. the weighted residual is equal to one previously obtained, as the prior information has been "forgotten". An extension of formula (16) was proposed in Sokolov et al. (2012). The coefficient $g^{(K-k)}$ was substituted for the function f(K):

$$\overline{Q_{\langle K \rangle}^{\langle \theta \rangle}} = \sum_{k=1}^K f(K - k) Q_{\langle k \rangle}^{\langle \theta \rangle}, \theta = 1,.. \Theta, \qquad (19)$$

Here f is a monotone decreasing (forgetting) function. It has the following properties:

$$f(a) > 0, f(0) = 1, \lim_{a \to ?} f(a) = 0, f(a) \ge f(a + 1), a = 0,1, ... \quad (20)$$

Now, the structural-adaptation algorithm is reduced to a search for the structure M_Θ , such that

$$\overline{Q_{\langle K \rangle}^{\langle \theta \rangle}} = \min_{\theta=1,\dots\Theta} \overline{Q_K^{\langle \theta \rangle}}, \tag{21}$$

Thus, it is necessary to calculate the quality measures (17) for all competitive structures $M_{\theta}, q = 1, \dots Q$ of CO models at each control cycle $k=1,\dots K$. All quality measures should be compared, and the structure M_θ with the best measure (minimal residual) should be chosen.

Another way to choose model $M_\Theta^{(l)}$ is through a probabilistic approach. In this case the following formula is used:

$$p_1\left(M_\Theta^l\right) = \frac{\sum_{p=k-d}^{k-1} J_i(M_i^{\langle p,L \rangle})}{\sum_{\Theta=1}^{\Theta^{\langle k-1 \rangle}} \sum_{p=k-d}^{k-1} J_i(M_\Theta^{\langle p,L \rangle})}, \tag{22}$$

where $J_i\left(M_i^{\langle p,l \rangle}\right)$ is a generalized quality measure value of model $M_i^{\langle p,l \rangle}$ functioning on previous time intervals, and d is a "forgetting" coefficient. It should be emphasized that the calculation of the quality measurement is expected to be carried out each time on the basis of a solution for the problem of multi-type selection. Thus, despite the random choice of the next model (multiple-model complex), the model that had the better measurement of generalized quality in the previous cycle control has the greater opportunity to be chosen.

The parametric adaptation of the model (Ivanov et al., 2010, 2012) should follow the structural one. It is important to determine a proper "forgetting" function under the perturbation influences $\vec{\xi}(t)$. The higher the noise levels in the CO, the slower the decrease of the function is implemented. However, if the CO changes its structure greatly then the function $f(a)$ has to be rapidly decreased in order to "forget" the results of the previous steps. It can be shown that the structural-adaptation algorithms based on model construction (synthesis) of atomic models (modules) are rather similar to the algorithms of CO structure-functional synthesis (Okhtilev et al.,2006). These algorithms are differing only in the interpretation of the results.

3.3   Example

This example demonstrates a structural adaptation of a CO containing the plurality of realization and multifunctional character. This CO was presented as original software for the remote sensing (RS) data processing for aboveground forest biomass estimation. The remote sensing data included airborne LiDAR data and hyperspectral imaging for the same forest territory in the Beskydy Mountains of the Czech Republic. The biomass assessment was performed for forest areas (over 60 plots of 500 m^2) where field measurements had been taken. The main goal was to find an appropriate model for aboveground forest biomass estimation for that region.

The CO consisted of four main blocks: input of airborne data (block 1), calculation of prediction metrics for a model (block 2), modeling (block 3), and models diagnostics (block 4). The perturbation influences were represented by the control model parameters, $\overline{w}^{(3)}$, that could be evaluated on the prediction metrics from airborne data available in the CO (block 2) and parameters that could be evaluated via simulation of various models for different scenarios of future events (block 3). The problem consisted of the structural adaptation of the CO models to perturbation influences. According to our algorithms it was necessary to choose a model for CO functioning under perturbation influences. Assessment of the model parameters from block 1 included:

— Characteristics of LiDAR data (points per square meter), spectral and spatial resolutions of the hyperspectral data;
— An area of the study forest site.

Assessment of model parameters from block 2 included:

— LiDAR metrics (mean of all plot returns; canopy density, tree height, standard deviations, average, maximum and minimum of each of the metrics);
— hyperspectral metrics (64 spectral bands in the 450-900 nm interval, atmospherically corrected and noise minimized; simple ration index, normalized difference vegetation index, red edge normalized difference vegetation index, Vogelmann red edge index, red green ratio).

Block 3 included retrieval of aboveground biomass models from the common modeling approaches as linear models (LM), support vector machines (SVM), nearest neighbor-based (NN), random forest (RF), and Gaussian processes (GP).

Block 4 included the model diagnostics based on field-measured data for forest plots for biomass estimation. Multiple mechanisms were used to avoid overfitting; Akaike Information Criterion, leave-one-out cross validation, k-fold cross validation, and bootstrapping.

The following global optimization problem was considered: $AD\left(w_1^{(3)}, w_2^{(3)}, ... w_n^{(3)}\right) \to min,$ (see formula 14) where the arguments $w_1^{(3)}, w_2^{(3)} ... w_n^{(3)}$ are model parameters from blocks 2-4 to be varied in order to receive the minimal (the best) value of the function AD. It was possible to find the biomass model due to the vector of program control based on the evaluated model parameters. The probability of the model parameter's choice was determined, based on the task's solution (formula 22, where J is the appropriate model for aboveground biomass assessment).

4   Conclusions

The proposed approach to structure dynamics description lets us adapt heterogeneous models to the changing environment.

Dynamic decomposition of control problems helps us to determine parameters of a model's adaptation.

There are the following advantages of the proposed approach: the joint use of diverse models (optimization, simulation, and heuristic) in the framework of poly-model systems allows us to improve the flexibility and adaptability of planning processes in a CO, as well as allowing us to compensate the drawbacks of one class of models by the advantages of another.

The adaptive plans (programs) of a CO' functioning include transition programs as well as programs of stable CO operation in intermediate multi-structural macro-states. One of the main opportunities of the proposed method of CO SDC program construction is that in addition to the vector of program control we receive a preferable multi-structural macro-state of the CO at the end. This is a state of reliable CO operation in the current (forecasted) situation.

Further investigations should include an analysis of the influence of external factors upon the convergence of the planning procedures. Attributes of Pareto's set of the multi-criteria problem (see formulas (11)-(12)) will be analysed.

Acknowledgments. The research described in this paper is supported by the Russian Foundation for Basic Research (grants 12-07-00302, 13-07-00279, 13-08-00702, 13-08-01250, 13-07-12120-ofi-m, 12-07-13119-ofi-m-RGD), Department of nanotechnologies and information technologies of the RAS (project 2.11), by Postdoc project in technical and economic disciplines at the Mendel University in Brno (reg. number CZ.1.07/2.3.00/30.0031), by ESTLATRUS projects 1.2./ELRI-121/2011/13 «Baltic ICT Platform» and 2.1/ELRI-184/2011/14 «Integrated Intelligent Platform for Monitoring the Cross-Border Natural-Technological Systems» as a part of the Estonia-Latvia-Russia cross border cooperation Program within the European Neighborhood and Partnership instrument 2007-2013. Grant 074-U01 from The Government of the Russian Federation, partially financially supported this work).

References

1. Ohtilev, M.Y., Sokolov, B.V., Yusupov, R.M.: Intellectual Technologies for Monitoring and Control of Structure dynamics of Complex Technical Objects, 410 p. Moscow, Nauka (2006) (in Russian)
2. Zaychik, E., Sokolov, B., Verzilin, D.: Integrated modeling of structure dynamics control in complex technical systems. In: 19th European Conference on Modeling and Simulation ESMS 2005, "Simulation in Wider Europe", June 1-4, pp. 341–346. Riga Technical University, Riga (2005)
3. Ivanov, D., Sokolov, B., Arkhipov, A.: Stability analysis in the Framework of decision Making under Risk and Uncertainty Network – Centric Collaboration and Supporting Frameworks. In: Camarinha-Matos, L., Afsarmanesh, H., OUus, M. (eds.) Network – Centric Collaboration and Supporting Frameworks, IFIP TC5WG 5.5 Seventh IFIP Working Conference on Virtual Enterprises, Helsinki, Finland, September 25-27. IFIP, vol. 224, pp. 211–218. Springer, Boston (2006)
4. Skurihin, V.I., Zabrodsky, V.A., Kopeychenko, Y.V.: Adaptive control systems in machine-building industry. Mashinostroenie, – M. (1989)

5. Rastrigin, L.A.: Modern principles of control for complicated objects. Sovetscoe Radio, – M. (1980) (in Russian)
6. Bellmann, R.: Adaptive Control Processes: A Guided Tour. Princeton Univ. Press, Princeton (1972)
7. Rastrigin, L.A.: Adaptation of complex systems. Zinatne, Riga (1981) (in Russian)
8. Fleming, W.H., Richel, R.W.: Deterministic and stochastic optimal control. Springer, New York (1975)
9. Moiseev, N.N.: Element of the Optimal Systems Theory. Nauka, – M. (1974) (in Russian)
10. Sowa, J.: Architecture for intelligent system. IBM System Journal 41(3) (2002)
11. Zypkin Ya. Z. Adaptation and teachning in automatic systems. Nauka, – M. (1969) (in Russian)
12. Bryson, A.E., Ho, Y.-C.: Applied optimal control: Optimization, Estimation and Control. Waltham Massachusetts, Toronto, London (1969)
13. Chernousko, F.L., Zak, V.L.: On Differential Games of Evasion from Many Pursuers. J. Optimiz. Theory and Appl. 46(4), 461–470 (1985)
14. Singh, M., Titli, A.: Systems: Decomposition, Optimization and Control. Pergamon Press, Oxford (1978)
15. Petrosjan, L.A., Zenkevich, N.A.: Game Theory. World Scientific Publ., Singapore (1996)
16. Roy, B.: Multi-criteria Methodology for Decision Aiding. Kluwer Academic Pulisher, Dordrecht (1996)
17. Fischer, M., Jaehn, H., Teich, T.: Optimizing the selection of partners in production networks. Robotics and Computer-Integrated Manufacturing 20, 593–601 (2004)
18. Huang, G., Zhang, Y., Liang, L.: Towards integrated optimal configuration of platform products, manufacturing processes, and supply chains. Journal of Operations Management 23, 267–290 (2005)
19. Kuehnle, H.: A system of models contribution to production network (PN) theory. Journal of Intelligent Manufacturing, 157–162 (2007)
20. Nilsson, F., Darley, V.: On complex adaptive systems and agent-based modeling for improving decision-making in manufacturing and logistics settings. Int. Journal of Operations and Production Management 26(12), 1351–1373 (2006)
21. Rabelo, R.J., Klen, A.A.P., Klen, E.R.: Multi-agent system for smart coordination of dynamic supply chains. In: Proceedings of the 3rd International Conference on Virtual Enterprises, PRO-VE, pp. 379–387 (2002)
22. Teich, T.: Extended Value Chain Management (EVCM). GUC-Verlag, Chemnitz (2003)
23. Wu, N., Mao, N., Qian, Y.: An approach to partner selection in agile manufacturing. Journal of Intelligent Manufacturing 10(6), 519–529 (1999)
24. Wu, N., Su, P.: Selection of partners in virtual enterprise paradigm. Robotics and Computer-Integrated Manufacturing 21, 119–131 (2005)

Electronic Computing Equipment Schemes Elements Placement Based on Hybrid Intelligence Approach

L.A. Gladkov, N.V. Gladkova, and S.N. Leiba

Southern Federal University, Rostov-on-Don, Russia
{leo_gladkov,nadyusha.gladkova77,lejba.sergej}@mail.ru

Abstract. The problem of electronic computing equipment (ECE) schemes elements placement within a switching field is considered in this article. It refers to the class of design problems that are NP-hard and NP-full. The authors suggested a new approach on the basis of genetic algorithm (GA) integration and a fuzzy control model of algorithm parameters. A fuzzy logical controller structure is described in the article. To confirm the method effectiveness a brief program description is reviewed.

Keywords: ECE, Design, Elements placement, Optimization, Genetic algorithm, Fuzzy logic.

1 Introduction

ECE design is a multilevel process and each level has its own specific software. Design problems are defined by high computational complexity due to the need of a search a huge number of different solutions. Moreover, to get the exact solution an exhaustive search is required to perform that is not possible. Design problems include partitioning problem, placement problem, routing problem etc. [1].

Nowadays evolutionary algorithms, genetic algorithms and other bioinspired methods are applied to increase effectiveness design problem solving [2].
A new development stage of the genetic algorithms theory became hybrid systems that are based on combination of different research areas, for example, genetic algorithms, fuzzy systems and neural networks. There are different hybridization methods. One of them is fuzzy genetic algorithms [3] in which fuzzy methods are applied to GA parameters setting. Also fuzzy operators and fuzzy rules are used for development of genetic operators with various characteristics; systems of fuzzy logical control of GA parameters in accordance with predetermine criteria; fuzzy stop criteria of genetic search. At this case the mathematical apparatus of fuzzy system theory is for encoding; setting of GA optimal parameters; setting of GA probability etc.

2 Problem Definition

In terms of the placement problem the initial data are represented by information on connection field configuration determined by installation requirements; quantity and

© Springer International Publishing Switzerland 2015
R. Silhavy et al. (eds.), *Intelligent Systems in Cybernetics and Automation Theory*,
Advances in Intelligent Systems and Computing 348, DOI: 10.1007/978-3-319-18503-3_4

geometrical dimension of design elements; net list; restriction on components mutual placement taking into account main features of the developing construction [1]. The placement problem reduces to finding such elements positions that a selected criterion is optimized. These elements positions should be provide favorable conditions for further electrical installation.

To evaluate the placement quality following criteria can be used:

1. summarized connections length;
2. length of the longest connection;
3. a number of connections that length is greater than predetermined one;
4. a number of bend connection;
5. a number of connections between elements in the nearest positions or in positions defined by developers;
6. chip area.

In the matter of placement algorithms the first criterion is most prevalent. This is because connection length reducing allows to improve schemes electrical characteristics that simplifies the further routing. Besides, this method is relatively simple to implement [4].

The optimization problem model can be defined by a 3-tuple which represented as <X,D,Q>, where X is a set of solutions (chromosomes); D is restrictions on the set X for the selection of feasible solutions; Q is a target function to find an optimal solution [5].

Let $E=\{e_i \mid i=1, \ldots, n\}$ be a set of components and $P=\{p_i \mid i=1, \ldots, c\}$ is a set of position within the connection field. To place all components it is necessary to satisfy the condition $c \geq n$. Random placement of components in positions is a permutation $P=P_{(1)}, \ldots, P_{(i)}, \ldots, P_{(n)}$, where $P_{(i)}$ is position number for component e_i. Depending on the selected criteria the target function F(P) is used to evaluate placement results. Thus, the placement problem consists in finding the optimal solution of the function F on the permutation set P.

3   Algorithm Description

To solve the placement problem the simple genetic algorithm can be used. An initial population is defined by the «shotgun» method; selection is implemented by a «roulette wheel» method; a single-point crossover operator and a two-point mutation operator is used [5, 6, 7].

The set of positions can be represented as regular structure (grid). Each position p_i has coordinates x_i, y_i. The first position has coordinates (1;1). It is essential to set up a one-to-one correspondence between the set of positions sequence numbers and the set of coordinates. For this purpose we order and enumerate positions in parameter $W_i = x_i + y_i$ ascending order. Since positions located on the same diagonal in the grid have the same value W_i they are ordered and numbered in parameter $Q_i = x_i - y_i$ descending order.

Each solution R_i is represented as a homologous chromosome H_i. The gene order number in a chromosome corresponds to the position order number in the connection field. The gene value conforms to the number of component that placed in the position corresponding to the gene order number. The genes number is equal to n in the chromosome. If the actual number of placed elements is less than the number of positions that "empty elements" is introduced so that the total number of elements be equal to the number of positions.

In the application a various sized elements, having a rectangular shape, are used because any geometrical shape element can be described as a rectangle.

Each element has a base point O^δ_i and base axes $O^\delta_i X^\delta_i$, $O^\delta_i Y^\delta_i$ with respect to which the contour description of the element a_i is define. We assume that the element a_i appointed to the position p_j, if its base point O^δ_i combined with point of connection field with coordinates (x_j, y_j).

For instance, the chromosome

$$<4\ 5\ 6\ 9\ 2\ 3\ 8\ 1\ 7> \tag{1}$$

define the placement represented in fig.1. The intersection points of the dashed lines correspond to element areas within the connection field.

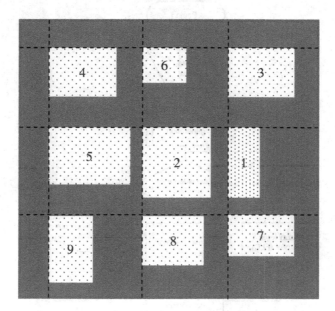

Fig. 1. Example of elements placement

To create the initial population is used the «shotgun» method. It assumes a random selection of alternatives from the search space [5].

To define the target function the total length of the connections between the points which are the gravity centers of the bounding rectangle is calculated.

$$F = \sum_{i=0}^{n}\sum_{j=0}^{n-i} k_{ij} \sqrt{((x_i + (l_i/2)) - (x_j + (l_j/2)))^2 + ((y_i + (h_i/2)) - (y_j + (h_j/2)))^2} \qquad (2)$$

where x_i, x_j, y_i, y_j are positions coordinates within the connection field in which elements i and j are placed; l_i, l_j, h_i, h_j are elements dimensions; k_{ij} is a number of connections between elements i and j; n is a total number of elements.

After the target function calculation genetic operators are applied to the solution population. Let us enumerate these operators. «Roulette wheel» selection is a simple and widely used method. To each element in the population corresponds the area on the roulette wheel, which is proportional to the target function value. Then, when the roulette wheel is rotated an element with larger value of target function has higher probability to select. Also an ordered crossover operator and multipoint mutation operator in which the number cut points proportional to the length of the chromosome are used [5].

Thus, taking into account the use of the fuzzy logic controller (FLC) block diagram of the genetic placement algorithm can be represented as follows (Fig. 2).

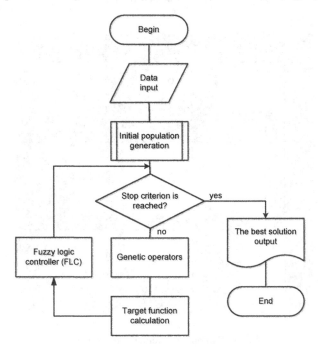

Fig. 2. Block diagram of the genetic placement algorithm

4   Algorithm Parameters' Control

Let us consider the work of fuzzy logic controller. It consists of the following blocks.

The Rule Base. Knowledge used for the correct functioning of the fuzzy control module, written in the form of fuzzy rules that represented as [8]

$R^k:$ **IF** $(x_1 \quad A_1^{\ k}$ **AND** ... **AND** $x_n \quad A_n^{\ k})$ **THEN** $(y \quad B^k)$.

It is also possible to present this knowledge in the form of fuzzy sets with membership function defined by the expression:

$$\mu_{R^k}(x,y) = \mu_{A^k \to B^k}(x,y) \tag{3}$$

The Output Block. The membership function particular form of a fuzzy set is dependent on the applicable T-norms, definitions of fuzzy implication and the representation of fuzzy sets Cartesian product.

The Fuzzification Block. As a results of fuzzification operation to the FLC input parameter $\bar{x} = (\overline{X_1}, \overline{X_2}, ..., \overline{X_n})^T \in X$ put into correspondence the fuzzy set $A' \subseteq X = X_1 \times X_2 \times ... \times X_n$.

The Defuzzification Block. The reflection problem of fuzzy sets $\overline{B_k}$ (or fuzzy set B') into a single value $\bar{y} \in Y$ is solved. This value represents the control action that affects on the FLC control objects. In the proposed algorithm the central average defuzzification is used [9].

The final stage in the FLC design is defining of the representation form of fuzzy sets $A_i^k, i = 1, ... , n; k = 1, ..., N$. For instance, that is a membership function

$$\mu_{A_i^k}(x) = exp\left(-\left(\frac{x_i - \bar{x}_i^k}{\sigma_i^k}\right)^2\right) \tag{4}$$

where parameter $\bar{x}_i^k$ is a centre; σ_i^k is a membership function (Gaussian function) width. These parameters can be changed in the learning process that allows to change fuzzy sets position and structure.

To improve the quality of the genetic search results expert information is included in the evolution circuit by the FLC creation that adjust evolution parameters on the basis of expert knowledge.

As input parameters are used the best, average and worst values of the objective function.

$$e_1(t) = \frac{f_{ave}(t) - f_{best}(t)}{f_{ave}(t)}; e_2(t) = \frac{f_{ave}(t) - f_{best}(t)}{f_{worst}(t) - f_{best}(t)}; \tag{5}$$

$$e_3(t) = \frac{f_{best}(t) - f_{best}(t-1)}{f_{best}(t)}; e_4(t) = \frac{f_{ave}(t) - f_{ave}(t-1)}{f_{ave}(t)}; \tag{6}$$

where t is a number of the current iteration; $f_{best}(t)$ and $f_{best}(t-1)$ are the best value of the target function at iterations t and (t-1) accordingly; $f_{worst}(t)$ is the worst value of the target function at the iteration t; $f_{ave}(t)$ and $f_{ave}(t-1)$ are average value of the target function at iterations t and (t-1) accordingly [9, 10].

The allowed value range for e1, e2, e3, e4 is within intervals: $e_1 \in [0; 1]$; $e_2 \in [0; 1]$; $e_3 \in [-1; 1]$; $e_4 \in [-1; 1]$.

The values of variables e1, e2, e3, e4, affect the probability of performing crossover and mutation operators

$$Pc(t) = Pc(t-1) + \Delta Pc(t), Pm(t) = Pm(t-1) + \Delta Pm(t), \tag{7}$$

5   Software Implementation and Experiments

To develop an application was selected Microsoft Visual Studio 2010. The experiments were carried out. FLC parameters were obtained as a result of the learning algorithm. The set of experiments reflects the dependence of the probability of crossover and mutation operators on the values of e_1 - e_4 that shown at fig. 4 (a–d).

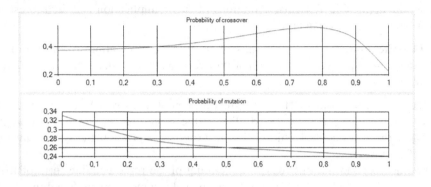

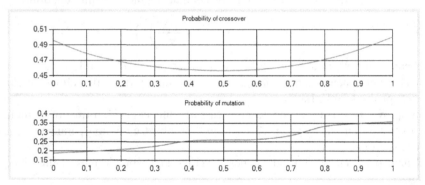

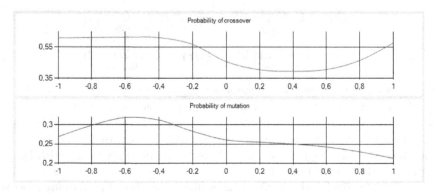

Fig. 3. Dependence of the probability of crossover and mutation operators on the e_1 - e_4

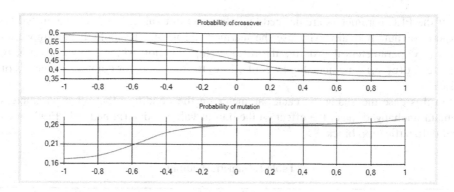

Fig. 3. (*Continued*)

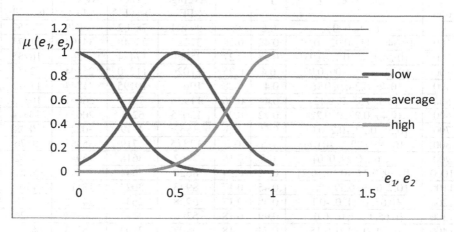

Fig. 4. The membership functions for e_1 and e_2 parameters

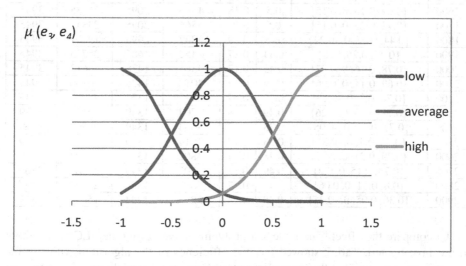

Fig. 5. The membership functions for e_3 and e_4 parameters

If the FLC parameters are defined by default output values need to be adjusted by experts or automatically with use the learning algorithm. By default the algorithm uses FLC parameters that define the membership function which shown in fig.5, 6, $\mu(e_1, e_2)$, $\mu(e_3, e_4)$ - membership level of e_1, - e_2 to fuzzy sets for default parameters of FLC.

In this case the change in each parameter on the condition that other parameters remain unchanged, has little effect on the output value is determined only FLC ratios and defuzzification block y^k.

Table 1. Numerical output

Itera ra-tion	With FLC					Without FLC	
	$E\{e_1, e_2, e_3, e_4\}$	P_c	P_m	Average FF	The best FF	Average FF	The best FF
0	{0, 0, 0, 0}	0.6	0.4	99404	78702	95079	74157
100	{0.34, 0.43, -0.02, -0.01}	0.42	0.28	37220	23748	34104	21471
200	{0.26, 0.45, 0.1, -0.03}	0.38	0.27	25069	17474	39262	16255
300	{0.41, 0.39, 0.09, 0.02}	0.4	0.25	21020	11290	25494	12257
400	{0.46, 0.28, 0, 0.16}	0.42	0.28	19664	10195	31244	11315
500	{0.5, 0.35, 0, 0.12}	0.44	0.26	11936	7919	26061	10314
600	{0.46, 0.27, 0, 0.22}	0.42	0.28	15085	6241	20653	10314
700	{0.56, 0.22, -0.2, 0.01}	0.55	0.3	8562	4719	30229	8808
800	{0.59, 0.29, 0, -0.01}	0.5	0.2	13315	4719	35385	8176
900	{0.64, 0.29, 0, 0}	0.5	0.19	7967	3618	15947	6052
1000	{0.59, 0.27, -0.19, -0.2}	0.52	0.24	8620	3044	20531	5721
1100	{0.7, 0.29, 0.39, -0.29}	0.45	0.17	8943	3027	18118	4560
1200	{0.67, 0.24, 0, -0.1}	0.48	0.17	6738	2634	24955	4560
1300	{0.65, 0.22, 0, -0.01}	0.49	0.18	5390	2193	24192	4326
1400	{0.74, 0.29, 0, 0.15}	0.74	0.18	6407	2193	15263	4326
1500	{0.79, 0.3, 0, 0.27}	0.50	0.18	8130	2105	14342	3945
1600	{0.69, 0.13, 0, 0.15}	0.5	0.18	4316	2095	21785	3945
1700	{0.72, 0.23, 0, 0.15}	0.5	0.18	5566	2095	23842	3945
1800	{0.44, 0.07, 0, -0.22}	0.49	0.19	5047	2095	19647	3945
1900	{0.77, 0.25, 0, 0}	0.44	0.17	4452	1685	25725	3945
2000	{0.78, 0.16, 0, 0.07}	0.42	0.17	9385	1438	14613	3945
2100	{0.69, 0.11, 0, 0.06}	0.49	0.17	4012	1438	20300	3945
2200	{0.74, 0.2, 0, -0.13}	0.37	0.17	6437	1354	25903	3945
2300	{0.71, 0.13, 0, 0.26}	0.49	0.18	7707	1349	29156	3869
2400	{0.77, 0.23, 0, -0.05}	0.38	0.17	5034	1349	26502	3869
2500	{0.73, 0.2, 0, -0.21}	0.33	0.17	3614	1349	19203	3869
2600	{0.79, 0.21, 0, 0.17}	0.48	0.17	4802	1349	20504	3869
2700	{0.73, 0.15, 0, 0.2}	0.49	0.18	4703	1349	19909	3869
2800	{0.8, 0.21, 0, 0.06}	0.41	0.17	7698	1340	28209	3867
2900	{0.36, 0.05, 0, -0.25}	0.48	0.21	2853	1334	15637	3867

To compare the effectiveness the test problems solved using the FLC and without it are investigated. Tab. 2 showed that the efficiency of the algorithm with use the controller is much higher than the efficiency of the algorithm without it. Efficiency of the controller is increased after the introduction of the training unit on the basis of an artificial neural network model. Coefficients parameters (x_{ik}, σ_{ik} и y_k) are given randomly or based on the direct search.

Table 2. Comparison

Number of research	Without FLC $N_{elem}=50$	With FLC $N_{elem}=50$	Without FLC $N_{elem}=100$	With FLC $N_{elem}=100$	Without FLC $N_{elem}=150$	With FLC $N_{elem}=150$
1	4585	3147	29658	21296	67953	48509
2	3870	3330	31145	23582	64311	51737
3	4245	2724	28192	23145	68989	50901
4	4056	3425	31632	23481	65576	50798
5	3774	2885	29761	21844	65184	48973
6	4896	2984	28487	23148	67925	49752
7	4129	2873	31845	22946	65427	52164
8	4812	3776	29145	21941	64964	48862
9	3981	3145	29411	22157	65817	50314
10	3876	3168	30491	22981	68482	50957
Average result	4222,4	3145,7	29976,7	22652,1	66862,8	50296,7
Increase quality of solution	25,6%		24,44%		24,78%	

6   Conclusion

FLC provides a probability value crossover and mutation operators based on the 4 input variables which characterize the process of search. In this case, the dependence of the output values of the input variables is determined by the parameters of the FLC. The number of these parameters depends on the number of input variables, the number of membership functions and parameters which define each membership function.

The studies were conducted for 4 input variables and 3 membership functions for each variable. In addition, each membership function defined by two parameters (center and width of the Gaussian function). Therefore, fuzzy logic controller, when determining the output value handles 28 parameters. Since in our case 2 output parameters were calculated (probability of crossover and mutation), the FLC behavior is determined by the 56 parameters. For an expert it would be problematic manually set these parameters so as to obtain the required behavior of FLC.

In this situation, it is necessary either to develop an interface that would allow to operate the linguistic variables to specify the desired behavior or develop an algorithm for neural network training. FLC parameters that were used in the study were obtained using a genetic algorithm learning. Training was carried out on the basis of statistical information on the dependence of the FLC parameters and the efficiency of the algorithm placement. After training NLC showed sufficiently high efficiency, while during the learning process were obtained depending on the values of the output parameters of the input variables, it would be almost impossible to do based on the experience of the expert.

Acknowledgment. This research is supported by grants of the Ministry of Education and Science of the Russian Federation, the project # 8.823.2014.

References

1. Shervani, N.: Algorithms for VLSI physical design automation, 538 p. Kluwer Academy Publisher, USA (1995)
2. Cohoon, J.P., Karro, J., Lienig, J.: Evolutionary Algorithms for the Physical Design of VLSI Circuits. In: Ghosh, A., Tsutsui, S. (eds.) Advances in Evolutionary Computing: Theory and Applications, pp. 683–712. Springer, London (2003)
3. Herrera, F., Lozano, M.: Fuzzy Adaptive Genetic Algorithms: design, taxonomy, and future directions. J. Soft Computing, 545–562 (2003)
4. Gladkov, L.A.: An integrated algorithm for solving the placement and the track-lock of the basis of fuzzy genetic methods. J. Izvestiya SFedU. Engineering Sciences 120, 22–30 (2011)
5. Gladkov, L.A., Kureichik, V.V., Kureichik, V.M.: Genetic algorithms. Fizmatlit, Moscow (2010)
6. Kureichik, V.M.: Modified genetic operators. J. Izvestiya SFedU. Engineering Sciences 12, 7–15 (2009)
7. Gladkov, L., Gladkova, N., Leiba, S.: Manufacturing scheduling problem based on fuzzy genetic algorithm. In: Proceeding of IEEE East-West Design & Test Symposium – (EWDTS 2014), Kiev, Ukraine, pp. 209–212 (2014)
8. Yarushkina, N.G.: Foundations of the theory of fuzzy and hybrid systems. Finance and Statistics, Moscow (2004)
9. Batyrshin, I.Z., Nedosekin, S.A.: Fuzzy hybrid systems. Theory and practice. Fizmatlit, Moscow (2007)
10. Gladkov, L.A., Gladkova, N.V.: New approaches to the construction of systems analysis and knowledge extraction based on hybrid methods. J. Izvestiya SFedU. Engineering Sciences. 108, 146–154 (2010)

Trends in the Sensor Development

Frantisek Hruska

Tomas Bata University, Faculty of Applied Informatics,
Nad Stranemi 4511, CS 76005 Zlin, Czech Republic
hruska@fai.utb.cz

Abstract. The current cyberspace also uses intensive development of sensors. The motivation are the requirements on control and informatics systems, security applications, monitoring and environmental protection, energy saving, health requirements etc. In the paper there are described the large-scale innovations of the sensor area for measuring the composition and concentration of the gas mixture, modern temperature measurement sensors and MEMS.

Keywords: Sensor, Transmitting, MEMS.

1   Introduction

Research and development in the field of sensors is highly developed. It is supported by research, materials, technology development, especially semiconductor technologies, the development of electronics etc. Development relates to standard sensors, e.g. for temperature, pressure, flow, etc. but also development of a fully new principles. Examples are sensors for non-contact temperature, sensors for measuring the concentration and composition of gas mixtures, ISFET sensors for measuring the concentration of dissociated ions and MEMS sensors. [1,2]

The basic principles used by the new sensors are known for their novelty and take on new material materials, new procedures for evaluating the output signals from the sensors, new technology. [3]

2   Temperature Sensor

As an example of the trends of new sensors is subsequently analyzed thermopile sensor (it enabled to extend the contactless measuring temperature), platinum temperature sensors (increasing the range and sensitivity), semiconductor temperature sensors and humidity sensors (extension for the indirect measurement of other parameters).

2.1   Subsection Heading

Semiconductor technologies facilitate to produce modern sensors at the thermocouple principle and expand the contactless temperature measurement. This sensors use a thermopile to absorb the passive infrared energy emitted from the object being

© Springer International Publishing Switzerland 2015

R. Silhavy et al. (eds.), *Intelligent Systems in Cybernetics and Automation Theory*,
Advances in Intelligent Systems and Computing 348, DOI: 10.1007/978-3-319-18503-3_5

measured and use the corresponding change in thermopile voltage to determine the object temperature. The thermopile voltage is digitized and reported with the temperature through output signals. The thermopile sensors are demonstrated in the fig.1.

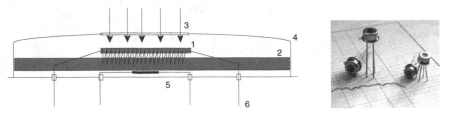

Fig. 1. Schema of thermopile (1-active part, 2-referency part, 3-window, 4-cover, 5- ambient temperature sensor, 6-output pins) and its real performance [4]

Infrared sensor is used in a wide range of applications. Low power consumption along with low operating voltage makes the device suitable for battery-powered applications. The low package height of the chip-scale format enables standard high-volume assembly methods, and can be useful where limited spacing to the object being measured is available. The sensor is cheap, it has the very low price, it can use by normal ambient temperature without cooling.

The figure 3 shows a thermopile sensor and application conditions during non-contact temperature measurement of the body surface. The energy electromagnetic radiation from the surface of the body (1) of its surface (2) is directed into the sensor (3) for contactless temperature measurement the angle by the optical system.   There it is the electromagnetic radiation energy (E) according to the temperature of the wall surface (T1) and the energy Pr, the energy reflected from the external environment (8). The sensor has at the input side the radiation optical element (4) which generates sensing geometrical conditions, i.e. the angle of the external sensing ($\varphi 1$) and the inner angle ($\varphi 2$) pointing towards the sensor (5). The incident radiation on the sensor is converted to an electrical signal which is processed in electronics (6) for indication of the surface temperature and performs with output (7) to other systems and devices. The fig.2 has a photo of exemplary devices at the measurement.

The performance of the radiant energy from the body surface in the direction of the sensor in the spatial angle ($\varphi 1$) is given by:

$$Q_t = A_t . \delta . T_t^{4} \varphi_1 . \varepsilon_t + P_r \tag{1}$$

where Q_t is the power of the radiation energy from the wall, power P_r reflected energy radiation emission, ε_t emission factor of area, A_t area of the surface radiating to the sensor, $\varphi_1 = A_t / l^2$ the spatial angle according to area A_t a distance l.

The energy passes through the optics of the sensor (4) and is absorbed in the face of the sensor (5) as follows:

$$Q_s = Q_t . p_4 . a_s \tag{2}$$

where is Q_s incidented and absorbed energy on the sensor, Q_t performance of radiant energy from the wall surface A_t, p_4 coefficient of penetration through the optics, a_s the absorption coefficient of the sensor surface.

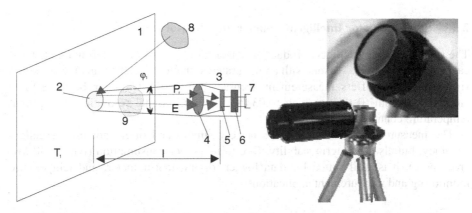

Fig. 2. Scheme of the thermopile sensor application and real view at equipment [4]

The sensor converts the incident electromagnetic radiation energy into electric voltage given its constant sensor units (V / W).

Next electronics circuits subsequently evaluate the surface temperature after putting all equations and adjusting to the formula:

$$T_t = \sqrt[4]{\left(\frac{U_{s,out}}{p_4 a_s k_s} - P_r\right)\left(\frac{1}{A_t.\delta.\varphi_1.\varepsilon_t}\right)} \qquad (3)$$

2.2   Metal Temperature Sensors

The metal sensor for the platinum is known very long period. The current demands are wider range of measurement and higher sensitivity. The standard Pt100 is changed to up to Pt1000 or more. There is used a new technology.

Table 1. Date of Pt sensors

Typ	TCR (ppm/K)	R_0 (Ω)	$\Delta R/K$ (Ω/K)
Pt100	3850, 3910 (USS), 3900 (BS), 3905 (GOST)	100	0,385
Pt1000	3850, 3910 (USS), 3900 (BS),	1000	3,85
Pt10000	3850, 3910 (USS), 3900 (BS),	10000	38,5
Pt100000	3850, 3910 (USS), 3900 (BS),	100000	385

The high sensitivity is possible by higher nominal resistance Ohm at 0°C. For example it is the Pt1000, Pt10000 or Pt100000 too. The downed table 2 shows some date.

2.3   Semiconductor Intelligent Temperature Sensor

The semiconductor intelligent product of sensor family consists of chip-integrated and calibrated temperature sensors with an integrated signal converter for analog or digital signal output. It offers measurement accuracy from ±0.5°C to far below ±0.1°C and also medium (0.1°C) to high (0.034°C, 0.004°C) signal resolution for optimal temperature control.

The measurement with the sensors is very simple and offers not only excellent accuracy, but also long-term stability. Due to its power consumption (typical 30 Microampere), it is well suited for data logger, digital thermometer, and temperature monitoring and measurement applications.

3   Sensors of Measuring the Composition and Concentration of the Gas Mixture

Area of sensors for measuring the concentration and composition of the gas mixture takes motivation of the requirements for air quality control interiors (Indoor Air Quality - IAQ), monitoring the concentration of CO_2 and VOC in the environment of interior and exterior. Indoor air quality is an important parameter of hygiene. Tracks on the one hand, concentration of CO_2, concentration mixtures of volatile hydrocarbons (Volatile Organic Compound-VOC) and on the other hand the concentration of toxic and hazardous gas mixtures (CO incomplete combustion, heating gas leak and explosion hazards related).

3.1   Catalytic Sensors

The group of sensors based on measuring the heat producing in the catalytic reaction is catalytic sensors. Measured gas (combustible and able to oxidize in the presence of a catalyst) during the catalytic reaction was oxidized and heat is generated. Schematic diagram of the catalytic cell and the pellistor is shown in Fig.3.

Performing a) contains the measuring chamber (1) with a measuring hot fiber (3) for the gas being measured. The second chamber (4) is filled with a reference gas and also contains a measuring fiber (2). Measuring fiber is preheating to the operating temperature according to the type of the measured gas is pure platinum or its alloy, which acts as a catalyst in oxidation e.g. H2, CO, CH4, and other oxygenated gases. The surface temperature must be high and reach a certain value according to the type of gas, e.g. H2 + 200 ° C, CO + 450 ° C, ethane, butane, propane + 500 to 550 °C, methane + 800 ° C. Another requirement is a sufficient supply of oxygen.

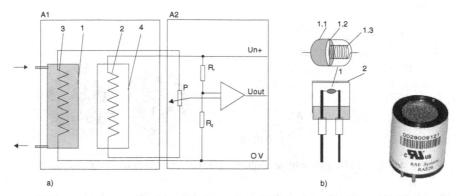

Fig. 3. Scheme of the electro catalytic sensor (1 – measuring cuvette, 2- hot catalytic wire, 3- measured gas) and real view on the sensor [9]

In addition to oxidation heat is generated. The electric resistance of the hot wire is changed according to the concentration of oxidized gas. Evaluating the measurement is performed in the transmitter (A2). Measuring bridge is formed by fibers (2, 3) and the resistors R1 and R2 and potentiometer P. The voltage at the diagonal is amplified into an corresponding output signal Uout.

Sensor according to the performance (b) is a version for a contact catalytic principle. The chamber has a bead (1) which has contact entering gas being measured through the sieve (2). The bead is consisted of the catalyst mass (1.1, ceramic separator (1.2) and incandescent 1.3). The measurement gas is oxygenated, it creates the heat, which changes the value of electrical resistance of incandescent. These changes are a measure of the concentration of combustible gas.

Electro catalytic sensor (pellistor in English) is produced as an embedded version, see the down image sensor (1). The content of combustible hydrocarbon gases or vapors of liquid hydrocarbons in the air can be measured in the ranges from the lower explosive limit. The measurement can be further CO content to 1000 ppm, the content of NH_3 in the range up to 5000 ppm, hydrogen concentration to 5000 ppm or H_2S content in the range of 100 ppm.

3.2   Spectral Analysis Sensor

Spectral analysis allows measuring the composition and properties of gases on the basis of changes in the absorption of electromagnetic radiation. Absorption of radiation in the non-ionized gas is in the wavelength band which is the gas able to emit in the ionized state. Absorption bands of the most gases are in the narrow strips predominantly in the ultraviolet (UV) and infrared (IR) range optionally in the visible (VIS) regions. The band VIS is used only a small number of gas to absorb the visible region of the spectrum.

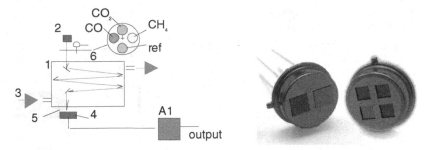

Fig. 4. Scheme of sensor with absorption of IR range, and photo of sensors [4]

Absorption monochromatic radiation in a gas is governed by Lambert equation:

$$e^{k.l.c} = \left(\frac{I}{I_{out}} \right) \tag{4}$$

where is k absorption coefficient depended on the wavelength, type of gas and concentration, l path length of the gas jet, c gas concentration, I intensity of input beam (W / m-2), I_{out} beam intensity at the output (W / m-2).

Scheme of the analyzer using the principle of absorption of electromagnetic radiation is shown in Fig.4. The radiation source (2) being according to the type of radiation is LED lamp with a small range of wavelengths corresponding to the intensity of the input radiation. The rays are guided into the chamber (1) and are typically multiple reflections on the walls. Their intensity is reduced at the gas and it transmits through the filter (5) to a radiation sensor (4). Signals from the sensor are evaluated in the evaluation unit (A1).

Most sensors for industrial devices use non-dispersive infrared radiation (NDIR). Method of NDIR sends the entire spectrum from the radiation source (1) into the measuring chamber (2) and further through the filter (3) on the sensor (4). In the method DIR (b) extends from a radiation source (1) via a prism (5) which disperses by sub wavelength. Only a specific range of wavelengths passes the measuring chamber then.

The NDIR absorption method utilizes the fact that certain gas molecules make absorption of radiation at the particular wavelength (so-called spectral line) which passes through it. With proper choice of radiation source or filters in the sensor, this method allows high selectivity identify and measure the composition of the particular gas in a gas mixture (qualitative method). The absorption band for each gas can be located in UV, VIS and IR wavelengths. This fact is shown for the IR range in fig.5.

Measurement accuracy class for this method is 1.5 according to the measuring range. Negative effects create pressure changes and gas flow in the cell, ambient temperature.

A

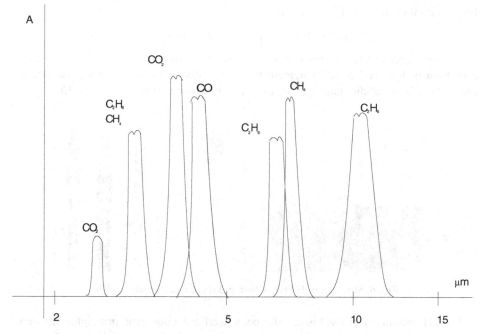

Fig. 5. Graphs of absorption electromagnetic radiation for the IR range for gases

4   Sensors of Liquid Composition and Properties

4.1   ISFET Sensor

Drinking water, surface and waste water, analysis in medicine, cosmetics, environmental, agricultural and various industrial technologies (e.g. in the production of iron
cyanide, calcium phosphate, paper, pharmaceuticals, food) need to determine pH in
liquid solutions.

The pH is determined by the concentration hydrogen ions. Clean water is dissociated to hydroxon (H3O +) and hydroxyl (OH-) ions according to the equation:

$$2H_2O=(H_3O^+)\ +(OH^-) \tag{5}$$

In the neutral environment there are the same number hydroxyl and hydroxon ions.
The solution of 1 liter at 25 ° C has as measured the number of gram-ions 1,004.10-7.
Furthermore, the concentration of both ions is equal to the constant:

$$C_{H+} \cdot C_{OH-} = K_v \tag{6}$$

where is $K_v=1,008.10^{-14}$ as a ion water coefficient.

For indicating the degree of acidity or basicity has been introduced calling "hydrogen exponent pH according to the formula:

$$pH=-\log C_{H+} \tag{7}$$

For a neutral environment is therefore:

$$pH(neutral)=-\log(C_{H+})=-\log(K_v/2)=-\log(1,004.10^{-7})=7 \qquad (8)$$

The above relationship means that an increase of the hydrogen ions reduces the concentration of hydroxyl ions. By agreement there is distributed environment for neural state (pH = 7), for acidic state (pH <7 to 0) and for alkaline state (pH> 7 to 15).

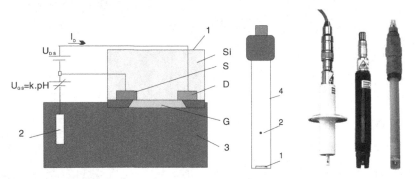

Fig. 6. Structure of ISFET sensor and its real performance [10]

For pH measurement has historically been used potentiometric principle. The new sensor is ISFET transistor. In fig.6 has the transistor (1) the standard electrode D, S, G. The electrode G is formed from SiO2, Si3N4, Al2O3 or Ta2O5, which is sensitive to H^+ ions. The pH change is manifested as a change in potential between the electrode S and the external reference electrode (2) and affects the conductivity of the DS transistor. The second part of the figure gives a realistic view at the electrode performance. The electrodes are shown in fig.7.

Fig. 7. Location the ISFET sensor in the pH electrode [10]

The advantage of the ISFET electrode is a plastic material which is mechanically very resistant and is operationally reliable, and can operate in a highly contaminated environment. The ISFET has a small time constant, always accurate in long time period, it is able sterilize, can be stored without water and resist for acids and alkalis.

The ISFET sensor can have other gate materials and so can have selectivity to other ions. The sensor is ion-selective. E.g. special material with Si for gate G gives selectivity to ions Na+, Ag+, with chalkogenits defined the selectivity to Pb_2^+, Cb_2^+.

4.2   Dissolved Oxygen Sensor

Reliable and continuous measurement of dissolved oxygen plays an important role especially in water management. The current measured values are necessary for safe monitoring and dynamic process control. [10]

The concentration of oxygen in water is a function of its partial pressure:

$$c_{(O_2)} = P_{(O_2)} \cdot \frac{k_a . M_{O_2}}{V_m} \tag{9}$$

where is $C_{(02)}$ oxygen concentration (mg/l), $p_{(02}$ partial pressure of oxygen (bar), k_a absorption coefficient (1/bar), M_{O2} the molecular weight of oxygen (32 g), Vm molar volume of oxygen (22,4 l). The table shows the oxygen content in water at $p_{vz} = 101$ kPa at 100% saturation.

Table 2. The oxygen content in water at $p_{vz} = 101$ kPa at 100% saturation

ta °C	0	5	10	15	20	25
mg/l	14,16	12,57	10,92	9,76	8,84	8,11

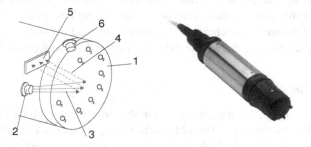

Fig. 8. Scheme of sensor with luminescence optic and view at electrode [10]

Sensors for dissolved oxygen content use electrode and new optical sensors working according to the principle of luminescence. Scheme is shown in fig.8. The oxygen molecules penetrate through the membrane (1) to the area where the rays of the LED (2) having a blue color (3). In contact with the O2 molecule the wavelength of radiation increases to red color. Red beams (4) are photoelectric sensed at sensor (5), followed by evaluation of the oxygen concentration. For calibration there are used a reference electrode (6).

5   MEMS Sensors

5.1   Accelerometer Sensor

Structure and function of the MEMS accelerometer is based on a variable capacity of three electrodes of capacitance. [11]There is used a known non-linear dependence of capacitance C according to the capacitor electrode separation x   (air gap). So if the one electrode has the floating movement it will depend on the operating acceleration, we got a capacitive accelerometer. With the actual MEMS structure is one of several parallel electrodes. The principle and the equivalent circuit is in Fig.9.

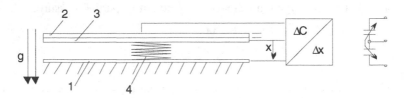

Fig. 9.   Function scheme of the MEMS accelerometer (1-base electrode, 2-secondary electrode, 3-moved electrode, 4-spring, g-acceleration, x-way of moved electrode, ΔC-change of capacitance, Δx-change of way of moved electrode)

Convert of the changes ΔC according to   Δx at ΔU is implemented in an ASIC electronics. The capacitance measurement uses the methods of switched capacitors controlled digital logic and switching the clock signal generator. It is then filtered again and linearized circuits with switched capacitors, and finally performs temperature compensation. The result is a linear, amplified and compensated voltage signal with a defined conversion constant - sensitivity disclosing how much needs to change the value of the measured acceleration, in order to change the output voltage of 1 V (value g / V). Everything is done separately for each axis sensing (main 3 channels).

5.2   Gyroscope Sensor

A gyroscope is a device for measuring or maintaining orientation, based on the principles of angular momentum.   The old mechanical gyroscopes typically comprise a spinning wheel or disc in which the axle is free to assume any orientation. The new gyroscopes based on electronic principles. Inexpensive vibrating structure gyroscopes manufactured with MEMS technology have become widely available. These are packaged similarly to other integrated circuits and may provide either analog or digital outputs. In many cases, a single part includes gyroscopic sensors for multiple axes. Some parts incorporate multiple gyroscopes and accelerometers (or multiple-axis gyroscopes and accelerometers), to achieve output that has six full degrees of freedom.

The sensor consists a resonating structure (1), the frames 2, 3,   which moves under its own mechanical resonance represented by springs (springs 5) in drive direction - perpendicular to the direction of rotation (see . Fig.10). This results in a Coriolis force proportional to the angular velocity of rotation, which compresses the outer frame. The fingers (6) are the Coriolis sense in the sensor and functioning as an electrode capacitors. The output is therefore proportional to the capacity change of angular velocity of rotation.

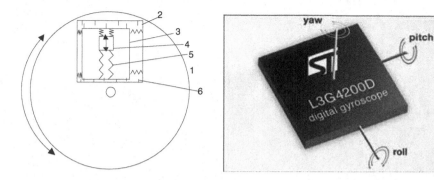

Fig. 10. Scheme of gyroscope principle and view at real sensor [12]

6   Conclusion

The paper shows the trends in the field of sensors choosing some examples of the new sensors. It pointed to specific parameters and application options. The selection is voted so that it was pointed out to significant developmental milestones of sensors.

References

1. Webster, J.G.: The measurement, instrumentation, and sensor handbook. CRC Press LLC; Springer-Verlag, New York (1999)
2. Fraden, J.: Handbook of Modern Sensors. Physics, designs, and Applications. Springer, New York (1996)
3. Hruska, F.: Senzory pro systémy informatiky a automatizace (Sensors for informatics and control system). UTB, Zlín (2007)
4. EG&G Heimann, http://www.heimannsensor.com/
5. Measurement Specialties, http://www.meas-spec.com/temperature-sensors/thermopile-infrared-sensors-modules/thermopile-infrared-ir-sensors.aspx
6. Excelitas Technologies, http://www.excelitas.com/downloads/DTS_A2TPMI.pdf
7. Innovative sensor technology, http://www.ist-ag.com/eh/ist-ag/
8. Sensirion, http://www.sensirion.com/
9. RAE Systems, http://www.raesystems.com/solutions/sensors
10. Hach-Lange, http://www.hach-lange.cz
11. Freescale Semiconductor, http://www.freescale.com
12. Analog Devices, http://www.analog.com
13. Bosch Sensortech, http://www.bosch-sensortec.com

Robust Stability Analysis for Families of Spherical Polynomials

Radek Matušů and Roman Prokop

Centre for Security, Information and Advanced Technologies (CEBIA – Tech)
Faculty of Applied Informatics, Tomas Bata University in Zlín,
nám. T. G. Masaryka 5555, 760 01 Zlín, Czech Republic
{rmatusu,prokop}@fai.utb.cz

Abstract. The families of spherical polynomials provide not very commonly utilized definition of uncertainty bounding set for systems with parametric uncertainty. The principal aim of this contribution is to present such spherical polynomial families, their description and related robust stability analysis. The illustrative example demonstrates an easy-to-use graphical method of robust stability investigation theoretically based on the value set concept and the zero exclusion condition and practically performed through the Polynomial Toolbox for Matlab.

Keywords: Spherical Uncertainty, Weighted Euclidean Norm, Robust Stability Analysis, Value Set Concept, Zero Exclusion Condition.

1 Introduction

Parametric uncertainty is frequently used tool for description of real plants as it allows using relatively simple and natural mathematical models for processes which true behaviour can be much more complex. The structure of the models with parametric uncertainty is considered to be fixed, but its parameters can lie within given bounds. Within this contribution, these bounds are going to be considered in not so commonly used way.

The "classic" approach assumes the bounds in the shape of a box. Here, the alternative approach, which employs the bounds in the shape of a sphere (ellipsoid), is going to be studied. The scientific literature contains much more works related to the standard "box" uncertainties than to the spherical ones. However, some basic information as well as possible extensions and various applications can be found e.g. in [1] – [5].

This contribution deals with polynomials with parametric uncertainty and spherical uncertainty bounding set. More specifically it is focused on description of a spherical polynomial family and tools for testing its robust stability while a special attention is paid to very universal graphical tool based on the combination of the value set concept and the zero exclusion condition [1]. The presented ideas are followed by an illustrative example supported by plots from the Polynomial Toolbox for Matlab [6], [5]. The previous version of this work has been already presented in the conference contribution [7].

© Springer International Publishing Switzerland 2015
R. Silhavy et al. (eds.), *Intelligent Systems in Cybernetics and Automation Theory*,
Advances in Intelligent Systems and Computing 348, DOI: 10.1007/978-3-319-18503-3_6

57

2   Uncertainty Bounding Set

The systems with parametric uncertainty can be described though a vector of real uncertain parameters (or just uncertainty) q. The continuous-time uncertain polynomial, which is a typical object of researchers' or engineers' interest, can be written in the form:

$$p(s,q) = \sum_{i=0}^{n} \rho_i(q)s^i \tag{1}$$

where ρ_i are coefficient functions.

Then, so-called family of polynomials combines together the structure of uncertain polynomial given by (1) with the uncertainty bounding set Q. Therefore, the family of polynomials can be denoted as:

$$P = \{ p(\cdot, q) : q \in Q \} \tag{2}$$

The uncertainty bounding set Q is typically given in advance by user requirements. It is supposed as a ball in an appropriate norm. The commonly used case employs L_∞ norm:

$$\|q\|_\infty = \max_i |q_i| \tag{3}$$

which means that a ball in this norm is a box. Practically, this box is defined componentwisely, i.e. by the real intervals which the uncertain parameters can lie within.

Another approach utilizes L_2 (Euclidean) norm:

$$\|q\|_2 = \sqrt{\sum_{i=1}^{n} q_i^2} \tag{4}$$

or more generally the weighted Euclidean norm:

$$\|q\|_{2,W} = \sqrt{q^T W q} \tag{5}$$

where $q \in \mathbf{R}^k$ and W is a positive definite symmetric matrix (weighting matrix) of size $k \times k$. Such definition of Q means that a ball in the norm can be referred as a sphere, or more generally as an ellipsoid. Under assumption of $r \geq 0$ and $q^0 \in \mathbf{R}^k$, the ellipsoid (in $\mathbf{R}^k$) which is centered at q^0 can be expressed by means of:

$$\left(q - q^0 \right)^T W \left(q - q^0 \right) \leq r^2 \tag{6}$$

or equivalently by:

$$\left\| q - q^0 \right\|_{2,W} \leq r \tag{7}$$

The ellipsoid can be easily visualized in two-dimensional space ($k = 2$) for:

$$W = \begin{bmatrix} w_1^2 & 0 \\ 0 & w_2^2 \end{bmatrix} \tag{8}$$

as it is shown in Fig. 1 [1].

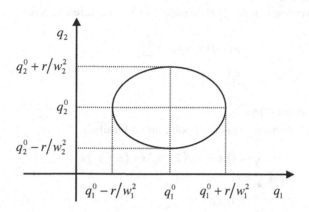

Fig. 1. Ellipsoid defined by weighted Euclidean norm

Decision on what type of norm should be used for uncertainty bounding set Q depends on several factors. In many engineering problems, the real uncertain physical parameters are independent on each other and thus Q should be a box naturally. However, according to [1], the ellipsoids could be useful and justifiable under "imprecise description" of the uncertainty bounds, i.e. if actual Q is located between some minimum and maximum and a suitable ellipsoid can interpolate them. The choice should respect also available tools for solving the specific problem. Besides, the mathematical models obtained on the basis of physical laws usually have Q in the shape of box, but the identification methods mostly lead to the ellipsoids [8].

3   Spherical Family of Polynomials

The family of polynomials given by (2) is called spherical one [1] if $p(s,q)$ has an independent uncertainty structure (i.e. all coefficients of the polynomial are independent on each other) and Q is an ellipsoid.

In fact, one can work with two basic representations of spherical polynomial families.

The first type assumes that polynomial is centered at zero:

$$p(s,q) = \sum_{i=0}^{n} q_i s^i$$

$$\left\| q - q^0 \right\|_{2,W} \le r$$

(9)

where W is a positive definite symmetric matrix, q_0 is the nominal and $r \ge 0$ means the radius of uncertainty.

In the second representation, Q is considered to be centered at zero:

$$p(s,q) = p_0(s) + \sum_{i=0}^{n} q_i s^i$$

$$\left\| q \right\|_{2,W} \le r$$

(10)

where moreover $p_0(s) = p(s,q^0)$.

As an example, suppose a spherical polynomial family:

$$p(s,q) = (1 + q_0) + (2 + q_1)s + (3 + q_2)s^2$$

$$\left\| q \right\|_{2,W} \le 1$$

$$W = \begin{bmatrix} 4 & 0 & 0 \\ 0 & 5 & 0 \\ 0 & 0 & 6 \end{bmatrix}$$

(11)

which can be centered on the vector:

$$\tilde{q}^0 = (1, 2, 3)$$

(12)

Then, the resulting spherical polynomial family, equivalent to (11) can be written as:

$$p(s,q) = \tilde{q}_0 + \tilde{q}_1 s + \tilde{q}_2 s^2$$

$$\left\| \tilde{q} - \tilde{q}^0 \right\|_{2,W} \le 1$$

$$W = \begin{bmatrix} 4 & 0 & 0 \\ 0 & 5 & 0 \\ 0 & 0 & 6 \end{bmatrix}$$

(13)

4   Robust Stability Analysis

The most critical feature of control loops is their stability. Under presence of parametric uncertainty, this term can be expanded to robust stability, which means that the whole family of closed-loop control systems must remain stable for all possible allowed perturbations of system parameters.

From viewpoint of practical testing, we are interested in the robust stability of the family of closed-loop characteristic polynomials in the form (2). This family is robustly stable if and only if $p(s,q)$ is stable for all $q \in Q$, i.e. all roots of $p(s,q)$ must be located in the strict left half of the complex plane for all $q \in Q$.

There are many results for robust stability analysis of systems with parametric uncertainty for Q in a shape of box. Their choice depends primarily on the complexity of the structure of investigated polynomial (or system). Doubtless, the most famous tool is the Kharitonov theorem [9] which is suitable for investigation of robust stability of interval polynomials (with independent uncertainty structure). Moreover, several modifications and generalizations of classical Kharitonov theorem are also available in literature [1], [10]. Other known tools are the edge theorem, the thirty-two edge theorem, the sixteen plant theorem, the mapping theorem, etc. [1]. Furthermore, it exists a graphical method which is applicable for wide range of robust stability analysis problems (from the simplest to the very complicated uncertainty structures, for various stability regions, etc.). This technique combines the value set concept with the zero exclusion condition [1], [11].

Robust stability analysis for systems affected by parametric uncertainty for the case of Q in a shape of ellipsoid is also relatively well developed and there are several methods available. The Soh-Barger-Dabke theorem [12], [1] represents analogical tool to Kharitonov theorem for spherical polynomial families. Furthermore, extensions are provided by the theorem of Barmish and Tempo [13], [1] based on the idea of the spectral set and the theorem of Biernacki, Hwang and Bhattacharyya [14], [1] which solves the robust stability for closed-loop system with affine linear uncertainty structure (e.g. a spherical plant family in feedback connection with a fixed controller).

Nonetheless, the very universal technique based on the value set concept and the zero exclusion condition, which is described in [1], is applicable also to the spherical polynomial families.

The value set at each frequency ω for a spherical polynomial family (2) supposed in the form:

$$p(s,q) = p_0(s) + \sum_{i=0}^{n} q_i s^i$$
$$p_0(s) = p(s,q^0) = \sum_{i=0}^{n} a_i s^i \quad\quad (14)$$
$$\|q\|_{2,w} \leq r$$
$$\deg p(s) \geq 1$$

is given [1], [15] by an ellipse centered at nominal $p_0(j\omega)$, with major axis (in the real direction) having length:

$$R = 2r \left(\sum_{i \ even} w_i^2 \omega^{2i} \right)^{1/2} \quad\quad (15)$$

and with minor axis (in the imaginary direction) having length:

$$I = 2r \left(\sum_{i \, odd} w_i^2 \omega^{2i} \right)^{1/2} \tag{16}$$

where W is a weighting matrix:

$$W = diag \left(w_1^2, w_2^2, \ldots, w_n^2 \right) \tag{17}$$

Moreover, for the special degenerate case of $\omega = 0$, the value set is just the real interval:

$$p(j0,Q) = \langle a_0 - r, a_0 + r \rangle \tag{18}$$

The practical visualization of the ellipsoidal value sets can be conveniently performed through the Polynomial Toolbox 2.5 [6], [15], [5] by using the "spherplot" command.

Then, the zero exclusion condition can be applied for robust stability investigation in the following way: The spherical polynomial family (2) with invariant degree and at least one stable member (e.g. nominal polynomial) is robustly stable if and only if the complex plane origin is excluded from the value set $p(j\omega,Q)$ at all frequencies $\omega \geq 0$, i.e. the spherical polynomial family is robustly stable if and only if:

$$0 \notin p(j\omega,Q) \quad \forall \omega \geq 0 \tag{19}$$

Generally, the detailed description, proofs and examples of the zero exclusion principle applications can be found in [1] or for instance in [8], [11].

5   Illustrative Example

Consider the spherical polynomial family defined by the uncertain polynomial:

$$p(s,q) = (1+q_4)s^4 + (2+q_3)s^3 + (3+q_2)s^2 + (2+q_1)s + (1+q_0) \tag{20}$$

and by the uncertainty bounding set:

$$\|q\|_{2,w} \leq 0.3$$
$$W = diag(5,4,3,2,1) \tag{21}$$

i.e.:

$$5q_0^2 + 4q_1^2 + 3q_2^2 + 2q_3^2 + q_4^2 \leq 0.3^2 \tag{22}$$

The polynomial (20) can be easily expressed in the form (14) as:

$$p(s,q) = s^4 + 2s^3 + 3s^2 + 2s + 1 + q_4 s^4 + q_3 s^3 + q_2 s^2 + q_1 s + q_0 \tag{23}$$

The nominal polynomial is stable and so the family fulfills the condition of at least one stable member. The value sets for the range of frequencies from 0 to 3 with step 0.01 was obtained with the assistance of the Polynomial Toolbox 2.5 for Matlab and its routine "spherplot" [6], [15]. They are plotted in Fig. 2.

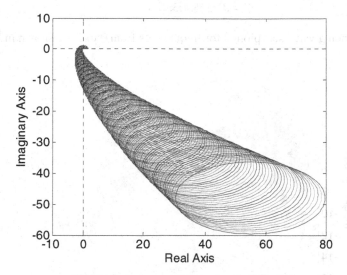

Fig. 2. The value sets for the family (20), (21)

The zoomed version of the same value sets, visualized in Fig. 3, provides better view of the neighborhood of the complex plane origin.

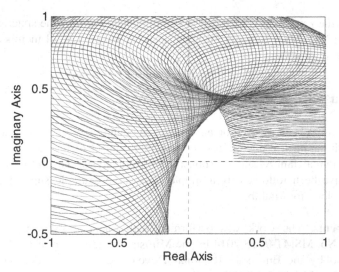

Fig. 3. The value sets for the family (20), (21) – detailed view near the point [0, 0j]

As can be observed, the origin of the complex plane is excluded from the value sets which means that the spherical polynomial family (20), (21) is robustly stable.

Now, suppose that the uncertainty bounding set (21) changes to:

$$\|q\|_{2,W} \leq 0.5$$
$$W = diag\,(5,4,3,2,1)$$
(24)

The corresponding value sets plotted for frequencies from 0 to 2 are shown in Fig. 4.

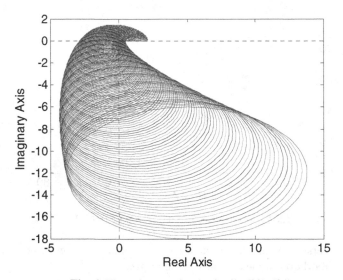

Fig. 4. The value sets for the family (20), (24)

In this case, the zero point is included in the value sets and thus the spherical polynomial family (20), (24) is robustly unstable. In other words, not all members of the prescribed family are stable.

6   Conclusion

The contribution has been focused on an alternative bounding of uncertain parameters in systems with parametric uncertainty, i.e. the main object of interest has been the spherical polynomial family and its robust stability analysis. The basic theoretical descriptions have been followed by a simple illustrative example supported by the Polynomial Toolbox for Matlab.

Acknowledgments. The work was performed with financial support of research project NPU I No. MSMT-7778/2014 by the Ministry of Education of the Czech Republic and also by the European Regional Development Fund under the Project CEBIA-Tech No. CZ.1.05/2.1.00/03.0089.

References

1. Barmish, B.R.: New Tools for Robustness of Linear Systems. Macmillan, New York (1994)
2. Tesi, A., Vicino, A., Villoresi, F.: Robust Stability of Spherical Plants with Unstructured Uncertainty. In: Proceedings of the American Control Conference, Seattle, Washington, USA (1995)
3. Polyak, B.T., Shcherbakov, P.S.: Random Spherical Uncertainty in Estimation and Robustness. In: Proceedings of the 39th IEEE Conference on Decision and Control, Sydney, Australia (2000)
4. Chen, J., Niculescu, S.-I., Fu, P.: Robust Stability of Quasi-Polynomials: Frequency-Sweeping Conditions and Vertex Tests. IEEE Transactions on Automatic Control 53(5), 1219–1234 (2008)
5. Hurák, Z., Šebek, M.: New Tools for Spherical Uncertain Systems in Polynomial Toolbox for Matlab. In: Proceedings of the Technical Computing Prague 2000, Prague, Czech Republic (2000)
6. PolyX: The Polynomial Toolbox, http://www.polyx.com
7. Matušů, R.: Families of spherical polynomials: Description and robust stability analysis. In: Latest Trends on Systems (Volume II) – Proceedings of the 18th International Conference on Systems, Santorini, Greece, pp. 606–610 (2014)
8. Šebek, M.: Robustní řízení (Robust Control). PDF slides for course "Robust Control", ČVUT Prague, http://dce.felk.cvut.cz/ror/prednasky_sebek.html or http://www.polyx.com/_robust/ (in Czech)
9. Kharitonov, V.L.: Asymptotic stability of an equilibrium position of a family of systems of linear differential equations. Differentsial'nye Uravneniya 14, 2086–2088 (1978)
10. Bhattacharyya, S.P., Chapellat, H., Keel, L.H.: Robust control: The parametric approach. Prentice Hall, Englewood Cliffs (1995)
11. Matušů, R., Prokop, R.: Graphical analysis of robust stability for systems with parametric uncertainty: an overview. Transactions of the Institute of Measurement and Control 33(2), 274–290 (2011)
12. Soh, C.B., Berger, C.S., Dabke, K.P.: On the stability properties of polynomials with perturbed coefficients. IEEE Transactions on Automatic Control 30(10), 1033–1036 (1985)
13. Barmish, B.R., Tempo, R.: On the spectral set for a family of polynomials. IEEE Transactions on Automatic Control 36(1), 111–115 (1991)
14. Biernacki, R.M., Hwang, H., Bhattacharyya, S.P.: Robust stability with structured real parameter perturbations. IEEE Transactions on Automatic Control 32(6), 495–506 (1987)
15. PolyX: The Polynomial Toolbox for Matlab – Upgrade Information for Version 2.5 (2001), http://www.polyx.com/download/OnLineUpgradeInfo25.pdf.gz

Algebraic Methods in Autotuning Design: Theory and Design

Roman Prokop, Jiří Korbel, and Libor Pekař

Tomas Bata University in Zlín, Faculty of applied informatics
Nad Stráněmi 4511, 760 05 Zlín, Czech Republic
prokop@fai.utb.cz

Abstract. The contribution presents a set of single input – output (SISO) principles for the design and tuning of continuous-time controllers for the utilization in autotuning schemes. The emphasis of the design is laid to SISO systems with time delays. Models with up to three parameters can estimated by means of a single relay experiment. Then a stable low order transfer function with a time delay term is identified. Two algebraic control syntheses then are presented in this paper. The first one is based on the ring of proper and stable rational functions R_{PS}. The second one utilizes a special ring R_{MS}, a set of RQ-meromorphic functions. In both cases, controller parameters are derived through a general solution of a linear Diophantine equation in the appropriate ring. The generalization for a two degree of freedom (2DOF) control structure is outlined. A final controller can be tuned by a scalar real parameter $m>0$. The presented philosophy covers a generalization of PID controllers and the Smith-like control structure. The analytical simple rule is derived for aperiodic control response in the R_{PS} case.

Keywords: Algebraic control design, Diophantine equation, Relay experiment, Autotuning, Pole-placement problem, Smith predictor.

1 Introduction

Feedback is the most powerful tool for the change of dynamic properties of systems. In process control, Proportional-Integral-Derivative (PID) controllers have been used for a wide range of problems for decades, they have survived changes in technology and they have been the most common way of using feedback in engineering systems [1], [9]. Yu [24] refers that more than 97 % of control loops are of this type and most of them are actually under PI control. The practical advantages of PID controllers can be seen in a simple structure, in an understandable principle and in control capabilities. It is widely known that PID controllers are quite resistant to changes in the controlled process without meaningful deterioration of the loop behavior. A solution for the qualified choice of controller parameters can be seen in more sophisticated, proper and automatic tuning of PID controllers, many modifications, unifications and generalizations of PID tuning and structure modifications can be found in [2], [3], [5], [8], [9]. Improvements can be seen in the change of the feedback structure (2DOF)

© Springer International Publishing Switzerland 2015
R. Silhavy et al. (eds.), *Intelligent Systems in Cybernetics and Automation Theory*,
Advances in Intelligent Systems and Computing 348, DOI: 10.1007/978-3-319-18503-3_7

and/or in the number of controller parameters, in robustness, in load disturbance attenuation, in predictive model principles, and in delay treatment and tuning.

In the field of modern controller design and tuning, algebraic notions and modules have successfully penetrated as useful engineering tools and methods [6], [7], [16], [21]. This approach considers a transfer function as an element of the field of fractions associated with an appropriate ring. Various design specifications can then be formulated through divisibility conditions and the control synthesis results in a general solution of the Diophantine equation in the appropriate ring [5], [6].

In this paper, new control design methods for autotuners are proposed and developed. Here, all autotuning principles are prepared for a combination of an asymmetrical relay identification test and a control design procedure performed in the ring of R_{PS} or R_{MS}. This factorization (fractional) approach proposed in [21] was generalized to a wide spectrum of control problems in [6], [11], [16], [26]. The pole placement problem in R_{PS} ring is formulated through a Diophantine equation and the pole is analytically tuned according to aperiodic response of the closed loop. The proposed method is compared by an equalization setting proposed in [4].

This contribution deals with two simplest SISO linear dynamic systems with a delay term. The first model of the first order (stable) plus dead time (FOPDT) is supposed in the form:

$$G(s) = \frac{K}{Ts+1} \cdot e^{-\Theta s} \tag{1}$$

Similarly, the second order model plus dead time (SOPDT) is assumed in the form:

$$G(s) = \frac{K}{(Ts+1)^2} \cdot e^{-\Theta s} \tag{2}$$

The contribution is organized as follows. Section 2 explains basic principle of algebraic syntheses in the special rings R_{PS} and R_{MS}. Basic derivations and tuning of controllers for first and second order systems is described in Section 3. Then the principle of the Smith predictor is introduced. Section 4 presents some facts about relay identification for autotuning principles. Then a Matlab program environment for design and simulations is described in Section 6. Finally, Section 7 presents simulation results and analysis for three types of SISO systems.

2   Algebraic Control Design

Algebraic notions, modules and tools were found to be useful in modern control theory for decades. Polynomials modules were firstly used in describing of linear systems continuous as well as discrete-time ones. The algebraic approach has dominated especially in the input-output properties of systems. The Diophantine equation in the ring of polynomials became a simple and powerful tool for the discrete control synthesis [5]. Later, rings and linear (Diophantine) equations penetrated into other fields of the control theory [6], [21], [26] as continuous-time, multivariable and nonlinear systems.

Apart from the ring of polynomials R_P, there are several rings used for control synthesis, e.g. the ring of stable and proper rational function R_{PS}, the ring of stable and proper meromorphic functions R_{MS} [6], [26].

Briefly speaking, a ring is a set equipped with two operations, addition and multiplication, that forms a (Abel) group with respect to addition and a semi group with respect to multiplication. It means that in the ring is not generally possible to divide. An element having a multiplicative inverse is called a unit. The ring is a commutative one if multiplication is commutative. If every non-zero element is a unit, the set is called a field. The Diophantine equation in a ring has the form

$$AX + BY = C \qquad (3)$$

where A, B, C are known elements and X, Y are unknown elements of given ring. Equation (3) is solvable if every common divisor A and B divides C. Without loss of generality, it can be supposed that A, B are coprime. Equation (3) has then an infinite number of solutions given by

$$\begin{aligned} X &= X_0 - AZ \\ Y &= Y_0 + BZ \end{aligned} \qquad (4)$$

where X_0, Y_0 are a particular solution of (3) and Z is an arbitrary element of a given ring.

A classical feedback, one-degree of freedom (1DOF), structure depicted in Fig. 1 and a two-degree of freedom (2DOF) structure depicted in Fig. 2 are supposed in this paper.

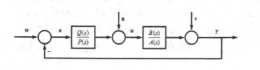

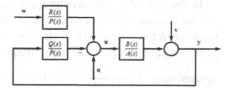

Fig. 1. One-degree of freedom (1DOF) control loop

Fig. 2. Two-degree of freedom (2DOF) control loop

In both cases, A, B represent a controlled plant, P, Q, R represent (unknown) controllers. Signals w, n, v, are reference, load disturbance and a background noise, respectively.

2.1   The R_{PS} Approach

The ring C represents a set of proper and Hurwitz stable rational functions analytic in $s \geq 0$ including $s = \infty$. Units of R_{PS} are rational functions of relative degree zero with stable numerator and denominator. The divisibility in this ring is defined through unstable zeros (including infinity). Traditional transfer functions as a ratio of two polynomials can be easily transformed into the fractional form in R_{PS} simply by dividing, both the polynomial denominator and numerator by the same stable polynomial of the appropriate order. Then all transfer functions can be expressed e.g. by the ratio:

$$G(s) = \frac{b(s)}{a(s)} = \frac{\dfrac{b(s)}{(s+m)^n}}{\dfrac{a(s)}{(s+m)^n}} = \frac{B(s)}{A(s)} \tag{5}$$

where $n = \max(\deg(a), \deg(b))$, $m > 0$

All feedback stabilizing controllers for the feedback system depicted in Fig. 1 are given by a general solution of the Diophantine equation (A, B coprime):

$$AP + BQ = 1 \tag{6}$$

which can be expressed with Z free in R_{PS}:

$$\frac{Q}{P} = \frac{Q_0 - AZ}{P_0 + BZ}, \quad P_0 + BZ \neq 0 \tag{7}$$

In contrast to the polynomial design, all controllers are proper and can be utilized. The Diophantine equation for designing the feedforward part in the structure 2DOF controller is:

$$F_w S + BR = 1 \tag{8}$$

where F_w is a denominator of the reference w.

In the case of 1DOF structure, asymptotic tracking is ensured by the divisibility of the denominator P in (7) by the denominator of the reference $w = G_w/F_w$. Asymptotic tracking in the case of 2DOF structure is achieved by the solution of the second Diophantine equation (8). The most frequent case for reference w is a signal with the denominator $F_w = \dfrac{s}{s+m}$; $m > 0$.

The similar conclusions can be also formulated for the load disturbance $n = G_n/F_n$. The divisibility in R_{PS} is defined through unstable zeros and it can be achieved by a suitable choice of rational function Z in (7), see [6], [16], [21] for details.

2.2   The R_{MS} Approach

This part utilizes a ring of stable and proper meromorphic functions R_{MS} omitting any approximation which was originally developed especially for delay systems by Zítek and Kučera [26] and later revised and extended to neutral systems in [10]. According to the original definition, an element of this ring is a ratio of two retarded quasipolynomials $y(s)/x(s)$.

A retarded quasipolynomial $x(s)$ of degree n means

$$x(s) = s^n + \sum_{i=0}^{n-1} \sum_{j=1}^{h} x_{ij} s^i \exp(-\vartheta_{ij} s), \quad \vartheta_{ij} \geq 0 \qquad (9)$$

where *retarded* refers to the fact that the highest s-power is not affected by exponentials. A more general notion of *neutral* quasipolynomials also can be used in this sense [10]. A quasipolynomial in the form of (9) is said to be asymptotically stable when it owns no finite zero s_0 such that $\text{Re}\{s_0\} \geq 0$.

A linear time-invariant delay system can be expressed as a ratio of two elements of the R_{MS} ring similarly as (5) but with quasipolynomials. The first order system with input-output time delay can be expressed in R_{MS} by

$$G(s) = \frac{\dfrac{K \exp(-\Theta s)}{s + m \ \exp(-\vartheta s)}}{\dfrac{Ts + 1}{s + m \ \exp(-\vartheta s)}} = \frac{\dfrac{b(s)}{m(s)}}{\dfrac{a(s)}{m(s)}} = \frac{B(s)}{A(s)} \qquad (10)$$

$$A(s), \ B(s) \in R_{MS}, \quad m_0 > 0$$

The traditional feedback loop 1DOF for the control design is displayed in Fig. 1. Generally, let model and controller transfer functions be expressed as

$$G(s) = \frac{B(s)}{A(s)}, \quad G_R(s) = \frac{Q(s)}{P(s)}, \quad A(s), B(s), Q(s), P(s) \in R_{MS} \qquad (11)$$

where $A(s)$, $B(s)$ and $P(s)$, $Q(s)$ are coprime pairs in R_{MS}.

Similarly, reference and load disturbance signals can be expressed by

$$W(s) = \frac{H_W(s)}{F_W(s)}, \quad D(s) = \frac{H_D(s)}{F_D(s)},$$

$$H_D(s), \ F_D(s), \ H_W(s), \ F_W(s) \in R_{MS} \qquad (12)$$

The aim of the control synthesis is to (internally) stabilize the feedback control system with asymptotic tracking and load disturbance attenuation.

The first step of the stabilization can be formulated in an elegant way in R_{MS} by the Diophantine equation

$$A(s)P_0(s) + B(s)Q_0(s) = 1 \qquad (13)$$

where $P_0(s)$ a $Q_0(s)$ is a (coprime) particular solution from R_{MS}. Since for stable and retarded systems, the R_{MS} ring constitutes the Bézout domain [10], the solution of (6) always exists. All stabilizing controllers can be expressed in a parametric form by

$$\frac{Q(s)}{P(s)} = \frac{Q_0(s) + A(s)Z(s)}{P_0(s) - B(s)Z(s)} \tag{14}$$

$$P_0(s) - B(s)Z(s) \neq 0$$

where $Z(s)$ is an arbitrary element of R_{MS}. The special choice of this element can ensure additional control conditions. Asymptotic tracking and disturbance attenuation result from expression for $E(s)$ which reads

$$E(s) = \frac{A(s)P(s)}{A(s)P(s) + B(s)Q(s)}W(s) - \frac{B(s)P(s)}{A(s)P(s) + B(s)Q(s)}D(s) \tag{15}$$

and they lead to the condition that both $F_w(s)$ and $F_D(s)$ divide $P(s)$. Details about divisibility in R_{MS} can be found in [10], [26]. It is clear that fraction (5) can be considered as a special case of (10) for $\vartheta = 0$. Then all feedback stabilizing controllers according to Fig. 1 are given by the same Diophantine equation (13), yet in the ring R_{PS}. Then all feedback controllers (14) can be utilized. In contrast to polynomial design, all controllers (14) are proper.

3 Basic Tuning Rules

3.1 First Order Systems

Suppose a first order system given by (1). The R_{PS} synthesis, the time delay term is omitted ($\Theta = 0$). Diophantine equation (6) then has the form:

$$\frac{(Ts+1)}{s+m}p_0 + \frac{K}{s+m}q_0 = 1 \tag{16}$$

It can be easily transformed to polynomial equation with general solution:

$$P = \frac{1}{T} + \frac{K}{s+m} \cdot Z, \quad Q = \frac{Tm-1}{TK} - \frac{Ts+1}{s+m} \cdot Z \tag{17}$$

where Z is free in the ring R_{PS}. Asymptotic tracking is achieved by the choice $Z = -\dfrac{m}{TK}$ and the resulting PI controller is in the PI form:

$$G_R(s) = \frac{Q}{P} = \frac{q_1 s + q_0}{s} \tag{18}$$

where parameters q_1 a q_0 are given by:

$$q_1 = \frac{2Tm-1}{K} \qquad q_0 = \frac{Tm^2}{K} \tag{19}$$

The feedforward part of the 2DOF controller follows from (8):

$$\frac{s}{s+m}s_0 + \frac{K}{s+m}r_0 = 1 \tag{20}$$

And the final feedforward part of the controller is given:

$$C_1(s) = \frac{R}{P} = \frac{r_1 s + r_0}{s} \tag{21}$$

with parameters

$$r_1 = \frac{Tm+m}{K}, \qquad r_0 = \frac{Tm^2}{K} \tag{22}$$

In all relations, the scalar parameter $m>0$ represents a tuning knob influencing control responses and behavior. A simple and attractive choice for the tuning parameter $m>0$ can be easily obtained analytically with the aim of aperiodic responses. For the system in Fig. 1, the closed-loop transfer function K_{wy} is for the first order system and PI controller (18) given in a very simple form:

$$K_{wy} = \frac{BQ}{AP+BQ} = BQ = \frac{(2Tm-1)s + Tm^2}{(s+m)^2} \tag{23}$$

The step response of (23) can be expressed by Laplace transform:

$$h(t) = L^{-1}\left\{\frac{K_{wy}}{s}\right\} = L^{-1}\left\{\frac{k_1 s + k_0}{s(s+m)^2}\right\} = L^{-1}\left\{\frac{A}{s} + \frac{B}{(s+m)} + \frac{C}{(s+m)^2}\right\}, \tag{24}$$

where A,B,C are calculated by comparing appropriate fractions in (24). The detailed derivation can be found in [19] and the result for the choice of the parameter $m>0$ is given by the inequality:

$$\frac{1}{2T} < m < \frac{1}{T} \tag{25}$$

Any positive parameter m from (25) ensures aperiodic response. It is a question for further investigation and simulation what choice from this interval is the best. The time constant is always estimation in the autotuning philosophy and then the mean value of (25) would be a reasonable choice in the form $m = \dfrac{3}{4 \cdot T}$. This result is in agreement with the equalization method developed by Gorez and Klán in [4].

The R_{MS} synthesis for FODPT supposes model (1) with a positive time delay term $\varTheta$ the coprime factorization in the R_{MS} ring can be then expressed by

$$\tilde{G}(s) = \frac{\dfrac{K\exp(-\Theta s)}{s+m}}{\dfrac{Ts+1}{s+m}} = \frac{B(s)}{A(s)} \tag{26}$$

where $m > 0$ is again a free (selectable) scalar parameter.

The control loop is considered as a simple feedback system (Fig. 1) with plant and controller transfer functions (11), respectively. Both external inputs w, n (Fig.1, Fig.2) are supposed as step functions.

The stabilizing Diophantine equation (13) reads

$$\frac{Ts+1}{s+m}P_0(s) + \frac{K\exp(-\Theta s)}{s+m}Q_0(s) = 1 \tag{27}$$

The choice $Q_0(s)=1$ gives the parameterized solution for $P(s)$ as

$$P(s) = P_0(s) - B(s)Z(s) = \frac{s+m-K\exp(-\Theta s)}{Ts+1} - \frac{K\exp(-\Theta s)}{s+m}Z(s) \tag{28}$$

In order to have $P(s)$ in a simple form satisfying $P_0(0)=0$, choose

$$Z(s) = \frac{s+m}{Ts+1}\left(\frac{m}{K}-1\right) \tag{29}$$

which gives the controller denominator and numerator by

$$P(s) = \frac{s+m(1-\exp(-\Theta s))}{Ts+1}, \quad Q(s) = \frac{m}{K} \tag{30}$$

according to (7). Thus, the final anisochronic controller structure reads

$$G_R(s) = \frac{m}{K}\frac{Ts+1}{s+m(1-\exp(-\Theta s))} \tag{31}$$

where m serves again as a tuning parameter. The denominator in (31) has infinite number of poles. The construction of this controller is more complex than usual PI or PID controllers.

3.2   Second Order Systems

The control synthesis for the SOPDT is based on stabilizing Diophantine equation (6) applied to the transfer function (2) without a time delay term. The Diophantine equation (6) takes the form:

$$\frac{(Ts+1)^2}{(s+m)^2}\cdot\frac{p_1s+p_0}{s+m} + \frac{K}{(s+m)^2}\cdot\frac{q_1s+q_0}{s+m} = 1 \tag{32}$$

and after equating the coefficients at like powers of s, it is possible to obtain explicit formulas for p_i, q_i:

$$p_1 = \frac{1}{T^2}; \qquad p_0 = \frac{3Tm - 2}{T}$$

$$q_1 = \frac{1}{K}\left[3m^2 - \frac{1}{T^2}(1 + 3m - \frac{2}{T})\right]; \qquad (33)$$

$$q_0 = \frac{1}{K}\left[m^3 - \frac{1}{T^2}(3m - \frac{2}{T})\right]$$

The rational function $P(s)$ has its parametric form (similar as in (18) for FOPDT):

$$P = \frac{p_1 s + p_0}{(s + m)} + \frac{K}{(s + m)^2} \cdot Z \qquad (34)$$

with Z free in R_{PS}. Now, the function Z must be chosen so that P is divisible by the denominator of the reference. The required divisibility is achieved by $Z = -\frac{p_0 m}{K}$. Then, the particular solution for P, Q is

$$P = \frac{s(p_1 s + (p_1 m + p_0))}{(s + m)^2}, Q = \frac{\tilde{q}_2 s^2 + \tilde{q}_1 s + \tilde{q}_0}{(s + m)^2} \qquad (35)$$

where

$$\tilde{q}_0 = q_0 + p_0 m, \quad \tilde{q}_1 = q_0 + q_1 m + 2Tp_0 m, \tilde{q}_2 = q_1 + T^2 p_0 m. \qquad (36)$$

The final (asymptotic tracking) controller has the transfer function:

$$G_R(s) = \frac{Q}{P} = \frac{\tilde{q}_2 s^2 + \tilde{q}_1 s + \tilde{q}_0}{s(p_1 s + (p_1 m + p_0))} \qquad (37)$$

Also the feedforward part for the 2DOF structure can be derived for the second order system. For asymptotic tracking the Diophantine equation takes the form:

$$\frac{s}{s + m}\frac{s_1 s + s_0}{(s + m)} + \frac{K}{(s + m)^2}r_0 = 1 \qquad (38)$$

The 2DOF control law is only dependent upon the rational function R with general expression $R = \frac{m^2}{K} - \frac{s}{s + m}Z$ also with Z free in R_{PS}. The final feedforward controller (according to Fig.2)

$$\frac{R}{P} = \frac{\frac{m^2}{K}(s + m)^2}{s(p_1 s + (p_1 m + p_0))} \qquad (39)$$

It is obvious that parameters of both parts of the controller (feedback and/or feed-forward) depend on the tuning parameter $m > 0$ in a nonlinear way. For both systems FOPDT and SOPDT the scalar parameter $m > 0$ seems to be a suitable „tuning knob" influencing control performance as well as robustness properties of the closed loop system. Naturally, both derived controllers correspond to classical PI and PID ones. Equation (18) represents a PI controller:

$$u(t) = K_P \cdot \left(e(t) + \frac{1}{T_I} \cdot \int_0^t e(\tau)d\tau \right) \tag{40}$$

and the conversion of parameters is trivial. Relation (37) represents a PID in the standard four-parameter form [1-4]:

$$u(t) = K_P \cdot \left(e(t) + \frac{1}{T_I} \cdot \int_0^t e(\tau)d\tau - T_D y_f'(t) \right) \tag{41}$$

$$\frac{T_D}{N} y_f'(t) + y_f(t) = y(t)$$

A second order model with dead time (SOPDT) has form (2) which can be expressed in R_{MS} as a ratio

$$\tilde{G}(s) = \frac{\dfrac{K\exp(-\Theta s)}{(s+m)^2}}{\dfrac{(Ts+1)^2}{(s+m)^2}} = \frac{B(s)}{A(s)} \tag{42}$$

Similarly as in (27) and (28) for the first order model, a stabilizing (non unique) particular solution of (13) can be obtained as

$$Q_0(s) = 1, \quad P_0(s) = \frac{(s+m)^2 - K\exp(-\Theta s)}{(Ts+1)^2} \tag{43}$$

and after the parameterization   the numerator and denominator result in

$$P(s) = \frac{s^2 + 2ms + m^2(1 - \exp(-\Theta s))}{(Ts+1)^2}, Q(s) = \frac{m^2}{K} \tag{44}$$

The final controller structure is then

$$G_R(s) = \frac{m^2}{K} \frac{(Ts+1)^2}{s^2 + 2ms + m^2(1 - \exp(-\Theta s))} \tag{45}$$

The program realization of the proposed controller is quite easy and it demonstrates the anisochronic structure of controller (45), see e.g. [12, 15, 19].

4   Conclusion

This contribution gives some control design rules for autotuning principles for a combination of relay feedback identification and a control design method.

Low order transfer function parameters supposed the estimation from asymmetric limit cycle data. The control synthesis is carried out through the solution of a linear Diophantine equation according to [6, 16, 21, 23, 26]. Two different algebraic control syntheses were studied and utilized. The first one utilizes a ring of proper and stable rational functions R_{PS}. The second approach utilized a ring of stable and proper meromorphic functions R_{MS} omitting any approximation for delay systems. Both control methodologies define a scalar positive tuning parameter which can be adjusted by various strategies. The generalization for a two degree of freedom (2DOF) control structure is outlined. A final controller can be tuned by a scalar real parameter $m > 0$. The presented philosophy covers a generalization of PID controllers and the Smith-like control structure. The analytical simple rule is derived for aperiodic control response in the R_{PS} case.

Acknowledgments. The work was performed with financial support of research project NPU I No. MSMT-7778/2014 by the Ministry of Education of the Czech Republic and also by the European Regional Development Fund under the Project CEBIA-Tech No. CZ.1.05/2.1.00/03.0089.

References

1. Åström, K.J., Hägglund, T.: PID Controllers: Theory, Design and Tuning. Instrumental Society of America, Research Triangle Park, NC (1995)
2. Åström, K.J., Hägglund, T.: The future of PID control. Control Engineering Practise 9, 1163–1175 (2001)
3. Åström, K.J., Murray, R.M.: Feedback Systems. Instrumental Society of America, Research Triangle Park, NC (2008)
4. Gorez, R., Klán, P.: Nonmodel-based explicit design relations for PID controllers. Preprints of IFAC Workshop PID 2000, pp. 141–146 (2000)
5. Kučera, V.: Discrete linear control, "The polynomial approach". Wiley, Chichester (1979)
6. Kučera, V.: Diophantine equations in control - A survey. Automatica 29(6), 1361–1375 (1993)
7. Matušů, R., Prokop, R.: Experimental verification of design methods for conventional PI/PID controllers. WSEAS Trans. on Systems and Control 5(5), 269–280 (2010)
8. Morari, M., Zafiriou, E.: Robust Process Control. Prentice Hall, New Jersey (1989)
9. O'Dwyer, A.: Handbook of PI and PID controller tuning rules. Imperial College Press, London (2003)
10. Pekař, L.: A Ring for Description and Control of Time-Delay Systems. WSEAS Trans. on Systems 11(10), 571–585 (2012)

11. Pekař, L., Valenta, P.: Algebraic 1DoF Control of Heating Process with Internal Delays. In: Advances in Mathematical and Computational Methods (Proceedings of the 14th WSEAS International Conference on Mathematical and Computational Methods in Science and Engineering), Sliema, Malta, pp. 115–120 (2012)
12. Pekař, L., Prokop, R.: Non-delay depending stability of a time-delay system. In: Last Trends on Systems, 14th WSEAS International Conference on Systems, Corfu Island, Greece, pp. 271–275 (2010)
13. Pekař, L., Prokop, R.: Control of Delayed Integrating Processes Using Two Feedback Controllers: RMS Approach. In: Proceedings of the 7th WSEAS International Conference on System Science and Simulation in Engineering, Venice, pp. 35–40 (2008)
14. Pekař, L., Prokop, R., Matušů, R.: Stability conditions for a retarded quasipolynomial and their applications. International Journal of Mathematics and Computers in Simulations 4(3), 90–98 (2010)
15. Pekař, L., Prokop, R.: Algebraic Control of integrating Processes with Dead Time by Two Feedback Controllers in the Ring RMS. Int. J. of Circuits, Systems and Signal Processing 2(4), 249–263 (2008)
16. Prokop, R., Corriou, J.P.: Design and analysis of simple robust controllers. Int. J. Control 66, 905–921 (1997)
17. Prokop, R., Korbel, J.: PI autotuners based on biased relay identification. Preprints of the 15th IFAC World Congress, Prague (2005)
18. Prokop, R., Korbel, J., Prokopová, Z.: Relay feedback autotuning – A polynomial approach. Preprints of 24th European Conf. on Modelling and Simulation, Kuala Lumpur (2010)
19. Prokop, R., Korbel, J., Líška, O.: A novel principle for relay-based autotuning. International Journal of Mathematical Models and Methods in Applied Science 5(7), 1180–1188 (2011)
20. Smith, O.J.M.: Feedback Control Systems. McGraw-Hill Book Company Inc. (1958)
21. Vidyasagar, M.: Control System Synthesis: A Factorization Approach. MIT Press, Cambridge (1985)
22. Vyhlídal, T.: Anisochronic first order model and its application to internal model control. Preprints of ASR 2000 Seminar, Prague, pp. 21–31 (2000)
23. Vyhlídal, T., Zítek, P.: Control System Design Based on a Universal First Order Model with Time Delays. Acta Polytechnica 44(4-5), 49–53 (2001)
24. Yu, C.C.: Autotuning of PID Controllers. Springer, London (1999)
25. Zhohg, Q.: Robust Control of Time-delay systems. Springer, London (2006)
26. Zítek, P., Kučera, V.: Algebraic design of anisochronic controllers for time delay systems". Int. Journal of Control 76(16), 905–921 (2003)

Algebraic Methods in Autotuning Design: Implementation and Simulations

Roman Prokop, Jiří Korbel, and Radek Matušů

Tomas Bata University in Zlín, Faculty of applied informatics
Nad Stráněmi 4511, 760 05 Zlín, Czech Republic
prokop@fai.utb.cz

Abstract. Autotuners represent a combination of a relay feedback identification test and some control design method. In this contribution, models with up to three parameters are estimated by means of a single asymmetrical relay experiment. Then a stable low order transfer function with a time delay term is identified by a relay experiment. Autotuning principles then combine asymmetrical relay feedback tests with a control synthesis. Two algebraic control syntheses then are presented in this paper. The first one is based on the ring of proper and stable rational functions R_{PS}. The second one utilizes a special ring R_{MS}, a set of RQ-meromorphic functions. In both cases, controller parameters are derived through a general solution of a linear Diophantine equation in the appropriate ring. A final controller can be tuned by a scalar real parameter $m>0$. The presented philosophy covers a generalization of PID controllers and the Smith-like control structure. This contribution deals with the implementation of proposed autotuners and presents some illustrative examples. A Matlab toolbox for automatic design and simulation was developed and various simulations performed and analyzed.

Keywords: Algebraic control design, Diophantine equation, Relay experiment, Autotuning, Pole-placement problem.

1 Introduction

The development of various autotuning principles was started by a simple symmetrical relay feedback experiment proposed by Åström and Hägglund [1] and the scheme is depicted in Fig. 1. The ultimate gain and ultimate frequency are then used for adjusting of parameters by the original Ziegler-Nichols rules [2, 3, 13]. The idea has gained wide popularity with control engineers and nowadays, many industrial controllers are equipped with automatic tuning abilities. During the period of three decades, many studies have been reported to extend and improve autotuners principles, see e.g. [5, 6, 7, 8]. The extension in relay utilization was performed in [8, 20, 21] by an asymmetry and hysteresis of a relay. System instability is dealt in [10] while time-delayed systems are solved in [9], aperiodic or robust features are outlined in [12, 18]. Over time, the direct estimation of transfer function parameters instead of critical

© Springer International Publishing Switzerland 2015
R. Silhavy et al. (eds.), *Intelligent Systems in Cybernetics and Automation Theory*,
Advances in Intelligent Systems and Computing 348, DOI: 10.1007/978-3-319-18503-3_8

values began to appear [5, 14]. Experiments with asymmetrical and dead-zone relay feedback are reported in [14, 15].

In this paper, new combinations for autotunig methods of PI and PID controllers are proposed and developed. Theory of the algebraic approach is studied in [19]. Here, all autotuning principles combine an asymmetrical relay identification test and a control design procedure performed in the ring of R_{PS} or R_{MS}. The pole placement problem in R_{PS} ring is formulated through a Diophantine equation and the pole is analytically tuned according to aperiodic response of the closed loop. Further details and aspects can be found in [16- 18]. Naturally, there exist many principles of control design syntheses which can be used for autotuning principles, see [22] as a monograph.

This contribution deals with two simplest SISO linear dynamic systems with a delay term. The first order (stable) plus dead time (FOPDT) model and the second order one plus dead time (SOPDT) are described by the transfer functions:

$$G_1(s) = \frac{K}{Ts+1} \cdot e^{-\Theta s}, \quad G_2(s) = \frac{K}{(Ts+1)^2} \cdot e^{-\Theta s} \tag{1}$$

Section 2 outlines the formulas for the estimation of transfer function parameters (1), section 3 deals with program implementation and section 4 illustrates some simulations and analysis.

2   Identification Procedure

The estimation of the process or ultimate parameters is a crucial point in all autotuning principles. The relay feedback test can utilize various types of relay for the parameter estimation procedure. The classical relay feedback test [1] was proposed for stable processes by symmetrical relay without hysteresis. Following sustained oscillation are then used for determining the critical (ultimate) values. The control parameters (PI or PID) are then generated in standard manner.

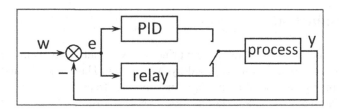

Fig. 1. Block diagram of an autotuning principle

Asymmetrical relays with or without hysteresis bring further progress [8, 14, 16, 21]. After the relay feedback test, the estimation of process parameters can be performed. A typical data response of such relay experiment is depicted in Fig. 2. The relay asymmetry is required for the process gain estimation K in (1) while a symmetrical relay would cause the zero division in the appropriate formula.

In this paper, an asymmetrical relay with hysteresis is used. This relay enables to estimate transfer function parameters as well as a time delay term. For the purpose of the aperiodic tuning the time delay is not exploited. The process gain can be calculated by the relation (see [21])

$$K = \frac{\int_0^{iT_y} y(t)dt}{\int_0^{iT_y} u(t)dt}; \quad i = 1,2,3,..$$ (2)

and the time constant and time delay terms are given by:

$$T = \frac{T_y}{2\pi} \cdot \sqrt{\frac{16 \cdot K^2 \cdot u_0^2}{\pi^2 \cdot a_y^2} - 1}$$

$$\Theta = \frac{T_y}{2\pi} \cdot \left[\pi - arctg \frac{2\pi T}{T_y} - arctg \frac{\varepsilon}{\sqrt{a_y^2 - \varepsilon^2}} \right]$$ (3)

where a_y and T_y are depicted in Fig. 2 and ε is the hysteresis.

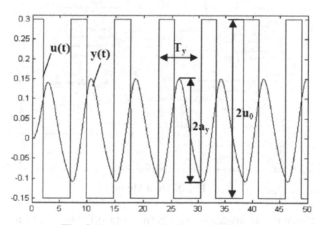

Fig. 2. Asymmetrical relay oscillation

The gain is given by (2), the time constant and time delay term can be estimated according to 21 by the relation:

$$T = \frac{T_y}{2\pi} \cdot \sqrt{\frac{4 \cdot K \cdot u_0}{\pi \cdot a_y} - 1}$$

$$\Theta = \frac{T_y}{2\pi} \cdot \left[\pi - 2arctg\frac{2\pi T}{T_y} - arctg\frac{\varepsilon}{\sqrt{a_y^2 - \varepsilon^2}} \right]$$

(4)

3   Simulation and Program System

A Matlab program system was developed for design, simulations and engineering applications of auto-tuning principles. This program enables a choice for the identification of the controlled system of arbitrary order. The estimated model is of a first or second order transfer function with time delay. The user can choose three cases for the time delay term. In the first case the time term is neglected, in the second one the term is approximated by the Pade expansion and the third case utilizes the Smith predictor control structure. The program is developed with the support of the Polynomial Toolbox. The Main menu window of the program system can be seen in Fig. 3.

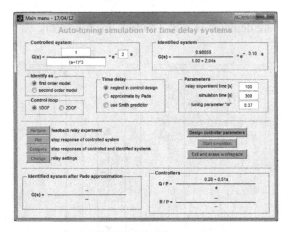

Fig. 3. Main Menu

In the first phase of the program routine, the controlled transfer function is defined and parameters for the relay experiment can be adjusted. Then, the experiment is performed and it can be repeated with modified parameters if necessary. After the experiment, an estimated transfer function in the form of (1) performed automatically and controller parameters are generated after pushing of the appropriate button. Parameters for experimental adjustment are defined in the upper part of the window.

The second phase begins with the "Design controller parameters" button and the actual control design is performed. According to above mentioned methodology and identified parameters, the controller is derived and displayed. The control scheme depends on the choice for the 1DOF or 2DOF structure and on the choice of the treatment with the time delay term.

During the third phase, after pushing the "Start simulation" button, the simulation routine is performed and required outputs are displayed. The simulation horizon can be prescribed as well as tuning parameter m, other simulation parameters can be specified in the Simulink environment. In all simulation a change of the step reference is performed in the second third of the simulation horizon and a step change in the load

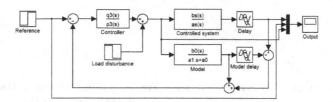

Fig. 4. Control loop in Simulink

is injected in the last third. A typical control loop of the case with the Smith predictor in Simulink is depicted in Fig. 4. Also the step responses can be displayed and the comparison of the controlled and estimated systems can be depicted. Further versions of the similar program systems were developed and they are referred in e.g. [16 - 18].

4   Examples and Simulations

The following examples illustrate the situation where the estimated model is in the form 19. The controllers are designed according to sections 3 of [19] with or without neglecting of the time delays.

Example 1: Controlled system with time delay with the transfer function:

$$G(s) = \frac{10}{(20s+1)(5s+1)(3s+1)} \cdot e^{-10s} \tag{5}$$

was identified by relay experiments as first and second order systems. The results give the following transfer functions:

$$\tilde{G}(s) = \frac{9.96}{27.85s+1} \cdot e^{-16.33s}, \quad \tilde{\tilde{G}}(s) = \frac{9.96}{(12.12s+1)^2} \cdot e^{-11.17s} \tag{6}$$

The first controller was designed for the identified system with neglecting of the time delay term and the tuning parameter $m = 0.03$ was derived from the aperiodic condition (25) in [19]. The PID for the second order estimation was designed for the tuning parameter $m = 0.06$. The final controllers are governed by the transfer functions:

$$G_{R1}(s) = \frac{\tilde{Q}(s)}{\tilde{P}(s)} = \frac{0.05s}{s}, \quad G_{R2}(s) = \frac{\tilde{\tilde{Q}}(s)}{\tilde{\tilde{P}}(s)} = \frac{0.46s^2 + 0.07s}{12.12s^2 + s} \tag{7}$$

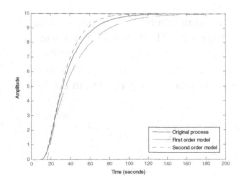
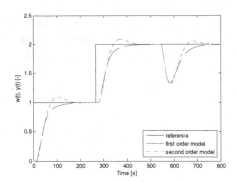

Fig. 5. Step responses of system (5) and estimations (6)

Fig. 6. Control responses 1DOF, first and second order controllers (7)

The original system $G(s)$ from (5) was controlled by (7) in two different control ways. The simple control response in the sense of 1DOF structure is depicted in **Fig. 6** for the first and second order controllers.

The same system was controlled also by anisochronic controller derived by the methodology mentioned in Section 2.3 of [19]. The relay identification yields a first order model approximation in the form:

$$\tilde{G}_1(s) = \frac{10}{23.45s + 1} e^{-16.91s} \tag{8}$$

The second order identified model has the transfer function:

$$\tilde{G}_2(s) = \frac{10}{(12.03s + 1)^2} e^{-11.04s} \tag{9}$$

Step responses of (5), (8) and (9) are shown in **Fig. 7**. The first order anisochronic methodology performed in R_{MS} gives the final controller:

$$G_{R1}(s) = 2.85 \cdot 10^{-3} \frac{23.45s + 1}{s + 2.85 \cdot 10^{-2} \left(1 - e^{-16.91s}\right)} \tag{10}$$

According to the mentioned algebraic control design in the R_{MS} ring, the second order controller in R_{MS} has the transfer function:

$$G_{R2}(s) = \frac{5.43 \cdot 10^{-4} (12.03s + 1)^2}{s^2 + 14.74 \cdot 10^{-2} s + 5.43 \cdot 10^{-3} \left(1 - e^{-11.04s}\right)} \tag{11}$$

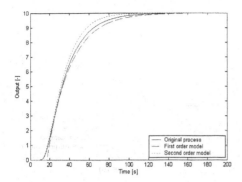

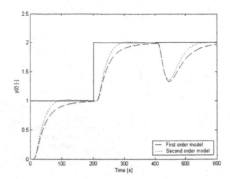

Fig. 7. Step responses of systems (5), (8) and (9)

Fig. 8. Control responses of anisochronic synthesis with 1DOF structure

Control responses for controllers(10) and (11) are compared in Fig. 8. In both cases the controlled plant was given by (5) and the same reference and load disturbance were assumed.

Example 2: Given a fifth order stable system with time delay with transfer function:

$$G(s) = \frac{3}{(2s+1)^5} \cdot e^{-5s} \tag{12}$$

The first and second order estimations result in the following transfer functions:

$$\tilde{G}(s) = \frac{2.99}{5.88s+1} \cdot e^{-10.35s}, \quad \tilde{\tilde{G}}(s) = \frac{2.99}{11.19s^2 + 6.69s + 1} \cdot e^{-8.49s} \tag{13}$$

Then controllers were designed for the identified models (13) with time delay terms neglected. The PI controller was derived for the value of $m = 0.13$ and the PID one was derived for $m = 0.22$. Both controllers in the 1DOF structure have the transfer functions:

$$G_{R1}(s) = \frac{0.17s + 0.03}{s}, \quad G_{R2}(s) = \frac{0.42s^2 + 0.23s + 0.03}{3.35s^2 + s} \tag{14}$$

The simulation control responses for the first order approximation and design are depicted in Fig. 9. Similar situation is shown in Fig. 10, where the second order approximation and synthesis were utilized. However, comparison of control responses shows that the first order synthesis is sufficient and the second order is redundant and needless complex.

In this case, the difference of responses between neglecting the time delay term and with the use of the Smith predictor is remarkably stronger. While the standard feedback control response is quite poor and oscillating then the response with Smith predictor in the loop is smooth and aperiodic.

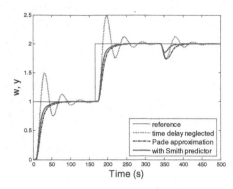
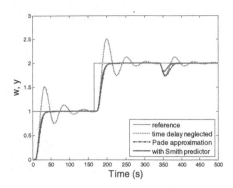

Fig. 9. Control responses 1DOF - first order approximation

Fig. 10. Control responses 1DOF - second order

Example 3: This example represents a case of a higher order system without delay which is approximated by a law order system with a time delay term. A higher order system (8th order) with transfer function $G(s)$ is supposed:

$$G(s) = \frac{3}{(s+1)^8} \tag{15}$$

Again, after the relay experiment, a first order and second estimation gives the following transfer functions:

$$\tilde{G}(s) = \frac{2.96}{4.22s+1} \cdot e^{-4.96s} \quad , \quad \tilde{\tilde{G}}(s) = \frac{2.96}{4.83s^2+4.40s+1} \cdot e^{-4s} \tag{16}$$

The step responses of systems (15) and (16) are shown in. Both step responses of the estimated systems (74) are quite different from the original system $G(s)$.

Again, PI controllers are generated from according to R_{PS} synthesis in [19] and the tuning parameter $m>0$ can influence the control behavior. Since the differences of controlled and estimated systems are considerable, it can be expected that not all values of and some of $m>0$ represent acceptable behavior. Some of control responses are shown in Fig. 12. Generally for the R_{PS} design, larger values of $m>0$ implicate larger overshoots and oscillations.

As a consequence, for inaccurate relay identifications, lower values of $m>0$ in interval can be recommended. The lower bound in $\frac{1}{2T} < m < \frac{1}{T}$ for the estimated $T=4.22$ in [19] gives $m = 0.118$ and the PI controller has the form

$$G_R(s) = \frac{0.17s+0.05}{s} \tag{17}$$

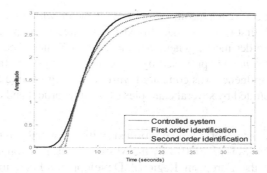

Fig. 11. Step responses controlled and estimated systems (15), (16)

The second order identification and synthesis of Example 3 for *m = 0.34* gives the PID controller:

$$G_R(s) = \frac{0.28s^2 + 0.23s + 0.05}{2.20s^2 + s} \tag{18}$$

The control responses of the combination (15), (18) are depicted in Fig. 13.

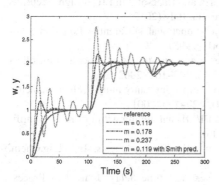

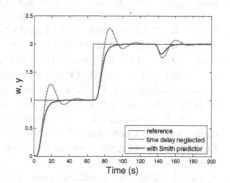

Fig. 12. Control responses 1DOF first order

Fig. 13. Control responses 1DOF second order

5   Conclusion

This contribution shows some simulation in autotuning principles with a combination of relay feedback identification and a control design method. The estimation of a low order transfer function parameters is performed from asymmetric limit cycle data. The control synthesis is carried out through the solution of a linear Diophantine equation and the methodology is described in [19]. Here, two different algebraic control syntheses were studied and utilized. The first one adopted ring of proper and stable rational functions R_{PS}. The second approach utilized a ring of stable and proper meromorphic functions R_{MS} omitting any approximation for delay systems. Both control methodologies define a scalar positive tuning parameter which can be adjusted by various strategies. A first order estimated model generates a group of PI-like controllers while a second order model generates a class of PID ones. The aperiodic tuning through the parameter $m>0$ is proposed by the analytic derivation. In both cases also the Smith predictor influence was compared with neglecting of time delay terms. The methodology is illustrated by several examples of various orders and dynamics.

Acknowledgments. The work was performed with financial support of research project NPU I No. MSMT-7778/2014 by the Ministry of Education of the Czech Republic and also by the European Regional Development Fund under the Project CEBIA-Tech No. CZ.1.05/2.1.00/03.0089.

References

1. Åström, K.J., Hägglund, T.: Automatic tuning of simple regulators with specifica-tion on phase and amplitude margins. Automatica 20, 645–651 (1984)
2. Åström, K.J., Hägglund, T.: PID Controllers: Theory, Design and Tuning. Instrumental Society of America, Research Triangle Park (1995)
3. Garpinger, O., Hägglund, T., Åström, K.J.: Criteria and trade-offs in PID design. Preprints of IFAC Conferencieon Advancesin Control, Brescia, Italy (2012)
4. Hale, J.K., Verduyn Lunel, S.M.: Introduction to Functional Differential Equations. Applied Math. Sciences, vol. 99. Springer, New York (1993)
5. Hang, C.C., Åström, K.J., Wang, Q.C.: Relay feedback auto-tuning of process con-trollers – a tutorial review. Journal of Process Control 12(6), 143–162 (2002)
6. Kaya, I., Atherton, D.P.: Parameter estimation from relay autotuning with asymmetric limit cycle data. Journal of Process Control 11(4), 429–439 (2001)
7. Leva, A., Bascetta, L., Schiavo, F.: Model based PI/PID autotuning with fast relay identification. Ind. Eng.Chem. Res. 45, 4052–4062 (2006)
8. Luyben, W.L.: Getting more identification from relay-feedback tests. Ind. Eng.Chem. Res. 40, 4391–4402 (2001)
9. Majhi, S.: Relay based identification of processes with time delay. Journal of Process Control 17, 93–101 (2007)
10. Majhi, S., Atherton, D.P.: Autotuning and controller design for unstable time delay processes. Preprints of UKACC Conf. an Control, pp. 769–774 (1998)
11. Morari, M., Zafiriou, E.: Robust Process Control. Prentice Hall, New Jersey (1989)
12. Morilla, F., Gonzáles, A., Duro, N.: Auto-tuning PID controllers in terms of relative damping. Preprints of IFAC Workshop PID 2000, pp. 161–166 (2000)
13. O'Dwyer, A.: Handbook of PI and PID controller tuning rules. Imperial College Press, London (2003)
14. Panda, R.C., Yu, C.C.: Analytical expressions for relay feedback responses. Journal of Process Control 13, 489–501 (2003)
15. Pecharromán, R.R., Pagola, F.L.: Control design for PID controllers auto-tuning based on improved identification. Preprints of IFAC Workshop PID 2000, pp. 89–94 (2000)
16. Prokop, R., Korbel, J.: PI autotuners based on biased relay identification. Preprints of the 15th IFAC World Congress, Prague (2005)

17. Prokop, R., Korbel, J., Prokopová, Z.: Relay feedback autotuning – A polynomial approach. Preprints of 24th European Conf. on Modelling and Simulation, Kuala Lumpur (2010)
18. Prokop, R., Korbel, J., Líška, O.: A novel principle for relay-based autotuning. International Journal of Mathematical Models and Methods in Applied Science 5(7), 1180–1188 (2011)
19. Prokop, R., Korbel, J., Pekař, L.: Algebraic methods in autotuning design, Part I: Theory and design. This Journal (2014)
20. Thyagarajan, T., Yu, C.C.: Improved autotuning using shape factor from relay feedback. Preprints of IFAC World Congress (2002)
21. Vítečková, M., Víteček, A.: Plant identification by relay methods. In: Dudas, L. (ed.) Engineering the Future, pp. 242–256. Sciyo, Rijeka (2010)
22. Yu, C.C.: Autotuning of PID Controllers. Springer, London (1999)

Extension of the Pole-Placement Shifting Based Tuning Algorithm to Neutral Delay Systems: A Case Study

Libor Pekař

Faculty of Applied Informatics, Tomas Bata University in Zlín, Czech Republic
pekar@fai.utb.cz

Abstract. In [1], a revised version of the Pole-Placement Shifting based controller tuning Algorithm (PPSA), a finite-dimensional model-matching controller tuning method for time-delay systems (TDS), was presented together with some suggestions about algorithm improvements and modifications. Its leading idea consists in the placing the dominant characteristic poles and zeros of the infinite-dimensional feedback control system with respect to the desired dynamics of the simple finite-dimensional matching model. So far, retarded TDS have been studied in the reign of the PPSA. This paper, however, brings a detailed case study on a more advanced and intricate neutral-type control feedback. Unstable controlled plant is selected in our example, in addition. The results indicate a very good applicability of the PPSA under some minor modifications of standard manipulations with the neutral-type delayed spectrum.

Keywords: Time delay systems, neutral—type delay, spectrum-shaping controller tuning, optimization, MATLAB, direct-search algorithms, model matching.

1   Introduction

Many parameters setting strategies for conventional proportional-integral-derivative (PID) controllers have been derived, proposed and presented so far. In the contrary, there may emerge various unusual control laws in modern control design for which a known tuning method can not be directly applied (e.g. nonlinear, hybrid, optimal control systems etc).

Time delay systems (TDS) and their control design strategies have been intensively studied during past decades, see e.g. [4-6]; however, many of them suffer from theoretical or computational complexity. This is mainly due to infinite spectra of TDS – this feature is well characterized by the characteristic quasipolynomial instead of polynomial containing exponentials yielding its transcendental nature. It is worth noting that we have investigated an elegant, relatively simple and attractive algebraic controller design methodology for TDS in a special ring [7], which has proved to be practically useful and applicable, in addition [8]; however, a sufficiently accurate plant model is required for this purpose and the mentioned unusual control low is obtained in most cases.

© Springer International Publishing Switzerland 2015
R. Silhavy et al. (eds.), *Intelligent Systems in Cybernetics and Automation Theory*,
Advances in Intelligent Systems and Computing 348, DOI: 10.1007/978-3-319-18503-3_9

Once the controller structure is designed, it is parameterized by the set of unknown adjustable parameters which have to be appropriately set. Unfortunately, there is a lack of engineeringly effortless controller tuning approaches for this class of systems and controllers or they are burden with an excessive mathematics. In [1], the PPSA (Pole-Placement Shifting based controller tuning Algorithm) (PPSA) based on the matching of the eventual closed-loop dynamic structure with the desired finite-dimensional model by spectral shaping was presented in the revised version. The procedure is relatively computationally simple to be implemented by means of modern programming languages. Up today, however, so-called retarded TDS have been controlled and tuned by using the PPSA. Hence, the challenging task is to applied it onto more advanced, tricky and intricate neutral-type TDS with a very complex spectral properties.

This paper is aimed at a particular case study on the use of the PPSA for neutral feedback delayed system with an unstable TDS plant. It is organized as follows: First, the specific properties of neutral TDS concerning the spectral ones and stability are concisely introduced to acquaint the reader with them. Consequently, the idea of the PPSA via the framework meta-algorithm is given. Third, a rather detailed example by means of MATLAB® and Simulink® is presented and, eventually, the results are concluded.

2   Preliminaries

2.1   Neutral TDS

Definition 1. A single-input single-output TDS governed by the delay-differential equation

$$d\dot{\mathbf{x}}(t) = \sum_{i=1}^{n_H} \mathbf{H}_i d\dot{\mathbf{x}}(t - \vartheta_i) + \sum_{i=0}^{n_A} \mathbf{A}_i \mathbf{x}(t - \vartheta_i) + \sum_{i=0}^{n_B} \mathbf{B}_i u(t - \tau_i),\ y(t) = \mathbf{C}\mathbf{x}(t) \tag{1}$$

where $\mathbf{x}, \dot{\mathbf{x}} \in \mathbb{R}^n$ stand for the state vector and its derivative, respectively, $u, y \in \mathbb{R}$, mean input and output, respectively, $0 = \vartheta_0 < \vartheta_1 < ... < \vartheta$, $0 = \tau_0 < \tau_1 < ... < \tau_{n_B}$ express internal and input-output delays, respectively, and $\mathbf{A}_i, \mathbf{B}_i, \mathbf{C}, \mathbf{H}_i$ are real-valued matrices of compatible dimensions, is said *neutral* if $\exists i : \mathbf{H}_i \neq \mathbf{0}$.

In the language of the Laplace transform, the corresponding transfer function reads

$$G(s) = \frac{b(s)}{a(s)} = \mathbf{C}(s - \mathbf{A}_H(s))^{-1} \mathbf{B}(s)$$

$$\mathbf{A}_H(s) = s\sum_{i=1}^{n_H} \mathbf{H}_i \exp(-s\vartheta_i) + \sum_{i=0}^{n_A} \mathbf{A}_i \exp(-s\vartheta_i), \mathbf{B}(s) = \mathbf{B}_0 + \sum_{i=1}^{n_B} \mathbf{B}_i \exp(-s\tau_i) \tag{2}$$

where $a(s)$, $b(s)$ are quasipolynomials, i.e. polynomials $s \in \mathbb{C}$ and $\exp(-s\cdot)$. Note that $a(s) = s^n + \sum_{i=0}^{n} \sum_{j=1}^{h_i} a_{ij} s^i \exp(-s\eta_{ij}), \exists a_{nj} \neq 0$ agrees with the characteristic quasipolynomial of the system.

Assume that $a(s)$ and $b(s)$ have no common root, then the roots ζ_i of $b(s)$ in the set Ω_Z are called system *zeros* and those σ_i of $a(s)$ in the set Ω_P stand for system *poles*.

Definition 2. The *spectral abscissa* $\alpha \in \mathsf{R}$ (of poles) is defined as $\alpha := \sup \mathrm{Re}\,\Omega_P$.

Lemma 1 [4]. System (1)-(2) is *exponentially stable* if and only if $\alpha < 0$.

Definition 3. The *associated characteristic equation* of a neutral system (1)-(2) is

$$a_n(s) := 1 + \sum_{j=1}^{h_n} a_{nj} \exp(-\eta_{nj}s) = 0 \tag{3}$$

Let the corresponding spectrum reads $\Omega_a := \{\zeta : a_n(\zeta) = 0\}$ with $\alpha_a := \sup \mathrm{Re}\,\Omega_a$.

Lemma 2 [10]. Introduce the polynomial $a_n(z) := 1 + \sum_{j=1}^{h_n} a_{nj} z^{\eta_{nj}}, z := \exp(s)$. Then

$$\alpha_a = -\ln|\zeta| \tag{4}$$

where ζ is the root of $a_n(z)$ with the minimum modulus.

Definition 4 [9]. System (1)-(2) is said to be *strongly stable* if

$$C := \lim_{\delta \to 0^+} \sup \mathrm{Re}\,\Omega_{a,\Delta\eta} < 0, \Omega_{a,\Delta\eta} := \{\zeta : a_n(\zeta, \eta + \delta\eta) = 0, \|\delta\eta\| < \varepsilon\}$$
$$\eta = [\eta_{n1}, \dots, \eta_{nh_n}] \tag{5}$$

Rephrasing the definition, strong stability means that roots of $a_n(s)$ remain in the complex left half-plane under infinitesimally small delay perturbations since C is continuous with respect to coefficients of $a_n(s)$ but not with respect to small delay perturbations [9].

Lemma 3 [11]. The system is strongly stable if and only if $\sum_{j=1}^{h_n} |a_{nj}| < 1$.

Lemma 4 [4, 9]. For system (1)-(2) it holds that:

i) Let $\exists N, \forall i > N : |\zeta_i| < |\zeta_{i+1}|, \zeta_i \in \Omega_a$. Then $\lim_{i \to \infty} \mathrm{Re}\,\zeta_i < \infty$, $\lim_{i \to \infty} \mathrm{Im}\,\zeta_i = \infty$.

ii) If there are poles of (1) in a bounded vertical strip on the complex plane, these poles converge for $i \to \infty$ to the infinite sequence defined in i).

iii) For any $\varepsilon > 0$, the system has only a finite number of poles in the right half-plane $\mathrm{Re}(s) > \alpha_a + \varepsilon$.

Note that some authors (e.g. in [9]) mean that it is not reasonably to deal with poles right from the vertical line $\alpha_a + \varepsilon$ (or rather $C + \varepsilon$) when shaping the infinite spectrum. The estimation of the upper bound on C was designed in [9].

3   PPSA – Pole-Placement Shifting Based Controller Tuning Algorithm

The PPSA [1, 12] serves as a procedure for tuning the infinite-dimensional spectrum of a feedback TDS (1)-(2). Typically, such a delayed feedback arises when designing a control loop for plants with internal and/or input-output delays [5-7], or even for delay-free plants with a delayed controller [13].

The algorithm framework is based on the matching a finite-dimensional model with the desired dynamic behaviour (for instance, a suitable maximum relative step response overshoot, the relative dumping factor, etc.) with the eventual infinite-dimensional system such that the dominant system poles and/or zeros are shifted to the model ones as close as possible while the rest of the spectrum is pushed to the "stable" half-plane as much as possible. The framework "meta-algorithm" can be characterized as follows:

Algorithm 1 (PPSA).

Input. Closed-loop reference-to-output transfer function $G(s)$.

Step 1. Select a suitable (admissible) finite-dimensional matching model structure of the reference-to-output relation, $G_m(s)$. Every substrategy has specific bounding conditions on degrees of polynomials in $G_m(s)$.

Step 2. Prescribe poles and zeros of the model.

Step 3. Place infinite-dimensional model roots into desired positions.

Step 4. If there roots are dominant (the rightmost), terminate the algorithm; else shift the dominant roots towards the desired ones, e.g. by using a quasi-continuous shifting algorithm [9, 14] followed, and the rest of both spectra push to the left as far as possible.

Step 5. If the shifting is successful, the algorithm is finished; otherwise, minimize the cost function reflecting the distance of dominant and prescribed roots and the spectral abscissa of the rest of both spectra by an iterative optimization [15].

Output. Positions of dominant poles (zeros) and controller parameters' values.

Three PPSA subalgorithms were provided to the reader in [12]; however, the presented versions suffer from many imperfections and are not entirely completed. Hence, two of the substrategies have been completely reformulated in [1], yet only retarded systems have been considered therein. The reader is kindly referred therein for further details.

A simulation study case example on PPSA for neutral TDS follows.

4   Study Case

Example 1. Consider the controlled plant governed by the transfer function

$$G_p(s) = \frac{s+4}{s+1-2\exp(-0.4s)}\exp(-s) \tag{6}$$

with poles

$$\Omega_{PP} = \{0.5836, -4.3826 + 11.0375j, -6.5569 + 26.9811j, -7.6892 + 42.8094j, ...\}$$

System (6) is of a retarded not neutral type; however, if it is subjected to any feasible linear continuous-time controller, a neutral feedback TDS is obtained. If, for example, the Two-Degrees-of-Freedom (2DoF) system [16] with the feedback and feedforward controllers

$$G_Q(s) = \frac{q_2 s^2 + q_1 s + q_0}{s^2 + p_1 s + p_0}, G_R(s) = \frac{r_2 s^2 + r_1 s + q_0 - p_0/4}{s^2 + p_1 s + p_0} \tag{7}$$

respectively, are used, it yields the following reference-to-output transfer function

$$G(s) = \frac{(s+4)(r_2 s^2 + r_1 s + q_0 - p_0/4)\exp(-s)}{(s+1-2\exp(-0.4s))(s^2 + p_1 s + p_0) + (s+4)\exp(-s)(q_2 s^2 + q_1 s + q_0)} \tag{8}$$

with the unit static gain where $\mathbf{K} = [p_1, p_0, q_1, q_0, q_2, r_2, r_1]^T \in \mathbb{R}^7$ are adjustable parameters.

Determine now the position of the rightmost vertical chain of poles and decide about strong stability. Since $a_n(s) = a_3(s) = 1 + q_2 \exp(-s)$, we have $a_3(z) = 1 + q_2 z$, the unique root of which is $\zeta = -1/q_2$. According to Lemma 2, the value

$$\alpha_a = -\ln\left|-1/q_2\right| \tag{9}$$

must be strictly negative, which results in the condition $|q_2| < 1$. Similarly, the same requirement stands for strong stability. This means that the poles should be found right from the vertical line $\text{Re}\, s = -\ln\left|-1/q_2\right|$; however, we made the following observation.

Observation 1. It is reasonable to deal with poles of a sufficiently small modulus left from the line C (or α_a) when shaping the infinite spectrum.

The proof is intuitive: In the light of Lemma 4, there exists a finite number of poles near the zero point of the complex plane which significantly determinates the system dynamics. These poles may lie right from the chain of poles introduced in Lemma 4, part i), yet left from C (or α_a), and, naturally, the moving of these poles influences α_a as well. The observation is utilized hereinafter.

Now let us follow Algorithm 1 with a specific substrategy called Poles First Independently introduced in [1]. Remark that the aim of the substrategy of the PPSA is based on placing the feedback poles to the desired positions first, and consequently, transfer function numerator parameters not included in the denominator serve as a tuning tool for inserting zeros to the desired loci. Simultaneously, rests of both the spectra is pushed to the left (stable) half-plane.

A possible finite-dimensional matching model can be governed by the transfer function

$$G_m(s) = \frac{b_1 s + b_0}{s^2 + a_1 s + a_0} = k \frac{s - z_1}{(s - s_1)(s - \bar{s}_1)} \tag{10}$$

where $k, b_1, b_0, a_1, a_0 \neq 0 \in \mathbb{R}$ stand for model parameters, $z_1 \in \mathbb{R}^-$ are the model zero and $s_1 = \alpha + j\omega \in \mathbb{C}_o^-, \alpha < 0, \omega \geq 0$, mean the model stable pole where $\bar{s}_1$ expresses its complex conjugate. Moreover, it must hold $b_0 / a_0 = 1, k = -|s_1|^2 / z_1$ to obtain the unit static gain. The analysis of the model dynamics with respect to zeros/poles loci was presented e.g. in [12, 16].

Select a pair of prescribed poles as $\{s_1, \bar{s}_1\} = -0.1 \pm 0.2j$ and the zero $z_1 = -0.18$. Note that the vertical strip of neutral poles is obtained left from the desired pair if the inequality $-\ln|-1/q_2| < -0.1$ must hold, according to (9), the solution of which reads $|q_2| < 0.9048$. Place a pair of feedback poles directly to the desired loci Then the initial spectrum calculated by means of [17] and corresponding values of denominator parameters, $\mathbf{K}_D = [p_1, p_0, q_1, q_0, q_2]^T \in \mathbb{R}^7$, read

$$\Omega_{P,0} = \left\{ \begin{matrix} 0.539 \pm 2.8461j, -0.1 \pm 0.2j, -0.2186 \pm 21.8116j, -0.2409 \pm 15.5849j, \\ -0.2418 \pm 9.0022j, \ldots \end{matrix} \right\}$$

$$\mathbf{K}_{D,0} = [0.821, 0.4329, -0.0722, -0.0445, 0.7664]$$

Obviously, the pair of poles at prescribed positions is not the dominant one, and moreover, the system is not exponentially stable (see Lemma 1) but it is strongly stable. In the accordance to Step 4 of Algorithm 1, the shifting is to be performed. The counter of poles currently shifted to these loci is initialized as $n_{sp} = 2$ while no root is pushed to the left – this value can be updated up to 3 during the shifting. Evolutions of the distance of the dominant pair, $\sigma_{1,2}$, from the desired one, $|\sigma_1 - s_1|$, the overall spectral abscissa, $\alpha(\mathbf{K}_D)$, and the spectral abscissa of the rest of the spectrum, $\alpha_r(\mathbf{K}_D)$, by means of the QCSA [9, 14] are displayed in Fig. 1 and the corresponding development of $\mathbf{K}_D$ can be seen in Fig. 2.

Faster changes in quality indicators (Fig. 1) and controller parameters (Fig. 2) are due to a much more effective shifting of the spectrum to the left where all three possible rightmost poles are pushed to the left. The course of q_2 implies that strong stability is preserved with $\alpha_a = -4.343$ during the QCSA of the PPSA.

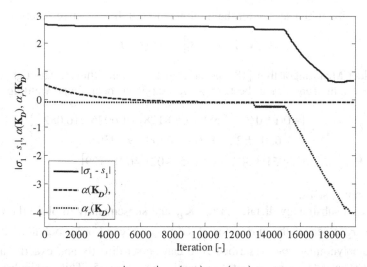

Fig. 1. The evolution of $|\sigma_1 - s_1|$, $\alpha(\mathbf{K}_D)$, $\alpha_r(\mathbf{K}_D)$ for (8) by using the QCSA

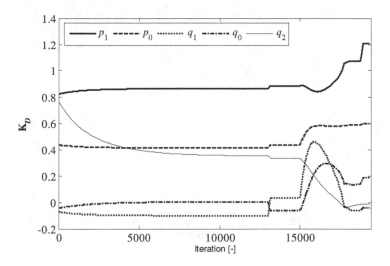

Fig. 2. The evolution of $\mathbf{K}_D$ for (8) by using the QCSA

The eventual results from the iteration process are

$$\Omega_{P,19400} = \left\{ \begin{array}{l} -0.0999 \pm 0.8432\mathrm{j}, -4.0358 \pm 2.3484\mathrm{j}, -4.0392 \pm 18.5881\mathrm{j}, \\ -4.3022 \pm 2.6475\mathrm{j}, -4.4912 \pm 24.7169\mathrm{j}, \ldots \end{array} \right\}$$

$$\mathbf{K}_{D,19400} = \begin{bmatrix} 1.2097, 0.5984, -0.043, 0.186, -0.013 \end{bmatrix}^T$$

which serve as the initial setting for the minimization of the objective function

$$\Phi(\mathbf{K}_D) = |\sigma_1 - s_1| + \lambda \alpha_r(\mathbf{K}_D)$$

via the Nelder-Mead algorithm [18], as in Step 5 of Algorithm 1. Through various simulations test, the suboptimal (best) value $\lambda = 0.05$ has been found yielding

$$\Omega_{P,opt,158} = \begin{Bmatrix} -0.1 \pm 0.2\mathrm{j}, -0.5884 \pm 3.8128\mathrm{j}, -1.6076 \pm 16.082\mathrm{j}, \\ -1.6302 \pm 22.1068\mathrm{j}, -1.6509 \pm 9.6930\mathrm{j}, \dots \end{Bmatrix}$$

$$\mathbf{K}_{D,opt,158} = [3.4051, 1.302, -0.9145, -0.2446, 0.1729]$$

after 158 iteration steps.

The selected substrategy dictates to fix $\mathbf{K}_D$ and subsequently to place the zero by means of r_1, r_2. Because the numerator of $G(s)$ is a second-order polynomial rather than a quasipolynomial, the prescribed zero can be set directly and exactly, and the second root is set arbitrarily yet not dominantly, e.g. $z_2 = -5$. This particular option results in $[r_1, r_2] = [-3.2815, -0.6335]$.

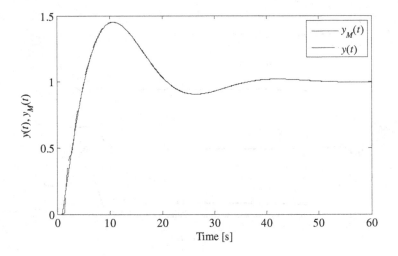

Fig. 3. The comparison of the desired $y_M(t)$ and real step response $y(t)$ from the PPSA for (8) – the global view

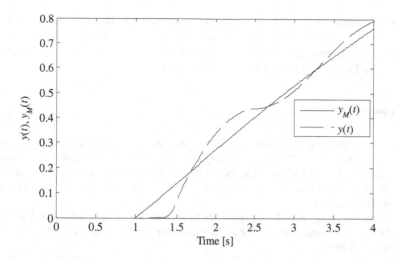

Fig. 4. The comparison of the desired $y_M(t)$ and real step response $y(t)$ from the PPSA for (8) – a detailed view

The final controller parameters then read

$$\mathbf{K} = [3.4051, 1.302, -0.9145, -0.2446, 0.1729, -0.6335, -3.2815]^T$$

The desired reference-to-output step response, $y_M(t)$, is compared that obtained using the PPSA, $y(t)$, is displayed in Fig. 3. Because of a very high similarity of both the courses in the figure, the decisive detailed view is provided to the reader in Fig. 4. They indicate a very good correspondence; however, there is a rather worse agreement near the zero due to high-frequency poles discrepancies.

5   Conclusion

The presented preliminary study has been aimed at attempting to use and extend the recently developed pole- and zero-shifting controller tuning PPSA algorithm for delayed systems to neutral ones. Up today, only simpler and less complex retarded systems were studied. As first, the paper has brought a concise overview of spectral properties of neutral TDS with respect to exponential and strong stability. Consequently, the idea of the PPSA has been brought to the reader's mind again, and finally, a study case utilizing a specific PPSA substrategy to neutral TDS has been presented. An interesting observation about the region of searched roots has been made. The presented results of the study have proofed a very good matching performance of the shifting algorithm. However, many aspects of the PPSA can be improved in the future; for instance, the effectively of the choice of the searching region, high-frequency prescribed/real zeros/poles agreement, etc.

Acknowledgements. The work was performed with the financial support of the research project NPU I No. MSMT-7778/2014 by the Ministry of Education of the Czech Republic and also by the European Regional Development Fund under the project CEBIA-Tech No. CZ.1.05/2.1.00/03.0089.

References

1. Pekař, L., Navrátil, P.: PPSA: A Tool for Suboptimal Control of Time Delay Systems: Revision and Open Tasks. In: Šilhavý, R., Šenkeřík, R., Komínková Oplatková, Z., Prokopová, Z. (eds.) Modern Trends and Techniques in Computer Science, 3rd Computer Science Online Conference 2014 (CSOC 2014). AISC, vol. 285, pp. 17–28. Springer, Heidelberg (2014)
2. Åström, K.J., Hägglund, T.: Advanced PID Control. ISA, Research Triangle Park (2005)
3. O'Dwyer, A.: Handbook of PI and PID Controller Tuning Rules. Imperial College Press, London (2009)
4. Hale, J.K., Verduyn Lunel, S.M.: Introduction to Functional Differential Equations. Applied Mathematical Sciences, vol. 99. Springer, New York (1993)
5. Richard, J.P.: Time-Delay Systems: An Overview of Some Recent Advances and Open Problems. Automatica 39, 1667–1694 (2003)
6. Sipahi, R., Vyhlídal, T., Niculescu, S.-I., Pepe, P.: Time Delay Sys.: Methods, Appli. and New Trends. LNCIS, vol. 423. Springer, Heidelberg (2012)
7. Pekař, L.: A Ring for Description and Control of Time-Delay Systems. WSEAS Trans. Systems, Special Issue on Modelling, Identification, Stability, Control and Applications 11, 571–585 (2012)
8. Pekař, L., Prokop, R.: Control of Time Delay Systems in a Robust Sense Using Two Controllers. In: Proc. 6th Int. Symposium on Communication, Control and Signal Processing (ISCCSP 2014), pp. 254–257. IEEE Press, Athens (2014)
9. Michiels, W., Vyhlídal, T.: An Eigenvalue Based Approach for the Stabilization of Linear Time-Delay Systems of Neutral Type. Automatica 41, 991–998 (2005)
10. Rabah, R., Sklyar, G.M., Rezounenko, A.V.: Stability Analysis of Neutral Type Systems in Hilbert Space. J. Differ. Equ. 214, 391–428 (2005)
11. Vyhlídal, T., Zítek, P.: Modification of Mikhaylov Criterion for Neutral Time-Delay Systems. IEEE Trans. Autom. Control 54, 2430–2435 (2009)
12. Pekař, L.: On a Controller Parameterization for Infinite-dimensional Feedback Systems Based on the Desired Overshoot. WSEAS Trans. Systems 12, 325–335 (2013)
13. Olgac, N., Sipahi, J., Ergenc, A.F.: 'Delay Scheduling' as Unconventional Use of Time Delay for Trajectory Tracking. Mechatronics 17, 199–206 (2007)
14. Michiels, W., Engelborghs, K., Vansevevant, P., Roose, D.: Continuous Pole Placement for Delay Equations. Automatica 38, 747–761 (2002)
15. Fletcher, R.: Practical Methods of Optimization. Wiley, New York (1987)
16. Šulc, B., Vítečková, M.: Theory and Practice of Control System Design. Czech Technical University in Prague, Prague (2004) (in Czech)
17. Vyhlídal, T., Zítek, P.: QPmR - Quasi-Polynomial Root-Finder: Algorithm Update and Examples. In: Vyhlídal, T., Lafay, J.-F., Sipahi, R. (eds.) Delay Systems: From Theory to Numerics and Applications, pp. 299–312. Springer, New York (2014)
18. Nelder, J.A., Mead, R.: A Simplex Method for Function Minimization. The Computer J. 7, 308–313 (1965)

Web Application for LTI Systems Analysis

Frantisek Gazdos and Jiri Facuna

Tomas Bata University in Zlin, Faculty of Applied Informatics
Nam. T.G. Masaryka 5555, 760 01 Zlin, Czech Republic
gazdos@fai.utb.cz

Abstract. The number of web-applications which employ external computing
tools to perform complex calculations on-line is rising. This contribution
presents one such application – web-interface for analysis of linear time-
invariant (LTI) systems. It is based on the interconnection of the MATLAB system with the latest web technologies and developer tools. It enables to analyze a
single input – single output (SISO) system in a user-defined form and a user-
friendly way on-line without the necessity to install the MATLAB environment
or similar computing tools. The paper explains motivation for development of
this site and gives also detailed description of the whole process including the
Web – MATLAB interconnection. The results are presented using selected
screen-shots of the site on a simple example of a LTI system analysis.

Keywords: Web-application, MATLAB, LTI systems, analysis.

1 Introduction

Providing and sharing information using the Internet is a standard nowadays. Simple
text information was soon supplemented with visual, audio and video data, forming
user-friendly environment for the information acquisition. Last decade has brought
interactivity into the web-pages, providing users with tools to better control the
process of information transfer. This opened the door to the development of applications which enable to process user-entered input-data and display corresponding results. As the requirements for complex calculations on the Web grow, there are rising
efforts to connect web-applications with effective computing tools such as The
MathWorks's MATLAB [1-2], Wolfram Research's Mathematica [3-4], Maplesoft's
Maple [5-6] and others which are more suitable for complex calculations and corresponding data processing. This paper presents one such web-application for LTI systems analysis which can be used widely by students, teachers or scientists interested
in control engineering or similar areas where linear time-invariant systems are used
for description of various processes. Users can choose different forms of data-input,
e.g. transfer function or state-space description, either continuous or discrete-time
with standard notation of the parameters as common in the MATLAB environment.
They can do various transformations, such as continuous to discrete-time or
state-space to transfer function conversions, and vice versa. Finally they obtain important information about the system, e.g. stability, gain, poles and zeros or selected

© Springer International Publishing Switzerland 2015
R. Silhavy et al. (eds.), *Intelligent Systems in Cybernetics and Automation Theory*,
Advances in Intelligent Systems and Computing 348, DOI: 10.1007/978-3-319-18503-3_10

characteristics, such as a step-response or nyquist plot. The presented application was designed in a bilingual (English/Czech) version within the scope of the Master's thesis [7] and further tuned by the authors. A similar application can be found at [8], however, with limited possibilities and only in the Czech language.

This paper is structured as follows: after this introductory part, main goals and motivation are outlined in the next section, followed by the employed software, hardware and Internet tools. Next part is devoted to the process of connecting the MATLAB system with web-applications and a brief description of the implemented functions for the LTI systems analysis follows. Further, the web-application user interface is presented on a simple example and described in detail, followed by some concluding remarks and perspective at the end of the paper.

2   Motivation and Goals

Nowadays, the academic standard for systems analysis and synthesis is the MATLAB environment [9] with its powerful computing possibilities and a great number of toolboxes for systems analysis, synthesis or modeling and simulation. However this system is not free and consequently not every student, teacher or scientist can afford it. Mainly students have problems of this type and often seek other possibilities how to use their favorite functions for free. The Octave [10-11] or Scilab [12-13] software can be the alternatives, however they are not as capable as the MATLAB system and have different notation. Other problem is that generally not every computer in the university network is equipped with the MATLAB software, and, often happens that you need to test something outside of the university network – at home, halls of residence, etc. These problems have led to the idea of the development of a simple, user-friendly web-application for basic analysis of widely used LTI systems. The site is primarily intended for students' purposes, however teachers or scientist can also gain advantages of using it from every place connected to the Internet.

For the web-application development, the main goals were formulated as follows: to develop a web-application which enables to analyze a user-specified linear time-invariant SISO system with the following requirements:

- Bilingual (English/Czech) version;
- Similar notation as in the MATLAB environment (e.g. for polynomials);
- Possibility to work either with continuous or discrete-time models;
- Possibility to work with both transfer function and state-space description;
- Possibility to do various transformations (e.g. continuous ↔ discrete-time, state-space ↔ transfer function);
- Provide important information about the system, such as poles, zeros, gain and stability;
- Display selected characteristics, such as various responses (step/impulse), bode or nyquist graphs.

Besides this, the web-application has to be user-friendly enough and provide help information when needed.

For the systems analysis, standard functions of the MATLAB SW are employed, mainly from the popular Control System Toolbox [14]. More information about the used functions are provided further in this contribution.

The designed web-application is intended to support pedagogical and research activities and one of the main advantages can be seen in the open access to the application without the need to install any kind of software on a user computer or to have a particular kind of operating systems. Apart from this, it can be administered and updated easily. The application is accessible directly at the URL: http://matserver.utb.cz/LTI [15].

3 Methodology

This section further explains used SW and HW tools, the process of MATLAB ↔ Web interconnection and main implemented functions for the LTI systems analysis.

3.1 Software and Hardware Tools

Web-applications are generally operating on web-servers – computers with a suitable operating system and a special application, web service enabling operation of the web-applications. The developed application runs on the PC Fujitsu Siemens Esprimo P5625 with the AMD Athlon 64 X2 5600+ processor with 4GB of RAM and two 500 GB hard-drives in the RAID 1 configuration. It hosts the Microsoft Windows Web Server operating system with the web service IIS.

Web-applications have usually several tiers and most common is the 3-tiered architecture with a presentation, application and storage tier.

In the developed application, the first, presentation tier accessible by a web browser was created using the HTML, CSS and AJAX technologies.

The middle application tier (logic tier) processing requests and data from the presentation tier and generating user-friendly interface dynamically was designed using the ASP.NET technology, a part of the Microsoft .NET Framework. For algorithms implementation, C# programming language and the Microsoft Visual Web Developer Express Edition software was used. Complex computations needed for the LTI system analysis were realized with the help of the MATLAB system. The following MATLAB components were employed for the required functions, corresponding data-processing and deployment to the Web:

- MATLAB;
- Control System Toolbox;
- MATLAB Builder NE [for .NET];
- MATLAB Compiler.

The last, third web-application tier is the storage one, also known as the data tier, which is accessed through the middle application tier and enables data-retrieving, storage and update. In the developed application, it is represented by the file-system of the server operating system.

3.2   MATLAB ↔ Web Interconnection

Up to the MATLAB version R2006b, there was a possibility to connect HTML web-pages with the MATLAB functions simply using the MATLAB Web Server component, where the functions were implemented directly in the form of MATLAB source codes (m-files). Next versions of the MATLAB system do not include nor support the Web Server component and the connection has to be realized in a different way. MATLAB functions can be implemented (deployed) in the form of so-called components – dynamic-link libraries (DLL-files) in the Microsoft Windows operating system. Development of a web-application connected to the MATLAB system can then be divided into the two independent parts: preparation of the source m-files with required MATLAB functions and components generation in the first step, and development of the web-application and implementation (deployment) of the generated components (DLL's) in the second step. Web-applications connected to the MATLAB system can be developed using various technologies such as Microsoft .NET Framework or JAVA. In this work the former tool was used.

An example of a source MATLAB m-file used for a web-application component is presented in Fig. 1. This function simply loads all the variables from a MATLAB binary data file (*.mat).

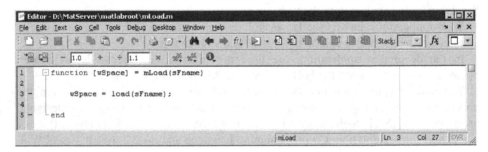

Fig. 1. Deployed m-file example

The resultant components for the web-application are then created from the m-files (and relevant toolboxes) using the MATLAB Compiler toolbox. This toolbox uses other supporting toolboxes for the compilation, depending on the chosen technology of web-application development. For the applications developed in the Microsoft .NET environment, the MATLAB Builder NE [for .NET] is needed which creates .NET components containing Microsoft .NET Framework classes. An example of a .NET component with the source m-file above is displayed in Fig. 2.

Fig. 2. Compiler toolbox

Components (DLL's files) created using the MATLAB Builder NE can be further used in the standard common way as other components of the Microsoft .NET Framework technology. It is only necessary to make a reference to the components in the web-application development environment. Then it is possible to use all the classes and functions from the source MATLAB m-files during the application development. An example of calling the function Load of the MatLoader object from the MatUtils component is given in Fig. 3. The component MatUtils.dll containing the MatLoaderClass was created from the m-file load.m, see Fig. 1, using the MATLAB Builder NE toolbox.

```
try
{

    // Create new matLoader and populate MWStructArray msaMatWrkSpace
    MatLoaderClass MatLoader = new MatLoaderClass();
    MWStructArray msaMatWrkSpace = (MWStructArray)MatLoader.Load(sFileName);
```

Fig. 3. Calling a function from an m-file

In web-application development environment, e.g. Microsoft Visual Studio, a reference is given to the MatUtils.dll file (see Fig. 4.) and then the class MatLoaderClass can be further used in the standard way as displayed in Fig. 3.

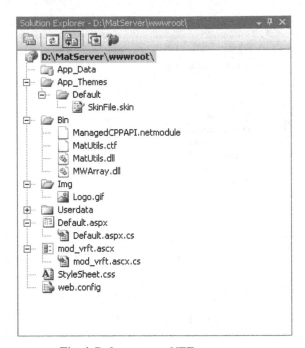

Fig. 4. Reference to a .NET component

3.3   Main Implemented Functions

All the used MATLAB functions (more than 40 programmed m-files) were divided
into 4 main categories: *transfer* (for various system transformations), *test* (to test
some system properties), *properties* (to determine certain system properties) and
graphs (to generate chosen characteristics). In the *transfer* class, these main
MATLAB functions were utilized:

- *ss*, *ssdata* for the state-space description and transformations;
- *tf*, *tfdata* for the transfer function description and conversions;
- *d2c*, *c2d*, *d2d* for the continuous ↔ discrete-time conversions and resampling.

The *test* class included implementation of these MATLAB functions:

- *size* to test dimensions (number of inputs/outputs) of the system;
- *isproper* to test if the system is proper.

The *properties* class was built on these standard MATLAB functions:

- *minreal* for the minimal realization of the system;
- *zero*, *pole* for zeros and poles of the system;
- *order* for the system order/degree;
- *isstable* to test stability;

- *ctrb*, *obsv* to test controllability and observability;
- *dcgain* for the system gain information.

The last class *graphs* contains implementation of these MATLAB functions:

- *step* for the step-response;
- *impulse* for the impulse response;
- *nyquist* for the nyquist graph;
- *bode* for the bode graph;
- *pzmap* for the pole-zero map of the dynamic system.

Detailed documentation for the above given standard functions of the Control Sytem Toolbox, part of the MATLAB system can be found in the MathWorks documentation, e.g. [14].

4   Results

A testing version of the developed application is accessible via the Internet at the following URL: http://matserver.utb.cz/LTI [15]. The application user interface (presentation tier) is designed as a guide and allows entering the required information for a given LTI system in several interconnected steps. Common control buttons such as "Next" / "Back" and "Help" are used to navigate throughout the application and provide corresponding help information when needed. The interface consists of these main parts:

- Start-up screen which provides basic information about the web-application, its possible usage and contact information; a language control button (Czech / English) and also the help button are accessible throughout the whole process of work with this site;
- In the next step, a user can choose the form of the system input – either as a transfer function or in the form of state-space description. For discrete-time systems, it is possible to give a sampling period. For both continuous and discrete-time systems also a time-delay can be easily entered;
- Further, coefficients of the transfer function or matrices of the state-space description can be entered in the same form as in the MATLAB environment, as illustrated in Fig. 5;
- The fourth step can be used for various transformations (continuous ↔ discrete-time, transfer function ↔ state-space description) and to find the so-called minimal realization of the system; default setting is no transformation. The entered system is also tested here – if it is SISO and if it is proper (if no, the user is warned and he/she can enter a new system). In case of a discrete system, the application also offers to resample it with a new sampling period;
- In the next screen, it is possible to choose which system properties should be determined (zeros, poles, order, stability, gain, controllability and observability);

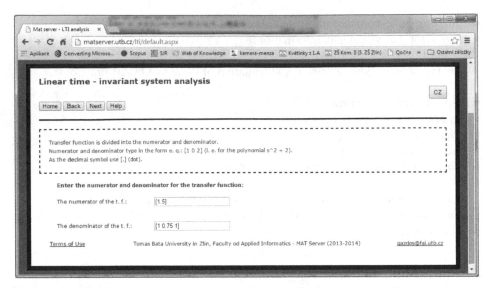

Fig. 5. Entering the system parameters (transfer function description)

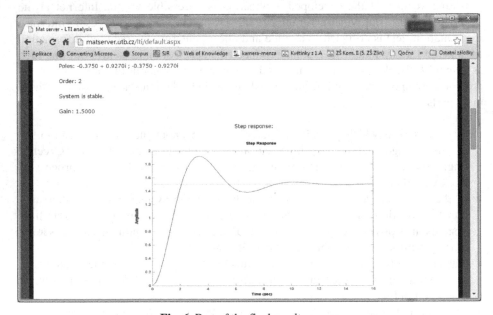

Fig. 6. Part of the final results screen

- The sixth step shows the selected properties form the previous step and enables to choose system characteristics to be displayed (step and impulse responses, bode and nyquist graphs, poles & zeros map in the complex plane);
- Finally, in the last step, all the entered information and performed transformation are summarized and displayed together with the required system properties and characteristics, as partly seen in Fig. 6. The graphs can be easily downloaded using

the right mouse click in the *.png image format. In every part of the application it is possible to return to previous steps to correct the entered information or directly back home for a new session.

5   Conclusion and Perspective

This paper has presented how it is practically possible to connect web-applications with the computing system MATLAB. The result, in this case, is the web-application for linear time-invariant systems analysis suitable mainly for students (and their teachers) of control engineering and similar fields of interest with limited access to the MATLAB system (or similar computing tools). The whole process of the application development with motivation and goals, SW and HW tools, utilized functions for the analysis and the process of MATLAB-Web interconnection has been explained in detail. Description of the resultant web-application user interface is also given. The presented application is still in the development and it is currently being tested; some improvements are being prepared, e.g. the possibility to handle also multi input – multi output (MIMO) systems in a certain way. The important thing is that this application has a modular structure, it is open for improvements and accessible from anywhere with the Internet connection.

References

1. Attaway, S.: Matlab: A Practical Introduction to Programming and problem Solving. Butterworth-Heinemann, Boston (2013)
2. MATLAB – The Language of Technical Computing, http://www.mathworks.com/products/matlab
3. Wolfram, S.: The MATHEMATICA Book. Cambridge University Press, Cambridge (1999)
4. Wolfram Mathematica: Definitive System for Modern Technical Computing, http://www.wolfram.com/mathematica
5. Garvan, F.: The Maple Book. Chapman and Hall/CRC, London (2001)
6. Maple – Technical Computing Software for Engineers, Mathematicians, Scientists, Instructors and Students – Maplesoft, http://www.maplesoft.com/products/Maple
7. Facuna, J.: Web-application for the LTI Systems Analysis. Master's thesis, Tomas Bata University in Zlin, Zlin (2014)
8. System analysis, http://winf230e-1.vsb.cz/crl002
9. Chaparro, L.: Signals and Systems using MATLAB. Academic Press, London (2014)
10. Hansen, J.S.: GNU Octave Beginner's Guide. Packt Publishing, Birmingham (2011)
11. GNU Octave, https://gnu.org/software/octave
12. Gomez, C.: Engineering and Scientific Computing with Scilab. Birkhäuser, Boston (1999)
13. Scilab, http://www.scilab.org
14. The MathWorks, Inc.: Control System Toolbox: User's Guide. Natick, MA, USA (2013)
15. Mat Server – LTI analysis, http://matserver.utb.cz/LTI

Predictive Control of Systems with Fast Dynamics Using Computational Reduction Based on Feedback Control Information

Tomáš Barot and Marek Kubalcik

Department of Process Control , Tomas Bata University in Zlín,
Faculty of Applied Informatics, nám. T. G. Masaryka 5555, 76001 Zlín, Czech Republic
{barot,kubalcik}@fai.utb.cz

Abstract. Predictive control is a method, which is suitable for control of linear discrete dynamical systems. However, control of systems with fast dynamics could be problematic using predictive control. The calculation of a predictive-control algorithm can exceed the sampling period. This situation occurs in case with higher prediction horizons and many constraints on variables in the predictive control. In this contribution, an improving of the classical approach is presented. The reduction of the computational time is performed using an analysis of steady states in the control. The presented approach is based on utilization of information from the feedback control. Then this information is applied in the control algorithm. Finally, the classical method is compared to the presented modification using the time analyses.

1 Introduction

Predictive control [1] is a modern method, which uses the principle based on computation on the future horizons in connection to the optimization problem solving. The future information about situation in the feedback control is predicted using a model of the identified system. The model is used as a predictor [2]. The unknown information is computed using optimization subsystem [3], in which are considered the requirements on the feedback control and all defined constraints.

In general, the algorithm of predictive control is suitable for many types of systems [4]. However, the category of systems with fast dynamics [5] could bring the problems. The disadvantage is based on the higher computational time of the algorithm with more difficult settings of the predictive controller, as can be seen in this contribution. These more demanding settings are the higher prediction horizons [6] and using of greater number of constraints [7] on variables in the control. The important assumption is solving prediction equations and the optimization problem in time of the sampling period.

The approach, when optimization in connection to predictions is solved in each sampling period, is referred to be online. The optimization part included more significantly time-demanding parts than the predictor; therefore the research is focused on the optimizer.

© Springer International Publishing Switzerland 2015
R. Silhavy et al. (eds.), *Intelligent Systems in Cybernetics and Automation Theory*,
Advances in Intelligent Systems and Computing 348, DOI: 10.1007/978-3-319-18503-3_11

The optimization task solved the minimization of the defined cost function, which is usually defined as the quadratic function. If the constraints are in the form of the linear inequalities [8], the task is defined as the quadratic programming problem [9]. The fast method used for this purposes is the Hildreth's method [10]. This algorithm uses the dual method [11] of the non-classical based extreme task [11].

For many types of controlled systems is this algorithm appropriate. However, the predictive control of systems with fast dynamics needs a modification of this principle. The modification [10] tries to remove all constraints in the optimization task in the current sampling period and then test the success of it. In this paper, the Hildreth's method with modification [10] is considered as the classical approach. The modified approach is based on using the information on the feedback control. If a steady state occured, the elimination of inactive constraints will be possible. The results are discussed in the final part of this contribution.

2   Model of Controlled System in Predictive Controller

In the predictive control, the controlled system can be mathematically described by a discrete transfer function (2) [5], which corresponds to the continuous representation [5] of this system (1) for given sampling period T. The linear variant of systems is considered in this paper. The future outputs of the system behaviour are determined using this model. The model is included in the predictor of the predictive controller.

In this paper, the systems with fast dynamics are considered. This category of systems is characterized by their roots with values in order of minus tens, in denominator of transfer function (1). In general, the sampling period is significantly low, in order of hundredths of a second.

$$G(s) = \frac{\overline{b}_1 s + \overline{b}_2}{\overline{a}_1 s^2 + \overline{a}_2 s + \overline{a}_3}, \quad G(z^{-1}) = \frac{b_1 z^{-1} + b_2 z^{-2}}{1 + a_1 z^{-1} + a_2 z^{-2}} \qquad (1), (2)$$

For the determination of future $N_1...N_2$ outputs (4), the difference equation (3) is used. The equation (4) is based on CARIMA (Controlled Autoregressive Integrated Moving Average) model [10], which includes N_u future increments of manipulated variable Δu instead the direct value u. In predictive control, the parameters N_1, N_u and N_2, determine the receding horizon window [6].

$$(1 - z^{-1})(1 + a_1 z^{-1} + a_2 z^{-2}) y(k) = (b_1 z^{-1} + b_2 z^{-2}).\Delta u(k) \qquad (3)$$

$$\begin{bmatrix} y(k + N_1) \\ \vdots \\ y(k + N_2) \end{bmatrix} = \boldsymbol{P}. \begin{bmatrix} y(k) \\ y(k-1) \\ y(k-2) \\ \Delta u(k-1) \end{bmatrix} + \boldsymbol{G}. \begin{bmatrix} \Delta u(k) \\ \Delta u(k+1) \\ \vdots \\ \Delta u(k + N_u - 1) \end{bmatrix} \qquad (4)$$

Matrices P and G contain coefficients of difference equations. They are determined recursively [10]. The part of the matrix equation (4) with matrix P includes the information from the past; whereas, the part with matrix G is related to the predicted situation. The future information is determined using optimization subsystem.

3   Control Law of Predictive Control

The control law has atypical form (5) in predictive control. Equation (5) is the solving of optimization task, where the sequence of N_u values of future values of manipulated-variable increments (the vector Δu) is determined. In this case, the quadratic cost function with constraints is used (6). This problem is then a quadratic programming task [9] and should be algorithmically solved by the fast Hildreth's method [10].

$$\Delta u = arg\left[min\left\{ \frac{1}{2}\Delta u^T.H.\Delta u + b^T.\Delta u \right\} \right],\ M.\Delta u \le K \qquad (5),(6)$$

The matrices P and G, from the predictor, are utilized for determination of vector b (7) and Hessian matrix H (8), where I is a unit matrix with dimension N_u and w is reference signal. Finally, the form of the cost function expresses the requirements on the feedback control [6].

$$b = 2(P[y(k) \quad y(k-1) \quad y(k-2) \quad \Delta u(k-1)]^T - $$
$$- [w(k+N_1)\cdots w(k+N_2)]^T)^T G \qquad (7)$$

Table 1. Rules for building matrices M and K for types of restrictions on control.

Restriction	M	K
u_{min}	$-T$	$-E^{N_u,1}.u_{min} + E^{N_u,1}.u(k-1)$
u_{max}	T	$E^{N_u,1}.u_{max} - E^{N_u,1}.u(k-1)$
y_{min}	$-G^{N_2-N_1,N_u}$	$-E^{N_2-N_1+1,1}y_{min} +$ $+ P.[y(k) \quad y(k-1) \quad y(k-2) \quad \Delta u(k-1)]$
y_{max}	$G^{N_2-N_1,N_u}$	$E^{N_2-N_1+1,1}.y_{max} -$ $- P.[y(k) \quad y(k-1) \quad y(k-2) \quad \Delta u(k-1)]$
Δu_{min}	$-I$	$-E^{N_u,1}.\Delta u_{min}$
Δu_{max}	I	$E^{N_u,1}.\Delta u_{max}$

$$H = 2.(G^T.G + I) \tag{8}$$

The form of matrices M and K can be constructed by rules, which can be seen in Table 1, where I is a unit matrix with dimension N_u, T is a lower-triangular matrix with dimension N_u and E is an ones matrix. Each type of constraint has a form, which has been derived for the variable Δu.

4   Classical Method

Subsystems of predictive controller – the predictor and the optimizer cooperate together by equations (5)-(6) and (4). The control law is enumerated in each sampling period in the predictive control. For higher setting of parameters in predictive controller, computations could be so time consuming, that they overload the time of sampling period. In case of controlled systems with fast dynamics, the higher value of parameter N_2 and the increasing number of constraints could cause a problem.

The modification of these cases was published in [10]. The main idea was to leave out all constraints (9) in computation of the optimization problem [9] by Hildreth's method.

$$\Delta u = -H^{-1}b \tag{9}$$

If the result passes the constraints condition (6) retrospectively, it is supposed to be the final result of the task. The non-classical based extreme task will be transferred to the multidimensional-free-extreme problem without consideration of all constraints. The time complexity can be decreased by this approach. However; this method has one disadvantage. This reduction of steps in optimization algorithm can not be applied in each sampling period in predictive control. This reduction is not naturally possible in all cases. The aim is to decrease the computational time in all steps of the discrete control.

5   Approach for Computation Reduction

The modified approach removes constraints, when the classical method with result (9) by condition (6) is not successful. The constraints in predictive control can be left out in case of stabilization of variables in the feedback control, when value y is steady. In this situation, all constraints can be tested for their reduction.

At first, the success of the classical method with constraints removing is tested in the optimization task. Otherwise; in steady state (10) in feedback control, the presented approach is applied and tested by (6). In other cases must be performed the whole optimization algorithm. The summary scheme can be seen in Fig. 1 and Fig.2.

$$\forall \left\{ \left| e(k+i\big|_{k-1}) = w(k+i\big|_{k-1}) - y(k+i\big|_{k-1}) \right| \leq \varepsilon, i = \left\{ N_1,...,N_2 \right\} \right\}$$
$$\varepsilon \approx 0, k \geq 1 \tag{10}$$

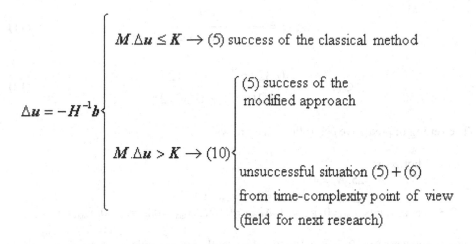

$$\Delta u = -H^{-1}b \begin{cases} M.\Delta u \le K \rightarrow (5) \text{ success of the classical method} \\ \\ M.\Delta u > K \rightarrow (10) \begin{cases} (5) \text{ success of the} \\ \text{modified approach} \\ \\ \text{unsuccessful situation } (5)+(6) \\ \text{from time-complexity point of view} \\ (\text{field for next research}) \end{cases} \end{cases}$$

Fig. 1. Inclusion of Proposed Approach in Whole Optimization Strategy

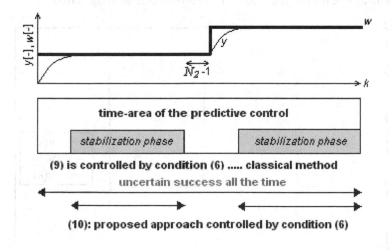

Fig. 2. Time-areas of Classical and Proposed Approaches

6   Results

Predictive control algorithm with optimization Hildreth's algorithm was realized in MATLAB environment. The time-measuring functions were implemented for the purposes of time analyses of presented approach in comparison to the classical method.

The controlled system (11) with fast dynamics is chosen and the corresponding discrete model (12) is determined for sampling period $T = 0.05$ s. The roots of the denominator are -10 and -14.

$$G(s) = \frac{s + 350}{s^2 + 24s + 140} \tag{11}$$

$$G(z^{-1}) = \frac{0.324z^{-1} + 0.1712z^{-2}}{1 - 1.103z^{-1} + 0.3012z^{-2}} \tag{12}$$

The setting of predictive controller is as follows:

- horizons:
  N_I=1, N_u=25, N_2=30
- reference signal values: w_{min}=0.5, w_{max}=1
- constraints: u_{min}=0, u_{max}= 1, Δu_{min} =0.02, Δu_{max} =0.2, y_{min} =0, y_{max} =1

The parameter ε in (10) is equal to 0 in this simulation example.
The simulation of the predictive control of system (12) can be seen in Fig. 3. On the axis k are ordinates of the discrete control that respect the sampling period T.

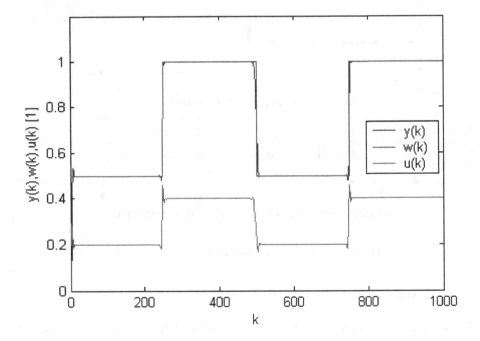

Fig. 3. Simulation of Predictive Control of System with Fast Dynamics

In Fig.4 and Fig.5, the time analysis can be seen. The time consuming operations of optimization subsystem were measured in MATLAB script using time-measuring function. Figures show the time T_c that was needed for executing of the optimization algorithm in each sampling period.

At first it is displayed the analysis for the classical method (Fig.4) and then for application of combination of both approaches (the classical one together with the proposed one) (Fig.5). The predictive control was realized hundred times and the average results are displayed for all sampling times.

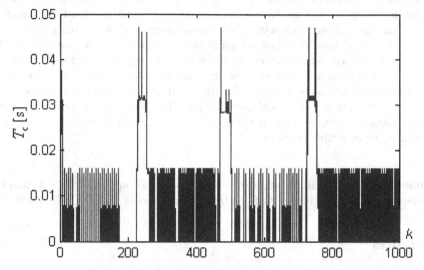

Fig. 4. Computing Time Analysis Only for Classical Method

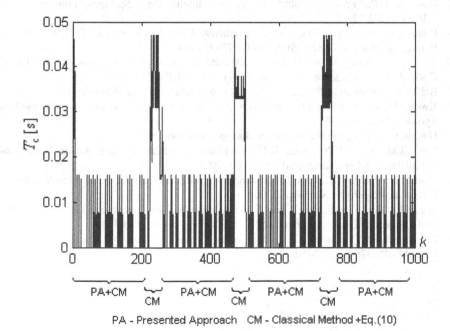

Fig. 5. Time Analysis for Proposed Approach Together with Classical Method

7   Conclusion

The improving of time-consuming approach was designed and realized for the predictive control of systems with fast dynamics. For higher horizons and many applied constraints it is needed to decrease the computational time in control algorithm. Without this modification, the performing of the algorithm can overload the sampling period. The classical approach tests, if the multidimensional free extreme task is appropriate for the optimization task of quadratic programming with constraints. This method may not pass the constraints conditions in all sampling periods of predictive control. The proposed approach uses the information from the control. In this situation some constraints can be reduced. The time-analyses confirmed, that better results were achieved using the presented modification. However, it is needed to eliminate the computational time in each sampling period of predictive control. This can be investigated in the further research.

Acknowledgment. The article was realized with financial support of IGA, Tomas Bata University in Zlín, Faculty of Applied Informatics number IGA/FAI/2015/006.

References

1. Camacho, E.F., Bordons, C.: Model Predictive Control. Springer, London (2007)
2. Corriou, J.-P.: Process control: theory and applications. Springer, London (2004) ISBN 1-85233-776-1
3. Rawlings, J.B., Mayne, D.Q.: Model Predictive Control: Theory and Design. Nob Hill Publishing, Madison (2009) ISBN 978-0-9759377-0-9
4. Kučera, V.: Analysis and Design of Discrete Linear Control Systems. Nakladatelství Československé akademie věd, Praha (1991) ISBN 80-200-0252-9
5. Balátě, J.: Automatické řízení, 2. vyd. BEN - technická literatura, Praha (2004)
6. Kwon, W.H.: Receding horizon control: model predictive control for state models. Springer, London (2005)
7. Huang, S.: Applied predictive control. Springer, London (2002)
8. Lee, G.M., Tam, N.N., Yen, N.D.: Quadratic Programming and Affine Variational Ine-qualities: A Qualitative Study. Springer (2005)
9. Dostál, Z.: Optimal Quadratic Programming Algorithms: With Applications to Variational Inequalities. Springer, New York (2009)
10. Wang, L.: Model Predictive Control System Design and Implementation Using MATLAB. Springer-Verlag Limited, London (2009) ISBN 978-1-84882-330-3.D.G
11. Luenberger, D.G., Yinyu, Y.E.: Linear and nonlinear programming. Springer, New York (2008) ISBN 978-0-387-74502-2

The Methods of Testing and Possibility to Overcome the Protection against Sabotage of Analog Intrusion Alarm Systems

Adam Hanacek and Martin Sysel

Tomas Bata University, Faculty of Applied Informatics,
Nad Stranemi 4511, 760 05 Zlin, Czech Republic
{ahanacek,Sysel}@fai.utb.cz
http://www.fai.utb.cz

Abstract. This paper deals with the testing of protection against sabotage in analog security systems. The first part of the paper is focused on the introduction to analog intrusion and hold-up alarm systems where there are described requirements of European standards and main parts of analog intrusion and hold-up alarm systems. In the second part, there are introduced two methods of testing the security and protection against sabotage of commonly used analog systems. In case of success one of the method many objects could be robbed. The third part of the paper describes the results of testing and the recommendation for future development in the field of analog security systems.

Keywords: Intrusion and Hold-up Alarm System, Detector, Control and Indicating Equipment.

1 Introduction

The main purpose of security systems lies in protection of life, health and properties. The European standard named EN 50 131-1 describes acronym "I&HAS" (Intrusion and Hold-up alarm system). It is possible to distinguish among intrusion alarm system and hold-up alarm system. Intrusion alarm system (IAS) is used in cases of finding out of intrusion of an intruder into protected object by the system. Hold-up alarm system (HAS) is used in cases of alarm activation intentionally.[1]

It is possible to choose among analogs, digitals or wireless control and indicating equipment for the protection of houses, cottages, garages or flats. The digital systems are the most appropriate for high level of protection against sabotage and interference; however, the price of digital systems is higher than the price of analog systems. The advantage of wireless connection is its simple installation. Nonetheless, the main disadvantages are its high susceptibility to interference, high costs and low protection against sabotage. Therefore, an analog system is very often used for protection of an object because of its lower price of each element. In addition, most signal sources especially the physical or chemical

© Springer International Publishing Switzerland 2015
R. Silhavy et al. (eds.), *Intelligent Systems in Cybernetics and Automation Theory*,
Advances in Intelligent Systems and Computing 348, DOI: 10.1007/978-3-319-18503-3_12

environment provides only an analog signal; however, it is possible to convert analog signal to digital and back using by A/D and D/A converters.[2],[3]

The main purpose of IAS includes discovering a presence of an intruder into secured area and sending an information about intrusion to a mobile phone or to an alarm receiving center. Sending an alarm message to a mobile phone is used the most often; however, an alarm receiving center is more appropriate for faster reaction. The first part of the work describes main elements of intrusion and hold up alarm system, the second part describes two methods of the testing of whole system and the last part is focused on results. Moreover, the last part describes possible threat of objects and subjects in case of using analog intrusion and hold up alarm system. The main parts of intrusion and hold-up alarm system contains:

− Control and indicating equipment
− Hold-up devices
− Detectors
− Keypad
− Output devices
− Alarm transmission system

Control and indicating equipment is elemetary component of I&HAS. It controls an activity of all connected components, provides the power for whole system, evaluates signals from keypad, saves an alarm history to a memory, receives alarm signals and sends an alarm information using by alarm transmission system to predetermined place.[4] The control and indicating equipment of I&HAS is divided on wireless or metallic. The main benefit of wireless connection between a control and indicating equipment and components consists in simply installation; however, the advantage of metallic connection consists in lower risk of false alarm caused by interference.

Hold-up device is facility which is used for inducing an alarm intentionally using by activation of public or hidden hold-up device.

The main purpose of detectors consists in guarding of predefined area and sending an alarm information to the control and indicating equipment in case of disturbing.

Intrusion alarm keypad is used for arming, disarming or programming of whole system.

Output devices have function as signalization equipment that usually represent optic or sound signalization and its purpose lies in frightening an intruder and in informing of surrounding area about object disruption.

Alarm Transmission System provides transfer of alarm messages from the secured object to alarm receiving center or to mobile phone by alarm transmission system.

Other used terms:

Periodic communication - "Periodic" means that at least one report should be realised in predefined interval to ensure that communication works.[5],[6],[7]

Expander - an electronic equipment used to extend the functionality of I&HAS.[8]

Interconnection - resources that help messages or signals with transmission between components I&HAS.[9]

Subsystem - part of an I&HAS located in a clearly defined part of the supervised premises capable of independent operation.[10]

Zone - assessed area where abnormal conditions may be detected.[11]

2   Methods of Testing of Analog Intrusion and Hold-Up Alarm Systems

This paper describes two procedures of testing of I&HAS to find out if it is possible to disconnect a detector without noticing by the control and indicating equipment. It is important to test the protection against sabotage in the field of security systems in order to ensure sufficient protection of life health and property. The effect of noise and climatic conditions on the detectors work is illustrated in [12],[13],[14].

2.1   Connection of Analog Control and Indicating Equipment to Each Component

Connection of each part to control and indicating equipment is shown in Figure 1. Each analog security system has 6, 8, or 16 analog zones and each detector is connected to a zone using by a circuit which has to be closed by terminating resistor to provide protection against sabotage. Depending on the design of security it is possible to connect more detectors to one circuit. Although, the possibility of exact identification of the alarm place decreases with increasing number of detectors connected to one zone. It is recommended to connect maximum seven detectors to one circuit. The principle of analog I&HAS consists in measuring the resistance of each circuit. Each detector has one alarm contact and one tamper contact connected to the circuit. The connecting of analog detectors to the CIE shows Figure 1. The contact "alarm" is connected in parallel with 1000Ω resistor. In case the ready state is activated, the resistance of whole circuit is roughly equal to closed resistor. Nevertheless, in case of intrusion of an intruder inside armed area is disconnected contact "alarm" and then the resistance of whole circuit is roughly equal summation of closed resistor with the resistor attached in parallel with the contact "alarm". The contact "tamper" is used for the detection of sabotage caused by opening the cover of detector. Each detector is also powered by the control and indicating equipment. The analog I&HAS uses 30% tolerance in case of measuring the resistance of the circuit due to different resistance depending on the temperature and different length of each circuit. Further, the analog control and indicating equipment does not react to alarm state for a short moment because of interference in the circuit; however, mentioned unmeasured time cannot be longer than 400ms. The most common uses unmeasured times are 100ms, 150ms, 250ms and 300ms. The main part of the work is focused on the testing of commonly used techniques of transmission messages in the security systems and necessity to digitize the transmission between analog components.

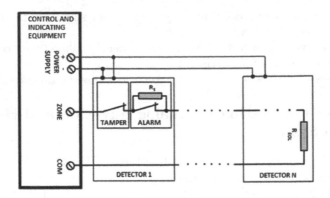

Fig. 1. Connecting of analog detectors to the I&HAS

2.2 The First Procedure of Testing of Analog Intrusion and Hold-Up Alarm Systems

The first procedure is based on the two conditions that have to be complied. The first condition is based on the fact that each control and indicating equipment has to have a time interval during which the resistance change is not recorded. Nevertheless, this time interval cannot exceed 400 ms. The second condition which has to be fulfilled lies in the need that the tolerance of the closed resistor is between 25-35%.

All testing was realized on the CIE named DSC PC 3000. The detector was connected to the circuit via connection described in the previous chapter.

The main parts of the first procedure of testing includes:

- Measuring of the voltage on terminating resistor
- Measuring of the current in circuit
- Connecting of alternate resistor
- Changing the switch to position "B"

The equipment used for testing of the first procedure are control and indicating equipment DSC PC 3000, keypad, voltmeter, ammeter, switch, detector, conductors and resistor.

Measuring of the Voltage on Terminating Resistor

It is necessary to insulate the conductor in the relevant circuit. In case of disconnecting the circuit, the control and indicating equipment sends an alarm message to an alarm received center or to a mobile phone. Connecting of a voltmeter shown in Figure 2 is identified by green color and by the number 2. There is pictured a parallel connection of a voltmeter for determining a voltage on the closed resistor. The most frequently used value of the closed resistor is 1000Ω. But there also exist control and indicating equipments that they are possible to change their ready state to 1100Ω, 2200Ω, 4700Ω or 5600Ω.

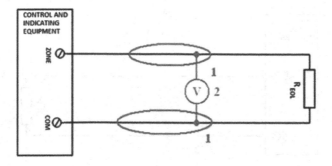

Fig. 2. Voltmeter connecting

Measuring of the Current in Circuit

Ammeter connecting is shown by light green color and by number 3 in Figure 3. The ammeter is connected to the switch between nodes "A" and "D". It is necessary to prevent disconnection of the circuit which the control and indicating equipment can note. The ammeter is connected with the switch to the circuit according to the Figure 3.

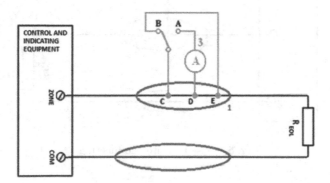

Fig. 3. Ammeter connecting 1

The next step of connecting the ammeter to the circuit consists in disconnecting the conductor between nodes "C" and "D". Further, it is necessary to put the switch to position "A". The situation is pictured in Figure 4. Once the switch is set to position "A", the ammeter displays the value of current which flows through the resistor. The range of ammeter is advised to be chosen in mA units, because lower ranges cause increasing of internal resistance of the ammeter and greater ranges cause higher inaccuracy of measuring. Once the switch is flipped to position "A", it is necessary to disconnect conductor between nodes "B" and "E".

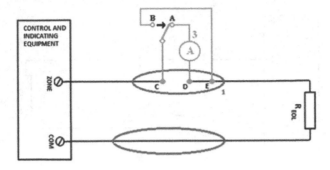

Fig. 4. Ammeter connecting 2

Alternate Resistor Connecting

Purple color in Figure 5 shows the resistor "RA" connected to the nodes "B" and "E" which replaces the terminated resistor. The value of the resistor "RA" was calculated from the current and voltage values that were measured in the previous points.

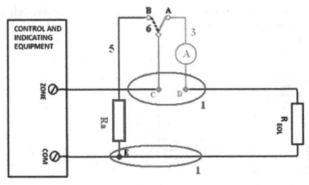

Fig. 5. Closed resistor connecting

Changing the Switch to Position "B"

Changing the switch to position "B" is pictured by blue color and by the number 6 in the Figure 5. This change causes that the current does not flow through the terminated resistor anymore, but flow through the resistor "RA". Switching has to be fast to the CIE does not note this short interruption. After this switching the CIE will not respond to any change in detectors connected to the circuit.

2.3   The Second Procedure of Testing of Analog Intrusion and Hold-up Alarm Systems

The second procedure of testing I&HAS capitalize on the tolerance of closed resistor between 25-35%.

The procedure is based on resistance reduction of the terminated resistor by the potentiometer connected in parallel and on resistance increase of the potentiometer connected in series.

The second procedure includes:

- Connecting of the first potentiometer in parallel with terminated resistor
- Connecting of the second potentiometer in series with an ammeter
- Reduction the resistance of the potentiometer connected in parallel with terminated resistor and increasing the resistance of potentiometer connected in series with terminated resistor

The equipment used for testing of the second procedure are control and indicating equipment DSC PC 3000, keypad, ammeter, two potentiometers, detector and conductors.

Connecting of the First Potentiometer in Parallel with the Terminated Resistor

The first step of the second procedure lies in connecting of the first potentiometer to terminated resistor in parallel. This step is shown in the Figure 6. The potentiometer has to be set to maximum resistance for ensuring that there is a minimal change of resistance in the circuit. The maximal resistance of the connected potentiometer in parallel is $22k\Omega$.

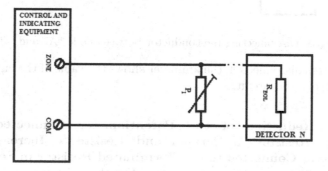

Fig. 6. Connecting a potentiometer in parallel with terminated resistor

Connecting of the Second Potentiometer in Series with an Ammeter

The second part of the second procedure is based on the connection of the second potentiometer with an ammeter to the circuit and then on the disconnection of a conductor.

This step is shown in Figure 7. The potentiometer and the ammeter are connected between nodes "A" and "B". Usually, the ammeter can't be connected in parallel; however, the voltage on the conductor between nodes "A" and "B" is too low for breaking the ammeter.

Despite of really low voltage between nodes "A" and "B", it is necessary to disconnect the conductor between nodes "A" and "B" as soon as possible. Once

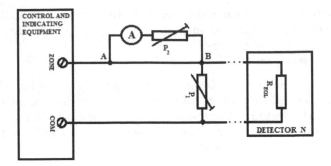

Fig. 7. Connecting of the second potentiometer with an ammeter in series

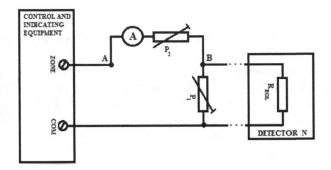

Fig. 8. Disconnecting the conductor between node "A" and "B"

the conductor is disconnected, the ammeter shows the value of the current which flows throughout the circuit.

Resistance Reduction of the Potentiometer Connected to the Terminated Resistor in Parallel and Resistance Increase of the Potentiometer Connected to the Terminated Resistor in Series

The last step of the second procedure is based on the resistance reduction of the potentiometer connected to terminated resistor in parallel and on the resistance increase of the potentiometer connected to terminated resistor in series. It is necessary to monitor and keep the same current flowing throughout the circuit.

3   Summary of Results and Future Recommendations

In the chapter "Methods of Testing of Analog Intrusion and Hold-up Alarm Systems", there are described two methods of testing the security of analog I&HAS. The procedures are effective if an intruder gains an access to the circuit. However, according to the standards the connection between the CIE and detectors does not has to be protected against a sabotage and also does not has to be placed in the armed zone and a many objects have connection between CIE and detectors placed in unarmed zone. The first procedure was applied 10 times while the CIE did not detect any disconnection of the detector 8 times. Procedure success is equal to 80%; however, for testing was used an ordinary switch. It is possible to use more suitable switches for which the switching takes less time. This way it could be possible to increase the efficiency of described procedure. The second procedure was also applied 10 times while the CIE did not detect any disconnection of the detector 9 times; therefore, the second procedure is more effective than the first procedure. This situation can cause the threat of many objects. In addition, the price of goods that may be stolen is just a part of detriment. Ordinarily the most important are data in notebooks or personal computers. It is recommended to accelerate the reaction of analog inputs and decrease the closed resistor tolerance to ensure data transmission security; however, mentioned problem can be also solved by transmitting messages using bus named controller area network(CAN). The other possible way how to solve mentioned problem lies in utilization of buses that are used in computational science. This enables data to be transmitted digitally and safety.

4   Conclusion

The paper introduced two methods of testing the protection against sabotage of commonly used analog security systems. The results are described in the chapter "Summary of Results and Future Recommendations". From mentioned chapter is apparently the possibility to overcome the analog security systems. Chapter "The First Procedure of Testing of Analog Intrusion and Hold-up Alarm Systems" and "The Second Procedure of Testing of Analog Intrusion and Hold-up Alarm Systems" shows the possibility to disconnect the detector without sending an alarm message and without activation alarm outputs by control and indicating equipment. In the chapter "Summary of Results and Future Recommendations", there are illustrated results of two methods of testing the security. Both methods were used 10 times and the success of the first method is equal 80% and the success of the second method is equal 90%. The burglary also depends on knowledge of circuits that connect the control and indicating equipment with detectors and on finding a way how to get to these circuits. However, European norm EN 50 131-1 ed.2 does not require arming connection among detectors and control and indicating equipment; therefore, it is possible to overcome a lot of object and gain access to the armed area. Analog security system cannot be recommended for building security. It is suggested to utilize already used digital systems despite their higher acquisition costs or to digitize the transfer between analog elements using by buses that are commonly used in computer science.

Acknowledgments. This work was supported by grant No. IGA/FAI/2014/043 from IGA (Internal Grant Agency) of Thomas Bata University in Zlin.

References

1. CSN. EN 50131-1 ed. 2, 40 s. Office for standards, Prague (2007)
2. Jaroslav, L., Drtina, R., Sedivy, J.: Special circuits of operational amplifiers for measurement purposes. International Journal of Circuits, Systems and Signal Processing 8 (2014), doi:1998-4464
3. Lonla, B., Mbihi, J., Nneme, L., Kom, M.: A novel digital - to - analog conversion technique using duty - cycle modulation. International Journal of Circuits, Systems and Signal Processing 8 (2013), doi:1998-4464
4. Design of a Battery - Considerate Uninterruptable Power Supply Unit for Network Devices. International Journal of Circuits, Systems and Signal Processing 7 (2013), doi:1998-4464
5. CSN. EN 50131-7, 48 s. Office for standards, Prague (2010)
6. CSN. EN 50131-1 ed. 2: A1, 12 s. Office for standards, Prague (2010)
7. CSN. EN 50131-1 ed. 2: Z2, 20 s. Office for standards, Prague (2011)
8. Horník, J.: Model of intelligent home security. Bachelor. Czech Technical University in Prague, Prague (2010)
9. Kindl, J.: Designing of security systems. Zlín, Learning text universities. Tomas Bata University (2007) ISBN 978-80-7318-554-1
10. Černý, J.J., Ivanka, J.: Systemization of security industry I. Zlín. Learning text universities. Tomas Bata University (2006) ISBN 80-7318-402-8
11. EN 50131-1. Alarm systems - Intrusion and hold-up systems. Brussels: rue de Stassart 35 (2006)
12. Bykovyy, P., Kochan, V., Sachenko, A., Markowsky, G.: Genetic algorithm implementation for perimeter security systems CAD. In: 4th IEEE International Workshop on Intelligent Data Acquisition and Advanced Computing Systems: Technology and Applications (IDAACS 2007), Dortmund, pp. 634–638 (2007)
13. Bykovyy, P., Pigovsky, Y.: Multicriteria synthesis of alarm perimeter security systems within undefined noise influence. In: Proceedings of 11th International Conference "Modern Information & Electronic Technologies", Odessa, vol. 1, p. 88 (2010) (in Ukrainian)
14. Bykovyy, P., Pigovsky, Y., Kochan, V., Vasylkiv, N., Karachka, A.: Assessment of probabilistic parameters of alarm security detectors taking uncertain noise into account. IEEE, Prague (2011) ISBN

Universal System Developed for Usage in Analog Intrusion Alarm Systems

Adam Hanacek and Martin Sysel

Tomas Bata University, Faculty of Applied Informatics,
Nad Stranemi 4511, 760 05 Zlin, Czech Republic
{ahanacek,Sysel}@fai.utb.cz
http://www.fai.utb.cz

Abstract. This work deals with the universal system which was basically developed for usage in the field of intrusion and hold-up alarm systems; however, it is also possible to connect analog components used for fire detection. The main advantage lies in resolving the problem with low protection against sabotage in analog systems and high cost of the digital systems. The introduction is focused on the current situation in the field of security systems and on the description of the bus named Controller Area Network which is used for the communication between connected devices. The main part of the work is focused on the description of the developed system which includes a method of devices connecting and a method of connecting a control and indicating equipment. Further, the communication protocol is described and the evolved system is tested in laboratory. The last part of the work contains the results and applicability of the developed system.

Keywords: Intrusion Alarm System, Fire Alarm System, Controller area network.

1 Introduction

Intrusion and hold-up alarm systems comply with standard CSN EN 50131 which specifies the terms alarm system for the detection of entry and alarm system for the detection of attack.[1] An Intrusion Alarm System (IAS) is system that was basically developed for the protection of properties; however, IAS also protects a life thanks to the possibility to inform the owner of the secured object using by short message service (SMS) or by activation of its outputs. Therefore, the next advantage of IAS lies in frightening possible intruder by activation of the object siren. On the other hand, a Hold-up Alarm System (HAS) cannot be activated automatically; therefore, a HAS is activated deliberately by pressing a switch. Thus, the siren of the object is not activated and an alarm is hidden. Unstable situation in the world shows necessity to improve the protection against sabotage in I&HAS and reduce the cost of security systems. Currently, it is possible to choose among analogs, digitals or wireless control and indicating equipment for object security. The digital systems are the most appropriate for high level

© Springer International Publishing Switzerland 2015
R. Silhavy et al. (eds.), *Intelligent Systems in Cybernetics and Automation Theory*,
Advances in Intelligent Systems and Computing 348, DOI: 10.1007/978-3-319-18503-3_13

of protection against sabotage and interference. However, the price of digital systems is higher than the price of analog systems. The advantage of a wireless connection between a CIE and detectors in comparison with a wire connection lies in simple installation and greater level of protection against sabotage. Nonetheless, the main disadvantage lies in greater susceptibility to interference and increased costs. Therefore, an analog system is very often used for the protection of objects because of the lower price of each element. In addition, most signal sources especially from the physical or chemical branches provides only an analog signal; however, it is possible to convert an analog signal to a digital and back using A/D and D/A converters.[2],[3] Mentioned analog communication is the most widely used for the protection of objects; however, this system has a trouble with the possibility to disconnect the detector without detection by an evaluation device.

1.1 Standard Connection of Alarm Circuits Used in Analog Security Systems

Fig. 1 illustrates the method of most commonly used connection of analog detectors to the control and indicating equipment (CIE). The connection is called NC(Normally Closed). An alarm can be activated by switching off one of contacts connected to the circuit.

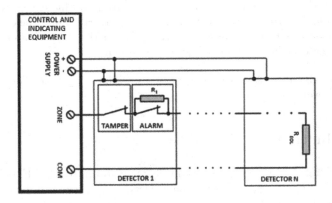

Fig. 1. Normally closed connection.[4]

Figure 1 shows connection of security detectors to the CIE. It is possible to connect standard fire detectors by the same way; however, fire detectors are usually connected by connection type NO (Normally Opened).

2   Current Situation and Proposed Solution

Current situation in the field of intrusion and hold-up alarm systems (I&HAS) shows an increasing ability of intruders to overcome security systems and gain access to an object. On the other side, security systems are currently very expensive for many subjects to protect their poverty; therefore, it is important to reduce the threat of sabotage by improving security systems. Further, it is also important to look for possible ways to reduce the cost of intrusion and hold-up alarm systems. Event so, researchers still have to find a way how to develop security system with sufficient protection against sabotage and with low cost of whole system.

The work deals with the possible solution of mentioned problem with low protection against sabotage in analog security systems and high price of digital systems using by adding a microprocessor named LPC11C24 with bus named controller area network (CAN) to each component that uses analog transmission. This enables data to be encrypted and transmitted by the protocol designed for mentioned CAN bus.

2.1   Controller Area Network

The bus named controller area network (CAN) was basically designed for usage in automotive industry, but currently it is often implemented in development kits. CAN bus is based on the CSMA/CR mechanism to prevent frame collisions during transmission messages on the bus.[5],[6] The CAN network protocol has been defined with the following features and capabilities:[7]

- Message priority
- Multi cast communication
- System-wide data format
- Error frame detection
- Detection of permanent failures in nodes and isolate faulty node

CAN is a high-integrity serial data communications bus for real-time applications and more cost effective than any other serial bus systems. Some another advantages are written below.[8],[9]

- Data rates of transmission up to 1 Megabit per second
- Length of CAN BUS up to 1 kilometer
- Is an international standard: ISO 11898

There are defined four types of messages in CAN protocol. [12]

- Data frame
- Remote frame
- Error frame
- Overload frame

The thorough description of layers, message frames and communication speed is explained in [11],[12]. Two wires used for transmission a message across the CAN bus are named CAN_L and CAN_H. Described bus has two states named dominant and recessive. The dominant state represents level log 0 transmitted by CAN bus and the recessive represent level 1 transmitted by CAN bus.

Further, the CAN bus supports a high speed and a low speed transmission. The low speed is used to long wire transmission. The recessive state represents zero voltage difference among CAN_L and CAN_H in case of using high speed transmission.[13],[14],[15]

3   Description of the Developed System

As it was mentioned in the introduction, the system was designed for usage of standard analog security components with the exception of adding a processor with a CAN bus to the control and indicating equipment and to each component. Then, the communication will be realized digitally and encrypted on the CAN bus by the created protocol. In Figure 2 is shown the method of connection of a detector to the designed system. The processor has to be located inside the detector to eliminate the possibility of a sabotage. Contacts named "tamper" and "alarm" are connected by evaluating circuit to the processor LPC11C24 which is further connected to the CAN bus. In the case of switching off one of the contacts, an alarm is transmitted to the processor LPC11C24 that send an alarm message to the CAN bus.

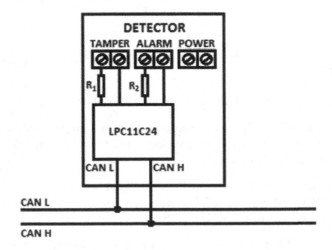

Fig. 2. Connection of detectors in the designed system

Figure. 3 provides the view of connection the CIE to the CAN bus. The circuits with closed resistor are connected to the CIE of intrusion and hold-up alarm system. Relay contacts are connected in parallel to the closed resistor. Switching relays is controlled by the processor LPC11C24 which is further attached to the CAN bus. The main purpose of relay contacts lies in resistance change in the evaluating circuit connected to the CIE. This way it is possible to send an alarm message from CAN bus to the CIE. The processor LPC11C24, relays and circuits

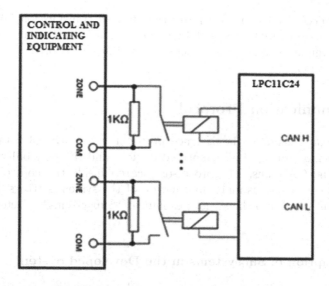

Fig. 3. Method of connection of the CIE to the bus

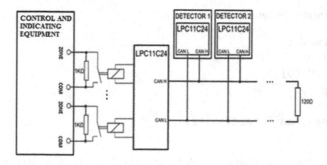

Fig. 4. Comprehensive view of the system

with closed resistor have to be placed inside the CIE of security system for the elimination of possible sabotage.

3.1   Transmission of Alarm Messages in the Designed System

In Figure 4 is depicted comprehensive view of the system. In the case of intrusion of an intruder into a protected object, an alarm message is sent to the processor LPC11C24 which is located inside the detector of secured area. The main purpose of mentioned processor lies in transmission of an alarm message to the another processor LPC11C24 which is located inside the control and indicating equipment of I&HAS. The maximum theoretical number of nodes connected to one CAN bus is 100; however, the maximum number depends on the speed of CAN bus and on the conductor used for connection. The designed protocol allow 60 connected nodes in order to prevent data transmission failure. Used CAN bus was configured for speed 125kbit/s and a length of 400m. It is necessary to terminate controller area network by 120Ω resistor to ensure correct data transmission.

4 Communication Protocol

The main requirements on the designed protocol are security of data transmission, regular monitoring of all connected devices and the possibility of system arming by the CAN bus. The processor located inside the control and indicating equipment contains main information of the system settings. Mentioned information is saved into the structures named "Subsystems", "Detectors" and "Timer".

4.1 Description of Subsystems in the Developed System

The evolved system allow four subsystems and main setting of each subsystem is stored in the structure named "Subsystems".

Subsystems in the designed protocol

```
typedef struct {
UINT32 Active;
UINT32 Det_ID[50];
UINT32 Output;
} Subsystems;
extern Subsystems Subsystem[4];
```

The command "extern Subsystems Subsystem[4];" creates four structures, where each structure has two variables named "Active" and "Output" and one field named "Det_ID". The first variable named "Active" contains the information about an activity of the subsystem and the first field named "Det_ID[50]" contains the information of detectors assigned to the subsystem. The second variable named "Output" contains the pin activation in case of alarm activation using by the controller area network.

4.2 Description of Connected Detectors

Information about each detector is saved in the structure named "Detector" which can be seen below.

The main information of detectors connected to the integrated system

```
typedef struct {
UINT8 ID;
UINT32 TimeM;
UINT8 Start_Key[24];
UINT8 Key[24];
} Detectors;
extern Detectors Detector[60];
```

Each detector has two variables named "ID" and "TimeM" and two field named "Start_Key" and "Key". The purpose of the first variable denominated "ID" lies in exact identification of the detector in the system by using modifier of the message. The variable "TimeM" is used for saving the time of last identification of the detector and the field "Start_Key" provides the basic encrypted key which is used to start the communication in case of power failure or in case of data read failure caused by electromagnetic interference. The second field of the structure "Detectors" is denominated "Key". In the mentioned field, there is saved actual key which encrypts data during a communication. Each detector changes its own encrypted key every 60 seconds. There are created 60 structures named "Detector" using by the command "extern Detectors Detector[60];" in the developed system; therefore, it is possible to connect 60 detectors to one CAN bus. It is important to set lower ID to I&HAS devices and higher ID for fire devices to ensure higher priority for I&HAS in case of utilizing the developed system for I&HAS and also for the detection of fire alarm.[16]

4.3   Usage of Time in the Developed System

In case of disturbing of an intruder into protected object, an alarm message is sent to the processor located inside the control and indicating equipment using by the technique described in the chapter 3.1. Then, the alarm is saved to a flash memory with the current time which is saved in the structure below the text.

Usage of time in the designed integrated system

```
typedef struct {
UINT32 Sec;
UINT32 Min;
UINT32 Hour;
} Timer;
extern Timer Time;
```

The structure contains variables named "Sec", "Min" and "Hour". The variable "Sec" contains the information about the number of second, variable "Min" contains the information about the number of minutes and variable "Hour" contains the information about the number of hours in actual time.

4.4 Sending a Message to the CAN bus

If an alarm is activated from the place of a processor input, the message is sent to the CAN bus by the technique which is described below.

Sending a message to the CAN bus

```
tdes_enc(&Data, &Can_Info.Alarm1, &Det.Key);
memcpy(&msg_obj.data, &Data, 8);
msg_obj.mode_id = DETECTOR_ID;
msg_obj.mask = 0x0;
msg_obj.dlc = 8;
(*rom)->pCAND->can_transmit(&msg_obj);
```

The message is encrypted by 3DES and saved in the field called "Data". Further, the data are copied into the data field of the message. The identifier of a message contains ID of the detector, followed by setting the mask, the message length is set to 8 bytes and command "(*rom)→pCAND→can_transmit(&msg_obj)" sends message to the CAN bus.

4.5 The Main Features of the Proposed Protocol Incudes:

- Regular monitoring of all connected devices every four seconds. An information about its activity is sent to a CAN bus by the processor located inside a detector. The identifier of each message contains network ID which is used for the identification in the network. The message is then identified by the development kit NXP OM13036. An alarm is switched on at the moment, when any detector interrupts sending the information about its activity.
- Save an alarm message to a flash memory already at each detector.
- The possibility of activation of multiple output devices already at each detector. In case of using the proposed system, it is possible to activate more output devices connected to each detector by using GPIO pins of the processor.
- Exact identification of the place of an alarm. One of main problems of analog systems lies in impossibility of exact identification of an alarm place. If described system is used in practice, it is possible to locate the place of an alarm exactly using by ID which every detector uses for reporting to the bus.
- Designed system can arm or disarm an intrusion and hold-up alarm system by the code keyboard connected to the CAN bus.
- Communication is encrypted by 3DES.
- Protection against possible attack is minimized by regular change of the encryption key every 60 seconds.

4.6 Testing of the Developed System

The developed system was tested in the laboratory environment at Faculty of Applied Informatics in Zlin. The name of control and indicating equipment was

DSC 3000. Detectors used for testing were passive infrared sensor PARADOX PRO+476, passive infrared sensor PARADOX - PRO+476 with animal immunity and 8x PARADOX - DG467. The testing also includes twelve development kits with the processor LPC11C24.

5   The Price Comparison

The minimum price of each detector in analog intrusion and hold-up alarm systems is approximately equal 12€ and the maximum price of each detector in analog intrusion and hold-up alarm systems is roughly equal 40€; however, digital detectors are much more expensive. Commonly price of digital detector which are currently uses in intrusion and hold-up alarm systems is from 40€ to 150€. Mentioned difference among analog and digital detectors is caused by essential of each producer to invest into development of own digital communication protocol and digital detectors. The price of each element in case of uses designed system is equal to the price of analog detector increased by the processor LPC11C24 with CAN bus which is added to each element. This can increase the price of each element by 8€. Despite of mentioned necessity to add processor LPC11C24, the finally price of each element in case of used descripted system is approximately equal one-third of the price of digital detector currently uses in I&HAS.

6   Conclusion

The paper clarifies the possible solution of low protection against sabotage in analog security systems and high price of digital systems using by developed system which combines the advantages of analog and digital communication in an intrusion and hold-up alarm systems. It is possible to choose from wide offer of analog detectors in comparison with limited selection of digital detectors and connect analog detectors to the developed system. Then, the communication between components and a CIE is digital and encrypted. Therefore, the created system provides better protection against sabotage than commonly used analog systems. Although, the currently uses digital systems are the most appropriate from the viewpoint of protection against sabotage, the developed system provides better protection against sabotage for the ability to not only encrypt all data, but also for the ability to change its own encrypted key every 60 seconds. According to the standards the communication does not have to be encrypted and the encrypted key does not have to be changed every 60 seconds; however, encryption is necessary to ensure the security of data transmission. One of the other benefits of described system lies in exact identification of the detector which activated the alarm. Moreover, the I&HAS system can be activated by the code keyboard connected to the CAN bus.

Acknowledgments. This work was supported by grant No. IGA/FAI/2014/043 from IGA (Internal Grant Agency) of Thomas Bata University in Zlin.

References

1. CSN. EN 50131-1 ed. 2, 40 s. Office for standards, Prague (2007)
2. Jaroslav, L., Drtina, R., Sedivy, J.: Special circuits of o perational amplifiers for measurement purposes. International Journal of Circuits, Systems and Signal Processing 8 (2014), doi:1998-4464
3. Lonla, B., Mbihi, J., Nneme, L., Kom, M.: A novel digital - to - analog conversion technique using duty - cycle modulation. International Journal of Circuits, Systems and Signal Processing 8 (2013), doi:1998-4464
4. Hanek, A.: Using Personal Computer as an Intruder Alarm System Control Unit. Zlín, Diploma thesis. Tomas Bata University (2012)
5. Xiao, H., Lu, C.-G.: Modeling and simulation analysis of CAN-bus on bus body. In: 2010 International Conference on Computer Application and System Modeling (ICCASM), October 22-24, vol. 12, pp. V12-205–V12-208 (2010), doi:10.1109/ICCASM.2010.5622238
6. Jiang, Y., Liang, B., Ren, X.: Design and implementation of CAN-bus experimental system. In: 2011 6th International Forum on Strategic Technology (IFOST), August 22-24, vol. 2, pp. 655–659 (2011), doi:10.1109/IFOST.2011.6021111
7. Ibrahim, D.: Controller Area Network Projects. Elektor International Media, United Kingdom (2011) ISBN 978-1-907920-04-2
8. Voss, W.: A Comprehensible Guide to Controller Area Network, 2nd edn. Copperhill Media Corporation, Canada (2005) ISBN 0976511606
9. Li, R., Wu, J., Wang, H., Li, G.: Design method of CAN BUS network communication structure for electric vehicle. In: 2010 International Forum on Strategic Technology (IFOST), October 13-15, pp. 326–329 (2010)
10. Vehicle Test Data Visualization and Processing. Prague, Diploma thesis. Czech Technical University in Prague (2011)
11. Spurný, F.: Controller Area Network. Automatizace 41(7), 397–400 (1998)
12. Device for transmitting messages on CANbus. Plzen, Diploma thesis. University of West Bohemia (2012)
13. Shweta, S.A., Mukesh, D.P., Jagdish, B.N.: Implementation of controller area network (CAN) bus (Building automation). In: Unnikrishnan, S., Surve, S., Bhoir, D. (eds.) ICAC3 2011. CCIS, vol. 125, pp. 507–514. Springer, Heidelberg (2011)
14. Understanding and Using the Controller Area Network Communication Protocol. Springer Science Business Media, London (2012) ISBN 978-1-4614-0313-5
15. Measurement system with CAN interface. Plzen, Diploma thesis. University of West Bohemia (2013)
16. Tan, X.-P., Li, X.-B., Xiao, T.-L.: Real-time analysis of dynamic priority of CAN bus protocol. In: 2010 International Conference on Electronics and Information Engineering (ICEIE), August 1-3, vol. 1, pp. V1-219–V1-222 (2010)

Embedded Supervisory Control and Output Reporting for the Oscillating Ultrasonic Temperature Sensors

A. Hashmi, M. Malakoutikhah, R.A. Light, and A.N. Kalashnikov

Department of Electrical and Electronic Engineering,
The University of Nottingham, Nottingham, NG7 2RD, UK
{eexah17,eexmm18,roger.light,
alexander.kalashnikov}@nottingham.ac.uk

Abstract. Ultrasonic temperature sensors can potentially outperform conventional sensors because they are capable of very fast sensing across the complete ultrasound pathway, whilst conventional sensors only sense temperature at a single point and have substantial thermal inertia. We report recent developments in electronic instrumentation for oscillating ultrasonic temperature sensors with the aim of achieving high accuracy and low scatter at a low cost.

Keywords: temperature sensor, ultrasonic instrumentation, ultrasonic NDE.

1   Introduction to Ultrasonic NDE Sensors

Ultrasonic sensors utilise ultrasonic waves for non-destructive or non-invasive probing of media or objects of interest. These sensors consist of at least one ultrasonic transducer to transmit and receive ultrasonic waves (or two separate transducers, one for reception and one for transmission) and supporting electronics [1]. Most applications of ultrasonic sensors are concerned with finding the voids or discontinuities from which the waves reflect in opaque objects or media. Examples include underwater sensors for locating fish and marine navigation; air sonars for range finding in construction and used as parking sensors; medical ultrasonic imaging and some other non-destructive testing, detection of obstacles, proximity sensing and imaging applications (e.g. [2]).

Another group of ultrasonic sensors is used to evaluate changes in the object or medium where the ultrasonic waves propagate; for example, non-destructive evaluation (NDE) for quality control or online process monitoring. In those types of sensors, changes in the ultrasound propagation parameters (amplitude and/or time of flight - TOF), sometimes across a range of frequencies for ultrasonic spectroscopy, are measured in order to evaluate the state of the wave propagation environment.

Compared to the majority of sensors that operate based on other physical principles, NDE ultrasonic sensors can sense the environment across the complete ultrasonic pathway instead of only a single point. This feature allows one to obtain "averaged" or "integrated" estimates using only one or two ultrasonic transducers without the need to install a number of conventional sensors, such as thermistors, to find the average temperature in a process vessel.

Another advantage of NDE ultrasonic sensors relates to the fact that the environment of interest is employed as part of the sensor itself without the need for any

© Springer International Publishing Switzerland 2015

R. Silhavy et al. (eds.), *Intelligent Systems in Cybernetics and Automation Theory*,
Advances in Intelligent Systems and Computing 348, DOI: 10.1007/978-3-319-18503-3_14

intermediation. Correct reading of most temperature sensors requires the sensor to first attain thermal equilibrium with the environment and that can take up to several seconds or even tens of seconds depending on the conventional sensor's thermal inertia. In contrast, ultrasonic waves propagate hundreds of metres in gases and thousands of metres in liquids and solids in just one second, enabling faster response to changing process conditions or potential thermal runaways.

Ultrasonic measurements frequently involve balancing between two contradictory requirements. On one hand, increasing the operating frequency of the transducer(s) decreases both the ultrasound wavelength and the time period, thus improving both the spatial and temporal resolutions. On the other hand, increasing the operating frequency leads to a rapid increase in the ultrasonic wave's attenuation, which reduces the signal-to-noise ratio (SNR) at the output of the ultrasonic receiver. Insufficient SNR could lead to substantial uncertainty of the measurement results [3]. In those cases, the ultrasonic pathway may need to be reduced in order to restore the SNR to an acceptable level. The cost of the transducer(s) is another important consideration that affects the selection of the operating frequency of an ultrasonic sensor. High frequency ultrasonic transducers (operating at tens of MHz and above) can cost over one-thousand dollars each, whilst mass produced devices operating at 25 or 32 or 40 kHz can be bought for a few dollars in large quantities.

Our research group aims to develop inexpensive ultrasonic sensors that can outperform their traditional counterparts (e.g. [4-7]). Cost requirements force the utilization of mass produced transducers operating in the 20 to 50 kHz frequency range. Section 2 discusses several electronic architectures for these sensors of which we believe the oscillating sensor architecture is most advantageous. In addition to the circuitry that is required to sustain the oscillations and keep the sensor within the desired operating conditions, oscillating ultrasonic sensors should be equipped with a microcontroller that provides supervisory control of electronics, measurement of the output frequency and the ability to communicate that measured frequency or the related process parameter to the data consumer. The output frequency should be measured with high resolution and accuracy (e.g. 100 ppm uncertainty may not be sufficient for some temperature measurements) using one out of several approaches discussed in Section 3. Section 4 summarises our experiences with various implementations for band pass filters (BPFs) that are required to keep the operating frequency of an oscillating sensor within a particular range. Section 5 presents recent developments for the amplifier, required to compensate for the energy losses in the sensor loop, related to the addition of the automatic gain control that enables scatter reduction of the output frequency of a sensor. The recent design of the phase shifter, needed to tune the output frequency of the sensor to the required value at the particular calibration point, is presented in Section 6. Section 7 concludes this paper.

2   Comparison of Electronic Architectures for Ultrasonic NDE Sensors

Ultrasonic NDE sensors can be used to measure and monitor various physical quantities using several arrangements of ultrasonic transducers [1,8]. More specifically, we focus our discussion on ultrasonic temperature measurements of water using two separate ultrasonic transducers fixed against each other at the boundaries of the water containing vessel (through transmission arrangement). Ultrasonic sensing utilises the ultrasound velocity that is dependent upon the temperature; for example, for water that dependency varies from 1482.36 m/s at 20°C to 1509.14 m/s at 30°C with a quoted measurement uncertainty of less than 0.02 m/s [9]. In order to measure temperature with an uncertainty and/or resolution of 0.1°C, one needs to achieve the relative uncertainty/resolution of measured TOF that can be estimated using the following equation:

$$\frac{(1509\text{m/s} - 1482\text{m/s})/(30°C - 20°C)}{(1509\text{m/s} + 1482\text{m/s})/2} \times 0.1°C \approx 1.8 \times 10^{-4}. \tag{1}$$

Let us consider several electronic architectures for the measurement of ultrasound TOF with the aim of determining how the above uncertainty can be achieved at low cost, assuming that we are interested in measurements for a typical process pipe with a 10 cm diameter where the expected TOF is around:

$$0.1\text{m} / 1500 \text{ m/s} \approx 60 \text{ μs}. \tag{2}$$

The first option relates to direct measurement of the TOF using the setup presented in Fig.1 (here, and thereafter, the amplifier is used to compensate for the propagation and energy conversion losses).

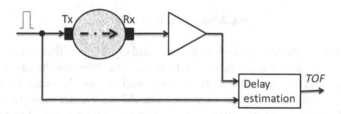

Fig. 1. Direct TOF measurement architecture

The delay estimation block measures the time interval between the instants of detection of the excitation pulse at input A and detection of the propagated pulse at input B. The time reference is provided by the clock oscillator. The delay estimator can be built to provide time resolution better than the period of the clock pulses. Examples of sub-sample delay estimation include cross-correlation processing (the shape of the pulse should not change much during its propagation, which holds in the being considered case), using the centre of gravity instants of both pulses to estimate the TOF ([10]) or by linearly interpolating samples of different signs to find the first zero crossing points for both pulses [11]. All these methods require considerable computing power, which would increase the cost of the sensor. A more affordable solution would simply involve counting the clock pulses between the detection of the two above mentioned pulses. In this case, the period of the clock oscillator should be smaller than the measured time interval by the inverse of the required resolution, i.e. the number of clock pulses counted during the measured time interval should be greater or equal to the inverse of the required resolution. In the considered case, this requirement translates to the clock oscillator period of:

$$appr.\ 60\ \mu s \times 10^{-4} \approx 6\ ns, \tag{3}$$

hence, the reference clock frequency needs to be around 150 MHz. Such a high frequency is difficult to use in low cost instrumentation; thus, this approach will only become feasible for pipes with larger diameters. Another potential problem with this approach is the potential jitter and uncertainty related to the pulse detections due to the additive noise presence.

TOF measurement can also be implemented by re-circulating a pulse (sending a pulse into water as soon as a pulse is detected at the receiver). The block diagram for an instrument implementing this "sing-around" architecture is presented in Fig. 2.

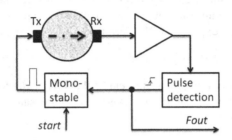

Fig. 2. Sing-around architecture

This arrangement enables one to measure the number of pulses that re-circulate over a known time interval and estimate the TOF as the ratio between the known measurement interval and the number of the re-circulated pulses. In order to achieve the required resolution, the number of pulses should be greater than the inverse of the required resolution. Consequently, the measurement time required to complete the measurement will be around:

$$appr.\ 60\ \mu s\ /\ 10^{-4} \approx 0.6\ s. \tag{4}$$

Although, in many cases, this measurement time is not prohibitive, consistent jitter-free detection of the received pulses may be complicated by the inevitable presence of additive noise.

Measuring the phase shift between the continuous sine wave supplied to the transmitter and the output wave at the receiver $\Delta\varphi$ (Fig. 3) could also be used to evaluate the TOF τ from the following equation:

$$\Delta\varphi = 2\pi f\tau \ => \ \tau = \Delta\varphi\ /\ (2\pi f), \tag{5}$$

where f is the frequency of the sine wave that is kept the same as the resonant frequency of the transducers in order to increase the SNR. For ultrasonic frequencies (f > 20 kHz) the phase shift in the considered case would be greater than 2π (20 kHz × 60 μs > 1) and the phase shift estimator could only evaluate the fractional part.

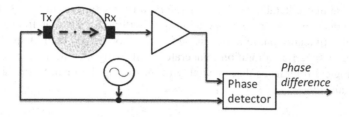

Fig. 3. Phase shift measurement architecture

Inexpensive phase shift estimators count reference pulses gated at the instants when the sine waves of interest cross the same amplitude level (e.g. zero crossing). To achieve the required resolution, the number of reference pulses for the complete period of the sine wave should be greater than the inverse of the resolution. For the lowest 20 kHz ultrasonic frequency, the period of the sine wave is 50 μs, which is even smaller than the TOF in the considered case. Therefore the frequency of the reference pulses should be even higher than that in the case of direct TOF measurement architecture.

The oscillating ultrasonic sensors are attractive by their potential simplicity (only a single amplifier is required to make the sensor work, Fig. 4) and their

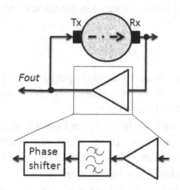

Fig. 4. Oscillating architecture

potentially shorter measurement time as compared to the sing-around ultrasonic sensors. The oscillating ultrasonic sensors, which have been developed in our group to date, oscillate at frequencies in the range 30-200 kHz, requiring less than

$$1/(30\text{ kHz} \times 10^{-4}) < 0.3\text{ s} \qquad (6)$$

in order to measure the sensor's output frequency to the required accuracy. In this case, a single amplifier does enable sustained oscillations, but very little control over the sensor's operation is possible. Moreover, most inexpensive ultrasonic transducers feature several resonances and they can start oscillations at different frequencies depending on, for example, the gain of the amplifier at start up, and some other factors, in a somewhat unpredictable fashion. Therefore, a robust design must include an electronic filter that reliably enables operation only within a particular frequency range. Inclusion of a tuneable phase shifter is desirable in order to obtain a specific output frequency at a specific calibration temperature, allowing electronic compensation of the technological tolerances that inevitably occur during the manufacture and mounting of the transducers.

3   Provision of Inexpensive but Accurate Measurements of the Oscillating Sensor's Output Frequency

Inexpensive but accurate frequency measurements can be achieved by using a digital counter and a reference clock oscillator with the frequency f_r. If the frequency of the reference oscillator is much higher than the frequency that is to be measured f_x, the counter is gated by a single period of the signal of interest counting N reference pulses. Then f_x is calculated as follows

$$f_x = f_r / N. \tag{7}$$

If the frequency of the reference oscillator is much lower than that is to be measured, the counter is gated by a single period of the reference oscillator and the input pulses are counted. The following expression becomes applicable for the frequency estimation:

$$f_x = f_r \times N. \tag{8}$$

The counter's output is accurate to a single pulse; thus, it can be inaccurate up to one count. Consequently, achieving the relative 10^{-4} resolution is possible if the number of pulses counted is no less than 10,000.

Sometimes the ratio between the frequencies is lower than the pulse count that ensures the required relative resolution. In this case, the counter needs to be gated not during a single period but over several periods of either f_x ($f_x < f_r$) or f_r ($f_x > f_r$) as appropriate [12]. In practice, frequency measurements with the required resolution can be achieved by employing two separate counters for the reference and input pulses; every time the lower frequency pulse counter increments, the value of the other counter is compared to the inverse of the required resolution. If the value of the counted high frequency pulses is higher, then the required resolution has already been achieved, the output frequency can be calculated and communicated and a new frequency measurement can be started by clearing both of the counters.

Another important factor in achieving the required resolution is the frequency stability and/or tolerance of the reference clock oscillator. Relatively inexpensive crystals, which cost a fraction of a dollar, can provide ±30–50 ppm or 0.3–$0.5_\times 10^{-4}$ frequency tolerances in 1–10 MHz range when connected to appropriate pins of a microcontroller, which could be just about enough for the considered application. A crystal oscillator with similar tolerances costs more (around a dollar), but does not require a microcontroller to be capable of using a wide range of crystals and can generate waveforms with low jitter. Temperature compensated crystal oscillators (TCXO) are a more accurate (a few ppms only) but more expensive (a few dollars) option for the reference oscillator. Higher accuracy oven controlled crystal oscillators (OCXO) are prohibitive for low cost instrumentation because they typically cost around one-hundred dollars or more.

4   Implementing BPFs for Oscillating Ultrasonic Sensors

Inclusion of a BPF into the signal loop of an oscillating ultrasonic sensor is essential if the sensor's ultrasonic transducers feature multiple resonances. This is frequently the case for low-cost, mass-produced transducers when they are securely attached to some supporting frame and/or are being submersed.

Dr Alzebda implemented a variable BPF using an LT1568 integrated circuit and digital potentiometers set by the supervisory microcontroller for ultrasonic oscillating temperature sensors operated above 300 kHz [4]. This relatively high operating frequency resulted in obtaining over 30 output frequency readings per second with the relative resolution of 10^{-4}, but the sensor could only operate at ultrasonic pathways of no more than about 30 mm—which is insufficient for the considered case.

Dr Popejoy developed an ultrasonic oscillating tilt sensor that operated with ultrasonic pathways of up to 500 mm at frequencies around 30 kHz [6]. As the operating frequency was well below the specified lowest operating frequency for the LT1568 parts, the variable BPF was built using two operating amplifiers and three digital potentiometers using the fliege BPF configuration [6,13].

Although the cost of the bill of materials (BOM) for this design was not too high, the adoption of specialised mixed-signal integrated circuits provides an opportunity to further reduce that cost. That approach consists of using programmable analogue and digital blocks, provided in addition to a fully featured microcontroller, in PSoC1 mixed signal microcontrollers manufactured by Cypress [14]. These devices include switched capacitor blocks that can be configured as various filters; additionally, they allow for adjustments of the filter properties at the run time by changing the values of the variable capacitors and/or changing the frequency at which the capacitors switch. There was a concern that using the switched capacitor principle would break the signal loop continuity, thereby disabling the sustained oscillations. However, PSoC1 band pass filters have been experimentally proven to be a viable low-cost option for implementing oscillating ultrasonic sensors, which on numerous occasions reliably operated for over fifty hours [15].

5   PSoC1-Based Amplifier Combining Both the Discrete and Continuous Gain Control

An oscillating ultrasonic sensor can function if and only if the energy conversion and propagation losses in the signal loop are fully compensated by an amplifier. The oscillations became sustained when the overall gain in the signal loop is greater or equal to unity; but, if it exceeds unity even slightly, then the output signal of the amplifier quickly saturates at the rail voltages.

Earlier oscillating sensor designs included a fixed gain amplifier built using one or two operating amplifiers with digital potentiometers at the input and output, which allowed for variation of the overall gain [4, 6].

PSoC1 devices can contain up to four programmable gain amplifiers (PGAs) with quite a wide range of available discrete gain settings. The PGAs can be cascaded and used for the loss compensation and adjustment of the signal loop gain. Fig. 5 presents the experimental results for the output frequency of an ultrasonic temperature sensor with ultrasonic pathway of around 50 mm. An amplifier that featured two cascaded PGAs and a band pass filter. The gain of the first stage G1 was fixed to the value shown in the figure legend, and the gain of the second stage G2 was varied to obtain all of the curves presented in Fig. 5.

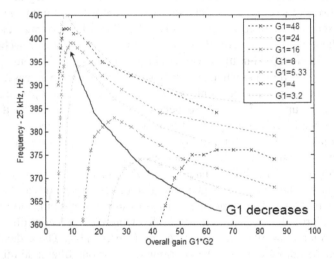

Fig. 5. Output frequency of an oscillating ultrasonic sensor, which was held at a constant temperature, depending on the overall gain in the signal loop and its composition

The results show that, even for the same overall gain, the output frequency could differ considerably depending on the gain composition, by around ±10 Hz for the higher overall gains or even more for the lower overall gains. For every curve there was a maximum point at which variations of the gain of the second stage affected the sensor output frequency to a lesser extent. For this reason, the gain for the amplifiers was selected at one of the maximum points that featured maximum flatness (G1=24 and overall gain of around 36 for the data presented in Fig. 5). It is important to note that this behaviour was observed without any involvement of the switched capacitor blocks present in PSoC; thus, it could not be attributed to the intermittent nature of their operation.

However, this selection could only be done once and at a single operating frequency. Temperature changes would affect both the ultrasound velocity and the gain of the amplifiers, to some extent, causing unwanted output frequency changes.

Additionally, ultrasonic transducers would age, thereby becoming less efficient in energy conversion, and in many situations the absorption of ultrasonic waves within the medium of interest could vary. These concerns called for the development of an amplifier that automatically adjusts its gain according to the changes of the signal losses in the signal loop.

The first design of the amplifier with supervisory gain control utilised PSoC1 comparators to detect whether the output voltage exceeded certain levels. If the upper level is exceeded by the output signal of the amplifier, then the gain of some of the PGAs is decreased. If the lower level is not exceeded, then the gain is increased. In practice, the output signal of the sensor was driven to saturation because digital control could not keep the overall gain exactly at unity because it was required to produce undistorted sine waveforms.

The first design of the amplifier with the analogue automatic gain control (AGC) included using an incandescent light bulb to set the gain of an operating amplifier, similar to [16]. Unfortunately, that design did not control the gain at the ultrasonic frequencies of interest (the reason for this remains unclear), despite the fact that the bulb itself featured nonlinear resistance and the operating amplifier provided enough current at a valid operating point. After examining, simulating and prototyping several other AGC circuits developed for audio processing, the best results overall were achieved using the circuit described in [17].

The final design of the amplifier featured both the supervisory and automatic gain controls and, in addition to a PSoC, it required one extra operating amplifier and one field effect transistor. The block diagram of this design is presented in Fig. 6. The first amplification stage (between points A and B) is implemented using a supervisory gain control to achieve the level of output voltage sufficient for the effective operation of the second amplifier with the AGC connected between points B and C. The circuitry between points C and D provides final amplification, frequency filtering and frequency measurement using another PSoC.

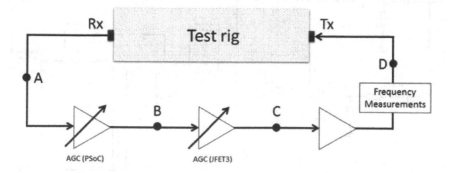

Fig. 6. Block diagram of the ultrasonic oscillating temperature sensor with supervisory and automatic gain controls

Experiments showed that this arrangement ensured the lowest scatter of the sensor output frequency at a constant temperature as compared to the previous designs. The downside of this arrangement is the increased BOM cost.

6   PSoC1-Based Tuneable Phase Shifter

Phase shifts at frequencies up to 1 MHz can be achieved at a low cost by employing RC circuits; the phase shift adjustment can be most conveniently implemented by controlling resistances in these circuits. In order to be deployed for an ultrasonic oscillating sensor, the phase shifter should operate across a range of frequencies and it should enable the setting of arbitrary phase shifts for flexibility.

Such a device can consists of several cascaded RC stages because a single RC stage cannot provide phase shifts of more than 90°. These stages would require buffering to reduce their influence on each other; compensation of the gain changes when tuning the phase; and quite an elaborated calibration to operate across a range of frequencies.

For this reason, we believe that a more robust approach is to create the required phase shift using the phasor diagram by adding in the phase and 90° shifted components with appropriate weights. The phase shifter that was previously developed utilised four digital potentiometers that set the required weights [13]. In the latest design, presented in Fig. 7 [18], these weights were set by altering gains of the programmable gain amplifiers (PGA1 and PGA2) and the values of the gain setting switched capacitors (SC) in SCBLOCK1. As the sign of the SCBLOCK1 input signal, coming from PGA1, can be altered programmatically, the output signal of this block can have a phase shift in the range from almost -90° to +90°, which may be sufficient in practice. The other components in the design were employed to extend this range to the full range of 360 angular degrees by outputting either the output signal of the SCBLOCK1 or its copy inverted by the low pass filter.

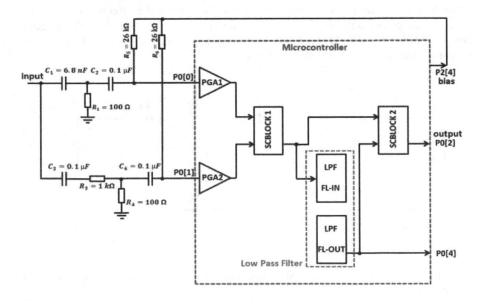

Fig. 7. Block diagram of the PSoC1 based tuneable phase shifter [18]

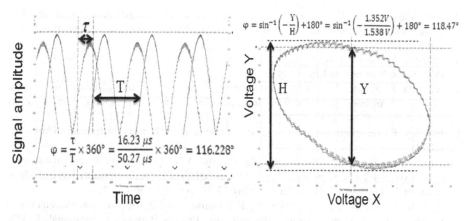

Fig. 8. Measurement of the actual phase shift using the direct oscilloscope method (left) and Lissajous figures (right) when the desired phase shift was set to 120° [18]

MATLAB simulation of the developed phase shifter showed that, by setting the correct values of the SCBLOCK1 capacitors, the resulting phase shift errors did not exceed ±1.5°, whilst the variation of the output amplitude (that should ideally stay the same) did not exceed ±1% [18]. In practice, some larger deviations were observed that depended upon the method of the phase shift measurement. These deviations occurred because the output signal was modulated by the switching frequency of the capacitors, which complicated the situation, resulting in somehow ambiguous readings. An example of an experimental measurement is presented in Fig. 8 [18].

7   Conclusions

Although some industrial applications of ultrasonic thermometers have been reported a long time ago (e.g. [19]), development of a reliable high accuracy ultrasonic sensor for cost-sensitive applications still remains an engineering challenge. Our development of oscillating ultrasonic sensors show that even though it can be simple to make a sensor oscillate, getting it to perform reliably to the required specification is not easy. In this paper, we reported our recent development that enabled better control of the sensor's operation and behaviour; this was achieved using a limited number of low-cost electronic components that were evaluated on their own and which showed notable improvements over previous designs. The experimental assessment of the developed module together will be carried out soon.

References

1. Asher, R.C.: Ultrasonic Sensors for Chemical and Process Plant (Sensors Series), 496 p. Taylor and Francis (1997)
2. Sonbul, O., Kalashnikov, A.N.: Electronic Architecture for Air Coupled Ultrasonic Pulse-Echo Range Finding with Application to Security and Surface Profile Imaging. Universal J. of Electrical and Electronic Eng. 2(3), 105–111 (2014) (open accesss)

3. Kalashnikov, A.N., Challis, R.E.: Errors and uncertainties in the measurement of ultrasonic wave attenuation and phase velocity. IEEE Trans. Ultrason., Ferroel. Freq. Control 52(2), 1754–1768 (2005)
4. Alzebda, S.: Low-cost oscillating sensing for ultrasonic testing and monitoring of liquids. PhD thesis, Nottingham University (2011)
5. Alzebda, S., Kalashnikov, A.N.: Ultrasonic sensing of temperature of liquids using inexpensive narrowband piezoelectric transducers. IEEE Trans. Ultrasonics, Ferroel., Freq. Contr. 57(12), 2704–2711 (2010)
6. Popejoy, P.: Development of an ultrasonic tilt sensor, PhD thesis, Nottingham University (2014)
7. Popejoy, P., Alzebda, S., Sonbul, O., Kalshnikov, A.N.: Linear angle measurement using continuous wave ultrasonic oscillator. In: Proc. 2012 IEEE Int. Instrumentation and Measurement Technol. Conf., Graz, Austria, May 13-16, pp. 733–736 (2012)
8. Afaneh, A., Alzebda, S., Ivchenko, V., Kalashnikov, A.N.: Ultrasonic measurements of temperature in aqueous solutions: why and how. Physics Research International, 156396 (2011) (open access)
9. Kaye, G.W.C., Laby, T.H.: Tables of Physical and chemical constants, Longman, 16th edn., sect. 2.4.1 "The speed and attenuation of sound"
10. Kalashnikov, A.N., Shafran, K.L., Challis, R.E., Perry, C.C., Unwin, M.E., Holmes, A.K., Ivchenko, V.: Super-resolution in situ ultrasonic monitoring of chemical reactions. In: Proc. 2004 IEEE Ultrasonics Symp., pp. 549–552 (2004)
11. Afaneh, A., Kalashnikov, A.N.: Embedded processing of acquired ultrasonic waveforms for online monitoring of fast chemical reactions in aqueous solutions. In: Haasz, V. (ed.) Advanced Distributed Measuring Systems: Exhibits of Application, pp. 67–93. River Publishers (2012)
12. Ess, D.V.: Measuring frequency., application note AN2283, http://www.cypress.com/?rID=2671 (accessed January 2015)
13. Popejoy, P., Alzebda, S., Hashmi, A., Harriott, M., Sonbul, O., Kalashnikov, A.N., Hayes-Gill, B.R.: Comparison of implementations of driving electronics for ultrasonic oscillating sensors. In: Proc. 2012 IEEE Ultrasonics Symp., pp. 2427–2430 (2012)
14. PSoC® 1product page, http://www.cypress.com/?id=1573 (accessed January 2015)
15. Hashmi, A.: PSoC1-based data logging instrumentation for overnight measurement of the output frequency of ultrasonic oscillating sensors versus the ambient temperature, MSc thesis, Nottingham (2013)
16. Wein bridge oscillator, http://en.wikipedia.org/wiki/Wien_bridge_oscillator (accessed January 2015)
17. Effective AGC amplifier can be built at a nominal cost, http://tinyurl.com/mqhz7rh (accessed January 2015)
18. Malakoutikhah, M.: Optimisation of the oscillating temperature sensor cell for differential calorimeter use, MSc thesis, Nottingham (2014)
19. Lynnworth, L.C.: Industrial Applications of Ultrasound - A Review: II. Measurements, Tests, and Process Control Using Low-Intensity Ultrasound. IEEE Trans. Sonics Ultrason. SU-22(2), 71–101 (1975)

Impact of Base Station Location
on Wireless Sensor Networks

Odeny Nazarius Koyi, Hee Sung Yang, and Youngmi Kwon

Department of Information and Communications Engineering
Chungnam National University, Daejeon, South Korea
odenyk@yahoo.com, yhshh001@naver.com, ymkwon@cnu.ac.kr

Abstract. Wireless sensor networks (WSNs) have attracted much attention in recent years due to their potential use in many applications such as surveillance, militant etc. Given the importance of such applications, maintaining a dependable operation of the network is a fundamental objective. In Wireless Sensor Networks, many algorithms have been devised to improve energy maintenance in a whole network. Most of them assume the location of the Base Station (BS) to be at the border of the network even though location of Base Station takes some role in overall performance of the network. So in this paper, we simulated WSN performance in energy consumption, throughput, packets delivery ratio and delay with different locations of BS. Simulation results showed the best performance when a Base Station is located in the center of the WSN field and the worst when a Base Station is in the corner of the WSN. Compared to the existing location assumption, with the best positioned BS, Cluster Heads consumed 69% and with the worst positioned BS, they consumed 127% in energy. When we build WSN, if we spend some higher cost for installing BS inside the network, its overall performance can be improved much.

Keywords: LEACH Protocol, Base Station Location, Energy Consumption.

1   Introduction

Wireless Sensor Networks [1] consist of hundreds and even thousands of tiny devices called sensor nodes distributed autonomously to monitor physical or environmental condition. Sensors have a very limited functionality, this normally consist of some type of short range wireless communication capabilities and some form of sensing, or actuator, equipment. WSNs are increasingly equipped to handle some of these complex functions, in-network processing such as data aggregation, information fusion, computation and transmission activities requires sensors to use their energy efficiently in order to extend network life-time. Sensor nodes are prone to energy drainage and failure, and their battery source might be irreplaceable, instead new sensors are deployed. Thus, the constant re-energizing of wireless sensor network as old sensor nodes die out and/or the uneven terrain of the region being sensed can lead to energy imbalances or heterogeneity among the sensor nodes. This can negatively impact the stability and performance of the network system if the extra energy is not properly

© Springer International Publishing Switzerland 2015   151
R. Silhavy et al. (eds.), *Intelligent Systems in Cybernetics and Automation Theory*,
Advances in Intelligent Systems and Computing 348, DOI: 10.1007/978-3-319-18503-3_15

utilized and leveraged. Several clustering algorithms such as LEACH have been pro-posed with varying objectives like load balancing, fault-tolerance, increased connec-tivity with reduced delay and network longevity in order to attain better performance. Base Station location plays an important role in wireless sensor networks lifetime because energy consumed by Cluster Heads to transmit data to Base Station doesn't only depend on the data bit rate, but also on the physical distance between Cluster Heads and Base Station. So it's important to understand the impact of Base Station location in Wireless Sensor Networks performance so that we can optimize topology during network deployment stage. In WSNs, the major source of node failure is bat-tery exhaustion and replacing this energy source in the field is usually not practical. Therefore the use of an energy efficient infrastructure can prolong the lifetime of the network and improve the overall network performance. This paper is organized as follows. In section 2 we have described LEACH protocol in detail. In section 3 we have explained how our proposed method improving lifetime of Wireless Sensor Network. In section 4 we present the analytical discussion of simulation and results. We have concluded our work in section 5.

2   LEACH Protocol

The LEACH protocol for sensor networks, proposed by Heinzelman et al. [2], mini-mizes energy dissipation in sensor networks. It partitions the nodes into clusters, and in each cluster, a dedicated node with extra privileges called a Cluster Head (CH) is responsible for creating and manipulating a time division multiple access (TDMA) schedule and sending aggregated data from the nodes to the Base Station where this data is needed using code division multiple access (CDMA), the remaining nodes are cluster members as shown in Figure 2. LEACH operation contains two phases, the setup phase and steady state phase. During setup phase, each node decides whether or not to become a CH for the current round. This decision is based on choosing a ran-dom number between 0 and 1, and if the number is less than a threshold $T(n)$, the node becomes a CH for the current round. The CH node sets up a TDMA schedule and transmits this schedule to all the nodes in its cluster, completing the setup phase, which is then followed by a steady-state operation.

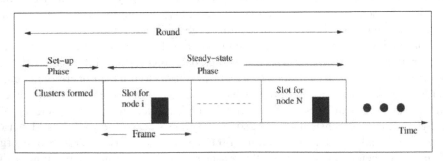

Fig. 1. LEACH Phases

The steady-state [3] operation is broken into frames, where nodes send their data to the CH at most once per frame during their allocated slot, shown in Figure 1. It assumes that nodes always have data to send, and they send it during their allocated transmission time to the CH. This transmission uses a minimal amount of energy. The radio of each non-CH node can be turned off until the node's allocated transmission time, thus minimizing energy dissipation in these nodes. The CH node must keep its receiver on to receive all the data from the nodes in the cluster. When all the data has been received, the CH node performs signal processing functions to compress the data into a single signal.

$$T(n) = \left\{ {0 \atop \frac{P}{1-P*\left(rmond\frac{1}{p}\right)}} \right. if \, n \in G \tag{1}$$

Where, P equals the suggested percentage of cluster-heads, r is the current round and G is the set of nodes that have not been Cluster Head in the last 1/p rounds. The node which has been selected as the Cluster Head will broadcast a message using CSMA MAC protocol and non Cluster Head nodes will choose nearest Cluster Head to join that cluster. Figure 2 shows the clustering architecture of LEACH protocol.

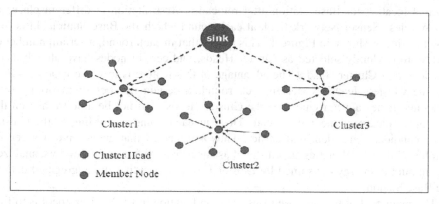

Fig. 2. Cluster formation in LEACH

This protocol is divided into rounds, where each round consists of setup phase and steady state phase. In this way, all sensor nodes take turns to serve as a Cluster Heads to balance the power consumption.

3   Improving Lifetime in Wireless Sensor Networks

Most of the existing LEACH-based papers assume the Base Station's location to be to the border of the network. But the location of the base station takes important role in overall performance of the network. So, in this paper, we compare the performances affected by the locations of the Base Station. The distance between Cluster Heads and Base Station is considered to be an important factor for lifetime in WSN. Once the distance between BS and CHs is minimized, it reduces extra transmissions of data in LEACH protocol. If the Base Station is located far from the Cluster Heads, direct transmission requires more energy since the distance from border to border is sufficiently large, this cause the energy in Cluster Heads to be drained faster and the overall life-time of the Wireless Sensor Network is shortened. On the contrary, if the Base Station is located close to the center of the Wireless Sensor Network field, the distance between CHs and BS is alleviated compared to the existing assumption and the CH's energy consumption can be improved because the transmission energy is proportioned to the square of distance. But to locate BS into the center of the huge network area is usually impractical. If we can get estimation for some practical cost to locate BS inside network in some degree and the effects that we can get from that expense, it would be practical suggestion to design an improved WSN. It is the goal of this paper.

4   Simulation and Results

In this chapter, we evaluate the performance of proposed approach of relocating the Base Station in Wireless Sensor Network field through simulations. We select an optimal location for Base Station after relocating Base Station in different places in the Wireless Sensor Network field, at each round which the Base Station shifts to a new location as shown in Figure 4. This means that in each round, a certain number of nodes are randomly selected as Cluster Heads, and certain nodes have already been considered as Cluster Heads. The advantage of this feature is low overhead of select-ing the Cluster Heads. Nodes in each round, a certain number of randomly self-generated nodes are introduced as the Cluster Heads, will not be able to take on the role of candidate in the next round. Thus, in each round, according to the Cluster Head candidates are identified earlier. Thus is expected that at the end of a certain number of rounds, all nodes are clustered as supervisor. After each round we analyzed the amount of energy consumed by Cluster Heads on sending the aggregated data to the Base Station.

The main goal of the proposed Base Station location in LEACH protocol is to find the best Base Station location (optimal location) where all selected Cluster Heads can spend minimal amount of energy during data transmission. By providing the optimal Base Station's location in LEACH protocol, we managed to minimize the high energy consumption in Cluster Heads, as well as to avoid network failure and finally prolong network lifetime.

4.1   Simulation Environments

As shown in Figure 3, we move the Base Station to different locations within and outside a Wireless Sensor Network field in order to find the Base Station location (BS-LOC) where Cluster Heads will spend minimum average amount of energy to transmit data to the Base Station. In this paper, we highlight the potential of careful positioning of the Base Station, which acts as a sink node for the collected data, as a viable means for increasing the dependability of WSN. We show that Base Station location can be very effective in optimizing the network functional and non-functional performance objectives. We analyze the effects of Base Station locations in a Wireless Sensor Network's lifetime and energy consumption in Cluster Heads. To find optimal Base Station location we relocate the Base Station to different locations (BS-LOC1 to BS-LOC9) within the network field and one location (BS-LOC0) outside the Wireless Sensor Network field which assumed to be original LEACH Base Station Location. Figure 3 shows ten different BS locations used in simulation. And Table 1 shows their matching coordinates in the 1500m x 1150m simulation field.

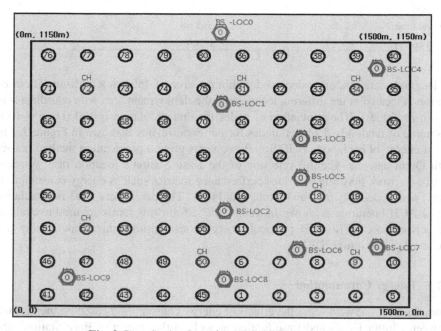

Fig. 3. Base Station Locations (BS-LOC) in simulation

Table 1. Base Station Coordinates of Each location

Base Station Location	Coordinates	Cluster Heads Energy Consumption (J)
LOC 0	(638, 1121)	51
LOC 1	(665, 795)	40
LOC 2	(653, 550)	35
LOC 3	(898, 654)	45
LOC 4	(1218, 928)	65
LOC 5	(906, 542)	38
LOC 6	(927, 247)	40
LOC 7	(1168, 217)	46
LOC 8	(703, 92)	49
LOC 9	(110, 88)	60

Table 2. Simulation Network Parameters and Values

Parameters	Values
Number of nodes	100
Simulation Area	1500m * 1150m
Base Station locations	10 locations
Radio Propagation	Two rays ground prop
Antennal	Omni antenna
Bandwidth	2e
Data size	512 bytes
Interval time	-0.2 ms
Initial energy	100 Joules

The nodes are randomly deployed within the area of 1500 m × 1150 m. The Base Station is located at ten different locations of the deployment area with coordinates as shown in Table 1. The population of nodes for this simulation is 100 (i.e. n = 100). A sample of randomly deployed nodes for our experiments is shown in Figure 3. The initial energy of each node is 100J and two ways ground propagation method is used with Omni antenna. Careful selection of the Base Station's location in a Wireless Sensor Network may affect various performance metrics such as energy consumption, delay, packet delivery ratio and throughput [4, 5]. The parameters used in simulation for LEACH operation is shown in Table 2. The following terms are used to evaluate the performance of LEACH protocol: energy consumptions, throughput, delay and packets delivery ratio (PDR).

4.2   Energy Consumption

From this section, we discuss the effects of energy consumption based on the distance between Cluster Heads and Base Station as well as the impact of Base Station location (BS-LOC) in a WSN's energy consumption. This section shows the energy comparison consumed by Cluster Heads in LEACH when the Base Station is located at different locations. BS in location 2 consumes less energy (35 Joules) than other locations. So by locating the Base Station to location 2, we can achieve the goal to save energy. By selecting an optimal Base Station location in a network field, we can influence the network performance as we have seen before. Compared to the existing original location assumption, with the best positioned BS, Cluster Heads consumed 69% and with the worst positioned BS, they consumed 127% in energy. For example, routing data to a Base Station that is far from the Cluster Heads usually involves numerous relay nodes which increases the aggregate delay, energy consumption and risks a packet loss due to link errors.

Figure 4 shows the effects of distance between the Cluster Heads and Base Station. As the Base Station moves toward the center of the Wireless Sensor Network field, it shows that Cluster Heads spend an average minimum amount of energy to transmit aggregated data to the Base Station. In LEACH Protocol, they assumed that Base

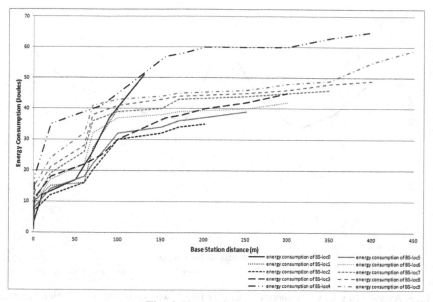

Fig. 4. Energy Consumption

Station is fixed and located far from the Wireless Sensor Network field [6]. Therefore, by introducing the optimal Base Station location (where the Cluster Heads and Base Station distance is minimal) we can improve the performance and efficiency of energy consumption in Wireless Sensor Networks, and hence prolong the network lifetime. Figure 5 shows that when the Base Station located at location 2 (BS-LOC2), location 3 (BS-LOC3) and location 5 (BS-LOC5), Cluster Heads consumed low average amount of energy during data transmission. At location 2 (BS-LOC2), the Base Station is located closest to the center of the network field as shown in Figure 3. So average energy consumed by Cluster Heads when the Base Station located at location 2 is the lowest compared to other locations as shown in Table 1.

Base Station at location 1 (BS-LOC1), location 6 (BS-LOC6), location 7 (BS-LOC7) and location 8 (BS-LOC8) is neither close to the center of the Wireless Sensor Network field nor at the edge of Wireless Sensor Network. Therefore, Cluster Heads consumed a bit high energy. This shows that average energy consumed by Cluster Heads keep increasing as the Base Station deviates from the center of the Wireless Sensors Networks.

Simulation result shows that when the Base Station is located far from the center of the Wireless Sensor Network field, the average energy consumed by Cluster Heads is higher than another cases, as seen at location 0 (normal LEACH), location 4(BS-LOC4) and location 9 (BS-LOC9).

By reducing the average energy consumption in the Wireless Sensor Network, we can prolong the network lifetime. LEACH based protocols can get better performance as the Base Station location is located close to the center of the Wireless Sensor Network.

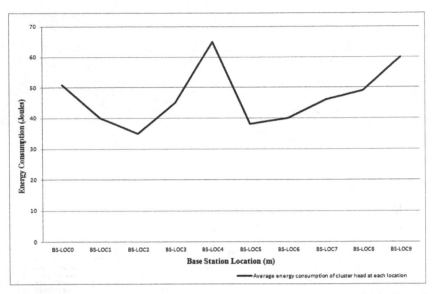

Fig. 5. Average Energy Consumption of Cluster Heads with different Base Station Location

4.3   Throughput

Generally throughput defines the average of successfully delivering data packets at the Base Station. Figure 6 shows that throughput in LEACH protocol is significantly greater when the Base Station is at optimal location (BS-LOC2) as compared to other locations in the network field.

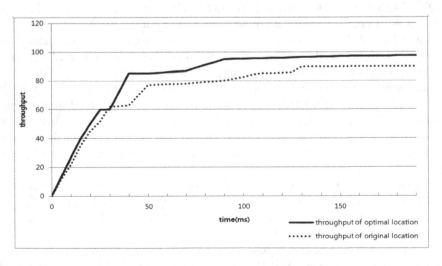

Fig. 6. Throughput Comparison according to different Base Station locations

From throughput comparison graph, we see that LEACH guarantees high packet delivery when the Base Station is located at minimum distance (closer to the center of the network) to all Cluster Heads.

$$Throughput = \frac{Packet\ Received}{Transmission\ Time} \qquad (2)$$

Original location (BS-LOC0) shows inefficient result in the case of throughput, the larger distance between Cluster Heads and Base Station causes degradation of throughput value. While at the Base Station's optimal location it shows high throughput value.

4.4   End to End Delay

Delay is one critical issue in Wireless Sensor Networks which causes high energy consumptions during data transmission. Delay occurred due to different reasons like queuing delay, propagation delay, processing delay etc.

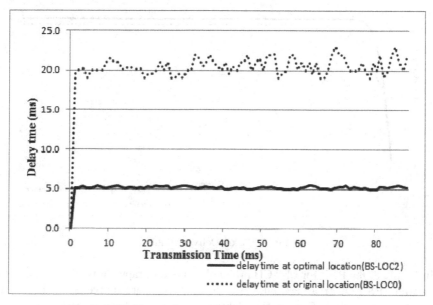

Fig. 7. Delay Comparison according to Base Station location

Average end-to-end delay (AED) is defined as the average time taken by the data packets to reach the intended destinations. End-to-End delay is calculated by considering the time taken by a packet to be transmitted from the source node to the destination node in the network field.

$$AED = \frac{\sum(Time\ received - Time\ sent)}{Total\ data\ packet\ received} \qquad (3)$$

Figure 7 shows the delay results obtained when the Base Station located closer to the center of the network, referred as our optimal location (BS-LOC2), and when the Base Station is outside the Wireless Sensors Network field. As the Base Station moved away from the center of the Wireless Sensor Network, the distance between the Cluster Heads and Base Station increase, as well as the data packet delay which increased from 5ms to 20ms. This means at location 2 (optimal location), LEACH protocol shows better performance compared to other locations.

4.5   Packet Delivery Ratio

Packet delivery ratio (PDR) is defined as the ratio of number of packets received by the destination to that of the generated packets. Figure 8 shows the results obtained by comparing the working of LEACH in a network with the Base Station located closer to the center of Wireless Sensor Network and when it is far away.

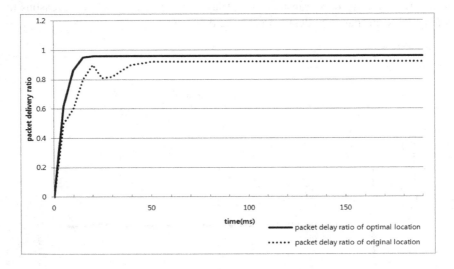

Fig. 8. Packet Delivery Ratio

In Figure 8, the PDR for LEACH protocol evaluates graph shows that at the optimal location (BS-LOC2), a maximum amount of packets are delivered to the destinations. A greater value of the packet delivery ratio means better performance of the protocol. When the Base Station is located far away from the center of the network the results is packet delivery rate drop which shows a reduction from 0.94 to 0.8 as shown in Figure 8.

$$PDR = \frac{\sum(Number\ of\ Received\ Packets)}{\sum(Number\ of\ packets\ sent)} \qquad (4)$$

Equation 4 gives us an idea of how successful the protocol is in delivering packets. A high value of packet delivery fraction indicates that most of the packets are being delivered to the higher layers and is a good indicator of the algorithm performance.

Overall Simulation results have shown that Base Station location in Wireless Sensor Networks has great impact in Cluster Heads energy consumption and general performance of the network. Therefore, the proposed approach is to locate the Base Station at an optimal location (in the center or near to the center of the WSN field) can prolong the network lifetime as long as it minimizes the amount of energy consumed by the Cluster Heads in Wireless Sensor Networks during data transmission.

5   Conclusion and Discussion

In this paper, we showed the effects of BS location in WSNs and its impact on energy consumption, network lifetime and general network performances. We focused on improving Wireless Sensor Network's performance in LEACH protocol through finding the optimal Base Station location. As LEACH Protocol is utilized, the Base Station is repositioned in different locations as shown in Figure 3, to show the BS location which provides a minimum distance between Cluster Heads and Base Station. We relocate the Base Station in different locations (BS-LOC 0 to BS-LOC 9) to observe the effects of BS location in WSNs performance. Simulation results show that the distance between Base Station and Cluster Heads is one of the factors affecting quick death of Cluster Heads in WSNs. When the BS is located far from the center of the Wireless Sensor Network field, Cluster Heads consume more energy on sending data to the Base Station as shown in Figure 5. At the optimal location (BS-LOC2), the Base Station is nearly at the center of Wireless Sensor Network field. At location 2, the Base Station can be accessed by all Cluster Heads within a minimum range of data transmission. This minimizes energy consumed by Cluster Heads on sending data to the Base Station. Impact of Base Station location in LEACH routing protocol for WSNs has been performed through Base Station relocation and shows its effects in Wireless Sensor Networks through comparison on parameters like throughput, delay in data transmission, energy consumption and packet delivery ratio in all Base Station locations.

References

1. Handy, M.J., Haase, M., Timmermann, D.: Low energy adaptive clustering hierarchy with deterministic cluster-head selection. In: Proceedings of the 4th International Workshop on Mobile and Wireless Communications Network, Stockholm, Sweden, pp. 368–372 (2002)
2. Heinzelman, W.R., Chandrakasan, A., Balakrishnan, H.: Energy-efficient communication protocol for wireless microsensor networks. In: Proceedings of the 33rd Annual Hawaii International Conference on System Sciences, p. 8020. HI, Maui (2000)

3. Heinzelman, W.B., Chandraskan, A.P., Blakrisshnan, H.: An Application-Specific Protocol Architecture for Wireless Microsensor Networks. IEEE Trans. on Wireless Communications. 1(4), 660–670 (2002)
4. Akkaya, K., Younis, M., Youssef, W.: Positioning of Base Stations in Wireless Sensor Networks. IEEE Communication Magazines (April 2007)
5. Mendis, C., Guru, S.M., Halgamuge, S., Fernando, S.: Optimized Sink Node Path using Particle Swarm Optimization. In: Proceedings of IEEE AINA 2006, Vienna, Austria, vol. 2 (April 2006)
6. Heinzelman, W.B., Chandraskan, A.P., Blakrisshnan, H.: An Application-Specific Protocol Architecture for Wireless Microsensor Networks. IEEE Trans. on Wireless Communications. 1(4), 660–670 (2002)

A New Implementation of High Resolution Video Encoding Using the HEVC Standard

Alaa F. Eldeken[1], Mohamed M. Fouad[1],
Gouda I. Salama[1], and Aliaa A. Youssif[2]

[1] Dept. of Computer Engineering, Military Technical College, Cairo, EGYPT
[2] Faculty of Computer & Information, Helwan University, Helwan, Cairo, EGYPT
{eldeken,mmafoad,drgouda,ayoussif}@ieee.org

Abstract. In this paper, the implementation method for encoding the high resolution videos using high efficiency video coding (HEVC) standard is introduced with a new approach. The HEVC standard, successor to the H.264/AVC standard, is more efficient than the H.264/AVC standard in the encoding high resolution videos. HEVC has been designed to focus on increasing video resolution and increasing the use of parallel processing architectures. Therefor, this approach merging all traditional configuration files used in the encoding process into only one configuration file without removing any parameters used in the traditional methods. Improvements are shown using the proposed approach in terms of encoding time as opposed to the traditional methods by reducing the access time by half which resulting from reducing the data exchange between the configuration files used in this process and without changing the rate-distortion (RD) performance or compression ratio.

Keywords: High efficiency video coding, HEVC implementation, rate-distortion.

1 Introduction

HEVC is a new video coding standard recently launched in 2013 in order to save the channel bandwidth and disk space as opposed to the standard H.264/AVC. It is also known as H.265 or MPEG-H Part-2 [1]. HEVC has been designed to focus on two key issues : increasing video resolution and increasing the use of parallel processing architectures. HEVC provides 50% more bit-rate reduction and a higher degree of parallelism when compared to H.264/AVC by adopting a variety of coding efficiency enhancement and parallel processing tools [2]. Typically, the H.264/AVC [3] divides a frame into 16×16 fixed size of macroblocks. However, this fixed size limits the ability of the H.264/AVC to encode/decode the high resolution videos. Contrarily, in HEVC, a frame is divided into coding tree units (CTU) of 16×16, 32×32 or 64×64. Each CTU can be further divided into smaller blocks, called coding units (CUs), using a quadtree structure. Each CU can be further split into either prediction units (PUs) or transform unit (TUs) using the quadtree structure. The size of each TU, used in the prediction error coding,

© Springer International Publishing Switzerland 2015
R. Silhavy et al. (eds.), *Intelligent Systems in Cybernetics and Automation Theory,*
Advances in Intelligent Systems and Computing 348, DOI: 10.1007/978-3-319-18503-3_16

is ranged from 4×4 upto 32×32 leading to larger transformations than that of the H.264/AVC that only uses 4×4 and 8×8 transforms [4]. In turn, the high resolution videos can be encoded using the HEVC more efficiently than that of the H.264/AVC standard [2].

The HM (High Efficiency Video Coding (HEVC) Test Model) software is the reference software for the HEVC project of the video sequences [5]. The HM software is written in C++ and is provided as source code. Since the HEVC project is still under development, the HM software is also under development and changes frequently. The HM software folders contains all files that are required for building and running the software(i.e., the folders bin and lib are created during building the software) [6]. A log file describing the changes (i.e., main changes) for HM software is the configuration file, which changes only the main changes in the encoding process but any other not existing in the configuration file is done directly into c++ source code and need to understand the all source code modules [7]. It is possible to build the HM software on a Windows 32 platform with microsoft visual studio.net and on a linux platform with gcc version 4. Since the HM software is written in C++, it should also be possible to build the software on other platforms, which provide a C++ compiler. All libraries are static libraries and all executable are statically linked to the libraries. The folder build contains a microsoft visual studio.net workspace and videoencdec.sln. In order to build the software, this workspace is opened with microsoft visual studio.net, and all project files are built [7]. This paper is organized as follows. The usage and configuration of the HM software and the proposed encoding method are shown in Section 2 and Section 3, respectively. Experimental results are shown in Section 4. Finally, conclusions are given in Section 5.

2   Usage and Configuration of The HM Software

In this section, information on usage the HM software and setup the configuration files in the HM software package are discussed. The libraries provided by the HM software [7] are descried as follows: (i)TAppCommon, provides classes that are used by both the encoder and decoder, as for example macroblock data structures, buffers for storing and accessing video data, or algorithms for deblocking. (ii)TLibEncoder, provides classes that are only used by the encoder. For example, it includes classes for motion estimation, mode decision, and entropy encoding. (iii)TLibDecoder, provides classes that are only used by the decoder. For example, it includes classes for entropy decoding and bitstream parsing. (iv)TLibVideoIO, provides classes for reading and writing NAL units in the byte-stream format as well as classes for reading and writing raw video data. Our work will focus on TAppCommon, TLibEncoder and TLibDecoder that is discussed in Section 2.1 and Section 2.2 respectively.

2.1   Encoder of The HM Software

The encoder can be used for generating HEVC bitstreams and reconstructed files [6]. The basic encoder file is illustrated in Fig. 1-(a) that represents the filename of the executable file (*i.e.*, TAppEncoder.exe) , the main configuration file (*i.e.*, encoder random-access-maim.cfg), the video configuration file (*i.e.*, Tennis.cfg), and the output file (*i.e.*, log-RA.txt), respectively. The -C parameter in the basic encoder file defines the configuration file to be used. Multiple configuration files may be used with repeated c options.

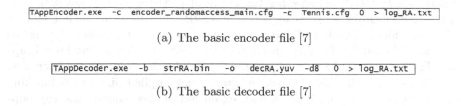

(a) The basic encoder file [7]

(b) The basic decoder file [7]

Fig. 1. The basic encoder and decoder files

In Fig. 2-(a) the main configuration file (*i.e.*, encoder random-access-maim.cfg) is shown that describes the method that is used in the coding process. This method can be one of three methods: All Intra method, random-access method, and low-delay method [6] as shown in Fig. 3-(a), Fig. 3-(b), and Fig. 3-(c), respectively.

In Fig. 3-(a) graphical presentation of all-Intra configuration is shown. Each picture in a video sequence will be encoded as instantaneous decoder refresh (IDR) picture. An encoder sends an IDR coded picture to clear the contents of the reference picture buffer. On receiving an IDR coded picture, the decoder marks all pictures in the reference buffer as unused for reference. All subsequent transmitted slices can be decoded without reference to any frame decoded prior to the IDR picture. The first picture in a coded video sequence is always an IDR picture. It is not allowed to change quantization parameter (QP) during a sequence within a picture [6].

In Fig. 3-(b) the random-access graphical presentation is shown. Each hierarchical bidirectional (B) structure will be used for encoding process. Intra picture will be inserted cyclically per about one second. The first intra picture of a video sequence will be encoded as IDR picture and the other intra pictures will be encoded as non-IDR intra pictures [6]. The pictures located between successive intra pictures in display order will be encoded as B-pictures. The second and third temporal layers consists of referenced B pictures, and the highest temporal layer contains non-referenced B picture only. QP of each inter coded picture will be derived by adding offset to QP of Intra coded picture depending on temporal layer.

In Fig. 3-(c) low-delay graphical presentation is shown. Only the first picture in a video sequence will be encoded as IDR picture. In mandatory low-delay (B low-delay) test condition (*i.e.*, two kinds of low-delay coding configurations, bidirectional (B) low-delay and predictive (P) low-delay), the other successive pictures will be encoded as generalized P and B-picture (GPB) [6].

Some of the parameters used in the main configuration file [7] for the HEVC will be discussed bellow: (1)OutputFile, specifies the filename of the bitstream to be generated. (2)ReconFile, specifies the filename of the coded and reconstructed input sequence. This sequence is provided for debugging purposes. It will be automatically created by the encoder. (3)BasisQP, specifies the basic quantization parameter. This parameter shall be used to control the bit-rate of a bitstream. (4)GOPSize, specifies the group of pictures (GOP) size that shall be used for encoding a video sequence. A GOP consists of an anchor picture and several hierarchically coded B pictures that are located between the anchor pictures. (5)LoopFilterDisable, specifies how the in-loop deblocking filter is applies. The following values are supported [8]: i) The deblocking filter is applied to all block edges. ii) The deblocking filter is not applied. iii) The deblocking filter is applied to all block edges with exception of slice boundaries. (6)LoopFilterTcOffset, specifies the Tc offset for the deblocking filter. (7)LoopFilterBetaOffset, specifies the beta offset for the deblocking filter. LoopFilterBetaOffset and LoopFilterTcOffset shall be the integer values in the range of -6 to 6 [9]. This parameter can be used to adjust the strength of the deblocking filter [10].

In Fig. 2-(b) the video configuration file (*i.e.*, Tennis.cfg) that gives all characteristics for the video used in the encoding process is shown. Some of theses characteristics are described bellow [7]: (1)InputFile, specifies the filename of the original raw video sequence to be encoded. The input files should have the format yuv. (2)SourceWidth, specifies the width of the input videos in luma samples. SourceWidth shall be non-zero. This parameter shall be present in each configuration file, since the default value of 0 is invalid. (3)SourceHeight, specifies the height of the input videos in luma samples. SourceHeight shall be non-zero. This parameter shall be present in each configuration file, since the default value of 0 is invalid. (4)FrameRate, specifies the frame rate of the input sequence in Hz. (5)FramesToBeEncoded, specifies the number of frames of the input sequence to be encoded for one view. In Fig. 2-(c) the output file (*i.e.*, log-RA.txt) that gives the the output results after encoding (*i.e.*, bit-rate, YUV values, and encoding time).

2.2 Decoder of The HM Software

The basic deccoder file is illustrated in Fig. 1-(b) that represents the filename of the executable file (*i.e.*, TAppDecoder.exe) , the bitstream file that specifies the filename of the bitstream to be decoded (*i.e.*, strRA.str), and the reconstructed file that specifies the filename for the reconstructed video sequence (*i.e.*, decRA.yuv), respectively. The -b and -o in the basic decoder file Specifies the output coded bit stream file and the output locally reconstructed video file, respectively [7]. The -d parameter is also specifies the luma internal bit-depth of the reconstructed YUV file (*i.e.*, internal bit-depth is equal to 8 or 10).

```
#======== File I/O ====================
BitstreamFile              : strRA.bin
ReconFile                  : recRA.yuv

#======== Unit definition =================
MaxCUWidth                 : 64        # Maximum coding
unit width in pixel
MaxCUHeight                : 64        # Maximum coding
unit height in pixel
MaxPartitionDepth          : 4         # Maximum coding
unit depth
QuadtreeTULog2MaxSize      : 5         # Log2 of maximum
transform size for
                                       # quadtree-based TU
coding (2...6)
QuadtreeTULog2MinSize      : 2         # Log2 of minimum
transform size for
                                       # quadtree-based TU
coding (2...6)
QuadtreeTUMaxDepthInter    : 3
QuadtreeTUMaxDepthIntra    : 3

#======== Coding Structure =============
IntraPeriod                : 32        # Period of I-Frame
( -1 = only first)
DecodingRefreshType        : 1         # Random Access
0:none, 1:CDR, 2:IDR
GOPSize                    : 8         # GOP Size (number
```

(a) The main configuration file (*i.e.*, encoder random-access-maim.cfg) [7]

```
#======== File I/O ===============
InputFile                  : d:/Tennis_1920x1080_24.yuv
InputBitDepth              : 8         # Input bitdepth
FrameRate                  : 24        # Frame Rate per
second
FrameSkip                  : 0         # Number of frames
to be skipped in input
SourceWidth                : 1920      # Input  frame width
SourceHeight               : 1080      # Input  frame
height
FramesToBeEncoded          : 3         # Number of frames to
be coded
```

(b) The video configuration file (*i.e.*, Tennis.cfg) [7]

```
HM software: Encoder Version [10.0][Windows][VS 1600][32 bit]

Input        File        : d:/video_coding/New_seq/Tennis_1920x1080_24.yuv
Bitstream    File        : str4.bin
Reconstruction File      : rec4.yuv
Real    Format           : 1920x1080 24Hz
Internal Format          : 1920x1080 24Hz
Frame index              : 0 - 9 (10 frames)
CU size / depth          : 64 / 4
RQT trans. size (min / max) : 4 / 32
Max RQT depth inter      : 3
Max RQT depth intra      : 3
Min PCM size             : 8
Motion search range      : 64
Intra period             : 32
Decoding refresh type    : 1
QP                       : 32.00
Max dQP signaling depth  : 0
Cb QP Offset             : 0
Cr QP Offset             : 0
QP adaptation            : 0 (range=0)
GOP size                 : 8
Internal bit depth       : (Y:8, C:8)
PCM sample bit depth     : (Y:8, C:8)
RateControl              : 0
Max Num Merge Candidates : 5

TOOL CFG: IBD:0 HAD:1 SRD:1 RDQ:1 RDQTS:1 RDpenalty:0 SQP:0 ASR:0 LComb:1 FEN:1 ECU:0
POC    0 TId: 0 ( I-SLICE, nQP 32 QP 32 )     267464 bits [Y 39.0962 dB   U 42.4387
```

(c) the output file (*i.e.*, log-RA.txt) [7]

Fig. 2. (a-c) The configuration files used in the coding process [7]

168   A.F. Eldeken et al.

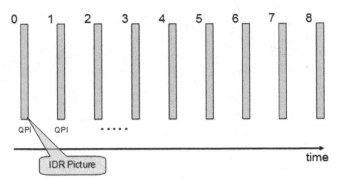

(a) Graphical presentation of all-Intra configuration [6]

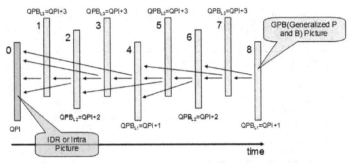

(b) Graphical presentation of Random-access configuration [6]

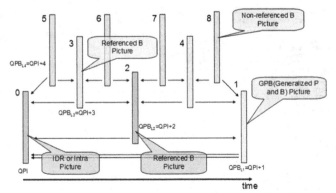

(c) Graphical presentation of Low-delay configuration [6]

Fig. 3. (a-c) Graphical presentation for the coding methods

3   The Proposed Approach

In the proposed approach the two configuration files (*i.e.*, random-access configuration file and Tennis configuration file) that present a collection of configuration parameters in the HEVC standard is called by the basic encoder file to start the encoding process. The first one describes the video coding method and the later describes the characteristics of the video that is used in the coding process. The access time for the basic encoder file to call and perform the two configuration files can be reduced by a half by merging the two configuration files into only one file. Each configuration file has it's own parameters which has a default values. So, when the configuration parameters is not present in the configuration file, the default values is taken instead. So, if we need to merge these two files into one file, we must first change these parameters calls in the C++ source code keeping all the functions related to theses parameters work well without any problems. Therefor, we modify all configuration parameters in the configuration files in the C++ source code to become suitable for calling from the new configuration file. The new configuration file is shown in Fig. 4.

```
TAppEncoder.exe  -c  encoder_randomaccess_main_Tennis.cfg   0 > log_RA.txt
```

(a) The new encoding file of the proposed approach

```
#======================= GENERAL
====================================================
InputFile             .\Tennis_1920x1080_24.yuv
OutputFile            .\strRA.bin
ReconFile             .\recRA.yuv
InputBitDepth             : 8          # Input bitdepth
FrameRate                 : 24         # Frame Rate per
second
FrameSkip                 : 0          # Number of frames to
be skipped in input
SourceWidth               : 1920       # Input   frame width
SourceHeight              : 1080       # Input   frame height
FramesToBeEncoded         : 3          # Number of frames to
be coded
#====================== CODING
====================================================
#======== Unit definition =================
MaxCUWidth                : 64         # Maximum coding unit
width in pixel
MaxCUHeight               : 64         # Maximum coding unit
height in pixel
MaxPartitionDepth         : 4          # Maximum coding unit
depth
QuadtreeTULog2MaxSize     : 5          # Log2 of maximum
transform size for
                                       # quadtree-based TU
coding (2...6)
```

(b) The new configuration file of the proposed approach

Fig. 4. The new configuration files of the proposed approach

Table 1. Description of data sequences used

Seq. #	Seq. Name	File size (MB)	Frame rate (fps)	# of Frames	Class	Resolution	Color format
1	PeopleOnStreet	555	30	150			
2	Traffic	563	30	150	A	2560×1600	
3	Train	1714	60	300			
4	BasketBallDrive	989	50	300			
5	Tennis	389	50	240	B	1920×1080	
6	ParkScene	475	24	240			
7	BasketBallDrill	20	50	200			
8	Keiba	97	30	500	C	832×480	4:2:0
9	BQMall	25	60	200			
10	RaceHorses	32	60	300			
11	BQSquare	64	60	600	D	416×240	
12	BasketBallPass	46	50	500			
13	Vidyo1	449	60	600			
14	SlideShow	88	20	500	E	1280×720	
15	Vidyo3	416	60	600			

Table 2. Encoding time comparison (hours)

Seq. #	Class	Seq. Name	Traditional approach (hours)	Proposed approach (hours)
1		PeopleOnStreet	5.33	5.21
2	A	Traffic	3.82	3.75
3		Train	8.72	8.64
4		BasketBallDrive	8.05	7.97
5	B	Tennis	6.80	6.76
6		ParkScene	3.23	3.12
7		BasketBallDrill	1.50	1.44
8	C	Keiba	1.53	1.51
9		BQMall	1.72	1.66
10		RaceHorses	0.27	0.25
11	D	BQSquare	0.40	0.37
12		BasketBallPass	0.41	0.36

In Fig. 4-(a) the new basic encoder file is illustrated that represents the file-name of the executable file (*i.e.*, TAppEncoder.exe), the main configuration and video file (*i.e.*, encoder random-access-maim-Tennis.cfg), and the output file (*i.e.*, log-RA.txt), respectively. In Fig. 4-(b) the new configuration file that is used in the encoding process contains both the encoding method besides the video characteristics.

4   Experimental Results

In this section data sequences, the implementation setup of experiments, and results are discussed. The data sets used in the experiments include five classes of real sequences [11]. Each class has three video sequences having different features with characteristics as shown in Table 1. Our implementation runs on Intel Core i5 with 4GB of RAM. The proposed approach (*i.e.*, referred to as *Proposed*) is compared to the traditional approach (*i.e.*, referred to as *traditional*) [6]. We use the HEVC standard software (HM10) [5] for encoding/decoding the data sets mentioned above. In this paper, the performance of competing approaches is evaluated by i) the rate-distortion (RD) (in dB/Kbps), ii) the Bjontegaard (BD) rate ratio [12], iii) the compression ratio between the decoded video sequence and its original using the two approaches for all data sets, and iv) the encoding time. The quantization parameter (QP) is set to 22, 27, 32, and 37 [13]. The group of picture (GOP) is set to 8. The coding method is random-access configuration (*i.e.*, class-E is skipped) [13] .

It worth noting that our implementation *Proposed* achieves the same result of the *traditional* when compared in terms of all metrics mentioned above at different QPs (22, 27, 32, 37), using the whole number of frames of all video sequences described above with the two competing approaches with using only one configuration file for the encoding process instead of using the two traditional configuration files for this process. Improvements are shown using the proposed approach in terms of encoding time as opposed to the traditional approaches by The Proposed approach surpasses the traditional approaches in terms of encoding time by a maximum decrease of 10% in class-D at QP equal to 32 as shown in Table 2.

5   Conclusions

In this paper, a new approach is introduced to implement the encoding process of high resolution videos using the HEVC standard with merging the two traditional configuration files into one configuration file. This modification is based on collecting all the parameters needed in the encoding process and the video characteristics into one file. Improvements are shown using the proposed approach in terms of encoding time as opposed to the traditional approaches by reducing the access time by half that resulting from reducing the data exchange between the configuration files. There is no change in the rate-distortion and compression ratio.

References

1. Bossen, F., Bross, B., Suhring, K., Flynn, D.: HEVC complexity and implementation analysis. IEEE Trans. on Cir. and Sys. for Video Tech. 22(12) (2012)
2. Sullivan, G., Ohm, J.-R., Han, W.-J., Wiegand, T.: Overview of the high efficiency video coding (HEVC) standard. IEEE Trans. on Cir. and Sys. for Video Tech., 22(12) (December 2012)

3. Wiegand, T., Sullivan, G.J., Björntegaard, G., Luthra, A.: Overview of the H.264/AVC video coding standard. IEEE Trans. on Cir. and Sys. for Video Tech. 13(7) (July 2003)
4. Hsia, S.-C., Hsu, W.-C., Lee, S.-C.: Low-complexity high-quality adaptive deblocking filter for H.264/AVC system. Signal Processing: Image Communication 27, 749–759 (2012)
5. Online (2012), http://hevc.hhi.fraunhofer.de/svn/svn_HEVCSoftware/branches
6. Kim, I.-K., McCann, K., Sugimoto, K., Bross, B., Han, W.-J.: High efficiency video coding (HEVC) test model draft 10 (HM 10) encoder description. Technical Report Doc. JCTVC-L1002, JCT-VC, Geneva, Switzerland (January 2013)
7. Bossen, F., Flynn, D., Suhring, K.: HM Software Manual. ITU-T SG16 WP3 and ISO/IEC JTC1/SC29/WG11, Geneva, Switzerland (January 2013)
8. List, P., Joch, A., Lainema, J., Björntegaard, G., Karczewicz, M.: Adaptive deblocking filter. IEEE Trans. on Cir. and Sys. for Video Tech. 13(7), 614–619 (2003)
9. Norkin, A., Björntegaard, G., Fuldseth, A., Narroschke, M.: HEVC deblocking filter. IEEE Trans. on Cir. and Sys. for Video Tech. 22(12) (December 2012)
10. Lou, J., Jagmohan, A., He, D., Lu, L., Sun, M.-T.: H.264 deblocking speedup. IEEE Trans. on Cir. and Sys. for Video Tech. 19(8) (2009)
11. Online (2003), ftp://hvc:US88Hula@ftp.tnt.uni-hannover.de/testsequences
12. Bjontegaard, G.: Calculation of average PSNR differences between RD-curves. VCEG-M33, Texas, USA (April 2001)
13. Bossen, F.: Common test conditions and software reference configurations. JCTVC-D600, Daegu, KR, U.S.A. (January 2011)

A Virtual Simulation of the Image Based Self-navigation of Mobile Robots

Mateusz Tecław, Piotr Lech, and Krzysztof Okarma

West Pomeranian University of Technology, Szczecin
Faculty of Electrical Engineering
Department of Signal Processing and Multimedia Engineering
26. Kwietnia 10, 71-126 Szczecin, Poland
{mateusz.teclaw,piotr.lech,krzysztof.okarma}@zut.edu.pl

Abstract. The paper concerns with the problem of fully visual self-navigation of mobile robots based on the analysis of similarity of images, acquired by the cameras mounted on the robot, with some previously captured images stored in a database. In order to simplify and speed-up the extraction of the necessary data from the image database it is assumed that the rough position of the robot is known e.g. based on the GPS module or some other sensors. Due to the application of the image analysis methods, the accuracy of the self-positioning of the robot can be significantly improved leading to fully visual self-navigation of autonomous mobile robots, assuming their continuous access to the image database. In order to verify the validity of the proposed approach, the virtual simulation environment based on the Simbad 3D robot simulator has been prepared. The initial results presented in the paper, obtained for synthetic images captured by the virtual robots, confirm the usefulness of the proposed approach being a good starting point for future experiments using the real images captured by the physical mobile robot also in various lighting conditions.

Keywords: machine vision, visual robot navigation, mobile robots.

1 Introduction

Due to rapid growth of availability of relatively cheap high quality cameras the application of machine vision algorithms in mobile robotics becomes more and more popular in recent years [2]. Similar trends can also be observed in industrial automation [4], mechatronics or transport applications, considering e.g. the development of the Intelligent Transportation Systems (ITS). Although the cameras are almost everywhere, their potential is not always fully utilized. One of the those areas where such possibilities of machine vision are still underused is the navigation of mobile robots, intelligent vehicles and Unmanned Aerial Vehicles (UAV). Information from cameras are often considered as supplementary for the infrared, ultraviolet, laser, sonar or some other types of sensors [3]. Because of low computational power in some embedded systems with limited memory, only some simple image analysis algorithms are used, often only for binary images.

© Springer International Publishing Switzerland 2015
R. Silhavy et al. (eds.), *Intelligent Systems in Cybernetics and Automation Theory,*
Advances in Intelligent Systems and Computing 348, DOI: 10.1007/978-3-319-18503-3_17

Another reason of such status quo corresponds with the main disadvantages of the vision systems related to the necessity of fast processing of bigger amount of data, higher memory requirements and the most important – sensitivity to changing lighting conditions. Nevertheless, the amount of information present on the images is much bigger and can provide more useful data improving the localization and navigation in comparison to e.g. simple line-following robot [8,9]. It is worth to notice that some active sensors, such as laser range finders or sonars, have their own limitations as well, e.g. related to limited angular resolution.

In some environments, e.g. larger open spaces, where the distance from the robot to the closest objects is relatively high, the advantages of the vision systems become obvious apart from their limitations. One of such examples may be the navigation based on the GPS which accuracy is limited to several meters. Considering some modern car navigation systems, the correction of the position of the vehicle is often made utilizing the additional analysis of the maps (e.g. assuming that the fast moving vehicle is not outside the road). Such an idea can be extended from the analysis of the 2D maps to the analysis of images representing the 3D surrounding of the mobile robot or vehicle, leading to the image based navigation with improved accuracy in comparison to the "pure" GPS. Such approach can be also used as supplementary solution for indoor environments although visual navigation in corridors is relatively simple [1] and therefore some dedicated systems have been developed similarly as for underwater navigation [5].

In our experiments it is assumed that some images of the surroundings have been captured previously from the known positions of the reference robot, similarly as for Google StreetView purposes. The task of the mobile robot is to find several images from the database taken for the nearby positions and calculated its more accurate position based on the analysis of their similarity with the images currently acquired by the cameras. Preliminary verification of the idea can be made using synthetic images utilizing the Java based 3D mobile robot simulation environment known as Simbad [6].

2   Virtual Environment for the Simulation of Self-localization of Mobile Robots

The main purpose of the publicly available Simbad environment, being a free Java 3D robot simulator, corresponds with usage of Artificial Intelligence and Machine Learning algorithms for autonomous robotics and autonomous agents. Due to the simplicity of the framework, some real-world issues are not supported e.g. there is no ray-tracer which would be helpful for the analysis of light changes. However, such an environment can be a convenient tool for preliminary testing of some control algorithms as well as the ideas requiring some further extensions and modifications. In addition to the color camera some other sensors are also available in Simbad such as contact sensors (bumpers) for collision detection and range sensors (infrared and sonars).

The assumed scenario of the experiments consists of several blocks with various colors and textures (used in further experiments) imitating the simplified

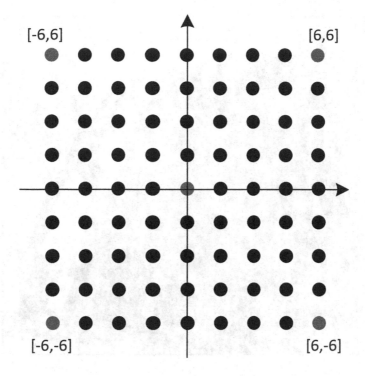

Fig. 1. The structure of the reference robot locations

urban area. In the first stage of experiments the database of images has been created which have been captured in four directions rotating the robot by 90 degrees in each of 81 predefined positions (a grid of 9×9 locations has been used). The reference positions of the robot have been equally distributed from -6 to 6 meters shifting by 1.5 m, both in horizontal and vertical direction, assuming the central location in the origin of the coordinate system as illustrated in Fig. 1. The perspective view of the 3D structure of the virtual scene (without textures) is shown in Fig. 2.

For each of 81 reference locations a set of four images has been captured and stored in the database together with the position of the robot. For testing purposes several different locations have also been chosen and the images acquired from them have been stored in the second part of the database. Considering the limited number of synthetic images, the first experiments have been conducted with the assumption that the most similar images are searched in the whole reference database consisting of 324 images. Obviously, such simplification is justifiable only in the closed virtual environment as in real scenarios for the images acquired from the camera there may be the situation where the images taken from distant locations are very similar to each other so additional rough location of the robot would be necessary to prevent the impact of images from distant locations. Since the "unknown" position of the robot is always located

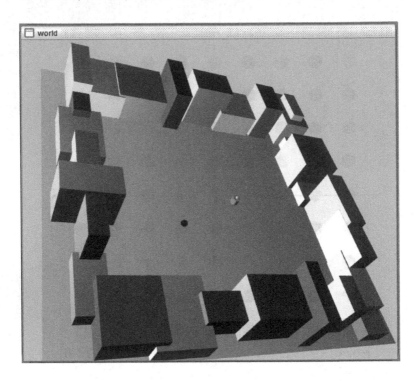

Fig. 2. The 3D structure of the virtual scene used in experiments

among four nodes of the 9×9 elements grid, we have set the minimum number of analyzed images to 4 in order to estimate the current position of the robot. Additionally, some further experiments utilizing higher number of the most similar images have been conducted as well.

A crucial element of the proposed approach in view of the accuracy of visual self-localization of mobile robots is the method used for determining the similarity factor between two images. Regardless of finding the most similar images an important issue is the estimation of the similarity level necessary for improvement of the localization accuracy between the grid nodes. The choice of the most suitable method is dependent on the image content and will be one of the most relevant topics of our further research. For the outdoor images containing many details some feature based methods utilizing Scale Invariant Feature Transform (SIFT) [11] or Speeded Up Robust Features (SURF) may be used. Promising results can also be obtained using some of the full-reference image quality assessment metrics such as Structural Similarity (SSIM) [13] or some of its modifications [7,14]. An additional application of the super-resolution algorithms may also lead to higher localization accuracy [10] as well as utilization of texture similarity metrics [12] also combined with some other features.

In order to verify the validity of the proposed approach in the assumed scenario constructed in Simbad simulator, our experiments have been conducted using the

SSIM index between two color images calculated as the mean values of the SSIM metric obtained for three RGB channels. For each channel the overall metric is defined as the average of the local SSIM index values calculated based on the local mean luminance, variances and the covariance for 11×11 pixels Gaussian window as:

$$SSIM = \frac{(2\bar{x}\bar{y} + C_1) \cdot (2\sigma_{xy} + C_2)}{(\sigma_x^2 + \sigma_y^2 + C_1) \cdot [(\bar{x})^2 + (\bar{y})^2 + C_2]} \ , \tag{1}$$

where the constants $C_1 = (0.01 \times 255)^2$ and $C_2 = (0.03 \times 255)^2$ are used as suggested by the SSIM authors in order to prevent the possible division by zero for flat and/or dark areas of the image.

Assuming the usage of at least four most similar images (one for each direction of observation), the current position of the robot can be estimated as weighted sum of the coordinates related to those images where the weighting coefficients are the values of the SSIM index calculated for the respective images. Nevertheless, much better results can be expected assuming the use of 8 best matched images by means of the SSIM index (two for each direction of observation), especially for locations relativelt far from the grid nodes. Both these approaches have been verified during experimental simulations.

The estimated position of the robot calculated using the average SSIM values for the 8 most similar images (two for each direction of observation) can be expressed as the weighted average of locations corresponding to those images according to the following formula:

$$[x \ y] = \frac{\sum\limits_{i=1}^{N} SSIM_i \cdot [x_i \ y_i]}{\sum\limits_{i=1}^{N} SSIM_i} \ , \tag{2}$$

where N is the number of considered most similar images.

3   Discussion of Results

The results of the experiments conducted assuming the usage of 4 and 8 most similar images for 13 different position of the virtual robot are shown in Table 1. The average error defined as the Euclidean distance between the estimated position of the robot and "ground truth" is equal to 0.4217 m using $N = 4$ images and 0.3308 m using $N = 8$ images. Considering the grid size of 1.5 m, a significant increase of the self-positioning accuracy can be noticed in comparison to the available data. In real outdoor scenarios, assuming the grid size comparable to the GPS accuracy, it should be possible to achieve the accuracy of even several centimeters instead of a few meters.

| Query images acquired by the virtual robot | Images from the database with the highest SSIM values | Images from the database with the second highest SSIM values |

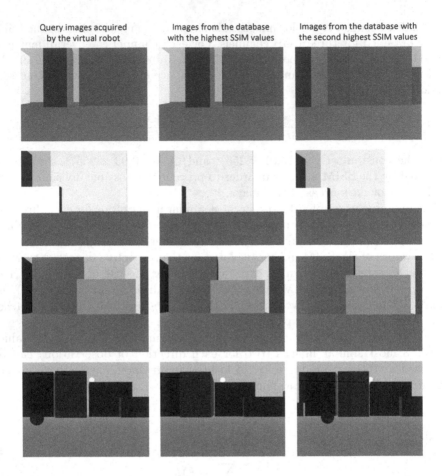

Fig. 3. Set of 4 images acquired for the robot's test position no. 2. (left column) and the most similar images found by means of the SSIM index for each observation direction

An exemplary set of four images acquired for the position no. 2 together with the reference images with the highest SSIM values found for those images are illustrated in Fig. 3 whereas corresponding set of images for the position no. 4 is shown in Fig. 4.

It is worth noticing that for some locations of the robot (4 out of 13) better accuracy can be obtained using only 4 best matched images instead of 8 (i.e. lower distance values presented in Table 1). This phenomenon is caused mainly by relatively small differences between the SSIM metric values used as weighting coefficients. In the situation when an image from the group of the second most similar images does not correspond to one of the four nodes closest to the robot's location, the error may increase in comparison to the use of only four images.

On the other hand, the presence of an image corresponding to further node in the group of four most similar images may increase the localization error more

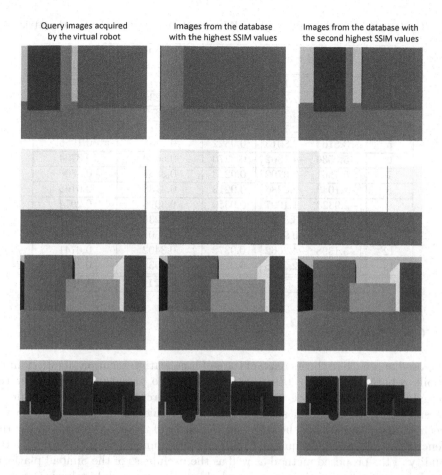

Fig. 4. Set of 4 images acquired for the robot's test position no. 4. (left column) and the most similar images found by means of the SSIM index for each observation direction

when only 4 images are used. It may be observed in in Fig. 4 where two of images with highest SSIM values are clearly acquired from more distant position than the closest grid nodes.

Since the experiments have been conducted for the images representing a simplified model of robot's surroundings without the use of textures, the dynamic range of the obtained SSIM values is relatively small so even small differences between them may influence the classification of the given image as the most similar to the query image captured by the robot's camera. It is also worth to notice that, due to relatively simple structure of compared images which contain many flat areas without textures, the influence of the constant values in the SSIM formula (Eq. 1) is relatively high, decreasing the dynamic range of obtained values.

Table 1. Obtained results of the self-localization based on the SSIM metric

Position	"Ground truth"	SSIM$_{max}$	Distance for 4 images	Distance for 8 images
1	[0.5621 ; −2.1804]	0.8589	0.3642	0.1189
2	[0.8975 ; −2.7104]	0.8750	0.3235	0.1626
3	[1.6589 ; −3.3684]	0.9149	0.4372	0.5452
4	[3.0278 ; −2.7214]	0.9636	0.7611	0.4653
5	[3.5164 ; −2.8015]	0.9522	0.2481	0.1955
6	[5.0536 ; −4.1802]	0.9270	0.6629	0.3764
7	[5.7661 ; −3.3702]	0.9223	0.3854	0.4806
8	[−1.1043 ; −2.6286]	0.9212	0.3738	0.6165
9	[−2.9323 ; 1.1997]	0.9359	0.3078	0.2974
10	[−3.2300 ; −3.5394]	0.9245	0.2235	0.3661
11	[−3.6786 ; −0.9097]	0.8688	0.6607	0.3715
12	[−5.1398 ; −4.2662]	0.9372	0.3402	0.0741
13	[−5.9762 ; 2.2628]	0.9119	0.3938	0.2297
Mean			**0.4217**	**0.3308**

4   Conclusions

Proposed approach to the increase of the self-localization accuracy of autonomous mobile robots using image based methods leads to promising results. For the virtual environment created using the Simbad 3D robot simulator a significant increase of the localization accuracy has been achieved in comparison to the specified distance between the reference locations of the robot for which the reference images have been acquired. Conducted experiments have allowed both the validity of the proposed method as well as the usefulness of the Simbad platform for rapid verification of the vision based algorithms for mobile robotics.

Further experiments conducted using more complex scene with additional textures imitating e.g. the buildings' facades have led to worse results. The reason is the specific character of the Structural Similarity metric which is calculated with the assumption of geometrical matching of images. Since the idea of our experiments is different and the images captured from the unknown position of the robot are shifted according to the images stored in the database, those differences are partially utilized for estimation of the robot's location but can be troublesome for the proper calculation of the SSIM metric. Possible solutions of this problem are related to the application of the multi-scale extension of the SSIM metric or some of its modifications not necessarily calculated in the pixel domain. Another possibility is the pre-matching of both compared images using some other methods e.g. based on the comparison of features. Both these approaches will be an important challenges for our future research, conducted also in real conditions.

References

1. Bonon-Font, F., Ortiz, A., Oliver, G.: Visual navigation for mobile robots: A survey. Journal of Intelligent and Robotic Systems 53(3), 263–296 (2008)
2. Chatterjee, A., Rakshit, A., Singh, N.N.: Vision Based Autonomous Robot Navigation – Algorithms and Implementations. SCI, vol. 455. Springer, Heidelberg (2013)
3. Desouza, G., Kak, A.: Vision for mobile robot navigation: a survey. IEEE Transactions on Pattern Analysis and Machine Intelligence 24(2), 237–267 (2002)
4. Domek, S., Dworak, P., Grudziński, M., Okarma, K.: Calibration of cameras and fringe pattern projectors in the vision system for positioning of workpieces on the CNC machines. In: Gosiewski, Z., Kulesza, Z. (eds.) Mechatronic Systems and Materials V, Solid State Phenomena, vol. 199, pp. 229–234. Trans Tech Publications (2013)
5. Garcia, R., Nicosevici, T., Ridao, P., Ribas, D.: Towards a real-time vision-based navigation system for a small-class UUV. In: Proceedings of the IEEE/RSJ International Conference on Intelligent Robots and Systems (IROS 2003), vol. 1, pp. 818–823 (2003)
6. Hugues, L., Bredeche, N.: Simbad: An autonomous robot simulation package for education and research. In: Nolfi, S., Baldassarre, G., Calabretta, R., Hallam, J.C.T., Marocco, D., Meyer, J.-A., Miglino, O., Parisi, D. (eds.) SAB 2006. LNCS (LNAI), vol. 4095, pp. 831–842. Springer, Heidelberg (2006)
7. Li, C., Bovik, A.: Three-component weighted structural similarity index. In: Proceedings of SPIE - Image Quality and System Performance VI, San Jose, California, vol. 7242, p. 72420Q (2009)
8. Okarma, K., Lech, P.: Monte carlo based algorithm for fast preliminary video analysis. In: Bubak, M., van Albada, G.D., Dongarra, J., Sloot, P.M.A. (eds.) ICCS 2008, Part I. LNCS, vol. 5101, pp. 790–799. Springer, Heidelberg (2008)
9. Okarma, K., Lech, P.: A fast image analysis technique for the line tracking robots. In: Rutkowski, L., Scherer, R., Tadeusiewicz, R., Zadeh, L.A., Zurada, J.M. (eds.) ICAISC 2010, Part II. LNCS, vol. 6114, pp. 329–336. Springer, Heidelberg (2010)
10. Okarma, K., Tecław, M., Lech, P.: Application of super-resolution algorithms for the navigation of autonomous mobile robots. In: Choraś, R.S. (ed.) Image Processing & Communications Challenges 6. AISC, vol. 313, pp. 147–154. Springer, Heidelberg (2015)
11. Se, S., Lowe, D., Little, J.: Vision-based global localization and mapping for mobile robots. IEEE Transactions on Robotics 21(3), 364–375 (2005)
12. Žujović, J., Pappas, T.N., Neuhoff, D.L.: Structural similarity metrics for texture analysis and retrieval. In: Proc. 16 th IEEE Int. Conf. Image Processing ICIP, pp. 2225–2228 (2009)
13. Wang, Z., Bovik, A., Sheikh, H., Simoncelli, E.: Image quality assessment: From error measurement to Structural Similarity. IEEE Trans. Image Processing 13(4), 600–612 (2004)
14. Wang, Z., Simoncelli, E., Bovik, A.: Multi-Scale Structural Similarity for image quality assessment. In: Proc. 37th IEEE Asilomar Conf. Signals, Systems and Computers, Pacific Grove, California (2003)

Implementation and Optimization of Stereo Matching Algorithm on ARM Processors

Peter Janků, Roman Došek, and Tomáš Dulík

Tomas Bata University in Zlin, Faculty of applied informatics,
Nad stráněmi 4511, 760 05 Zlín, Czech republic
{pjanku,dosek,dulik}@fai.utb.cz
http://www.utb.cz/fai

Abstract. This paper analyses possibility of implementation of stereo matching algorithms on ARM based processors with special attention on the final algorithm performance. Semi global block matching and Block matching algorithms were chosen as a base of this research. First, the technologies used in the implementation are described, then the optimization approach is discussed. The main part of this paper deals with algorithms performance depending on the chosen optimization.

Keywords: stereo matching, computer vision, OpenCV, ARM.

1 Introduction

Computer vision methods have recently become very popular in robotic systems. One area of the computer vision deals with stereo vision and depth map acquisition. It represents one of the methods to gain information about the environment for systems such as autonomous robots or self-driving cars. While the depth measurement precision is not as high as with laser range finders or lidars, the stereo camera is much cheaper. Precision of a stereo camera depth map is higher than that of ultrasound sensors, which are susceptible to reflections so direction of obstacles can be detected improperly. [1]

Although the stereo vision has been around for a long time, high computing power is still required for a real-time depth map computing. Therefore, several approaches are commonly employed to build a real-time embedded stereo vision systems. For example, the whole system may be implemented by one powerful FPGA, or combination of FPGA with set of other processors. In most of these cases, depth map computing is performed inside FPGA. [2,3]

The ARM based processors are very popular in robotic systems thanks to their high performance and small power consumption; moreover, they are able to run standard operating systems such as Linux or Windows. Despite this, the real-time depth map computing on ARM devices is problematic. It requires a good algorithm, solid implementation and many optimizations.

There are already several papers quantifying performance of the block matching algorithm under various conditions. The most comprehensive paper written by Szeliski and Scharstein compares performance impact depending on selected

© Springer International Publishing Switzerland 2015
R. Silhavy et al. (eds.), *Intelligent Systems in Cybernetics and Automation Theory*,
Advances in Intelligent Systems and Computing 348, DOI: 10.1007/978-3-319-18503-3_18

block size.[4] However, the issue of running stereo vision algorithms directly on ARM processors is not discussed much.

The rest of this paper is organized in the following manner. Section 2 describes basic optimization strategies available on ARM based processors. Section 3 contains description of test environment and methodology of measurement. Section 4 then contains results of measured performance of selected variants of algorithms.

2   Proposed Optimization on the ARM Platform

Optimization of stereo matching algorithms can take place on various levels. The simplest approach is to split image into separate parts and apply stereo matching algorithm on each of these parts separately.

2.1   Thread Based Parallelization

The main requirement for this approach is to have multiple independent processing cores. This requirement is satisfied on most of modern high performance ARM platforms such as Nvidia Tegra T30, Freescale i.MX6 or Qualcomm Snapdragon 800. [5]

Most of stereo matching algorithms require rectified image and the disparities are sought inside horizontal lines, so the best approach is to split image horizontally to get multiple subimages. To attain best performance, the number of subimages should match the number of processing cores.

As the algorithm runs independently on multiple parts of image, the performance is directly proportional to the number of cores. However, depending on implementation, it may be necessary to create overlapping subimages, because the algorithm does not operate on areas near image edges. Furthermore, the resulting depth map has to be reconstructed back from depth maps of subimages.

2.2   SIMD Parallelization

As name SIMD (Single Instruction Multiple Data) implies, SIMD instructions basically works by applying the same operation on multiple data at the same time. On x86 & x86_64 platforms, there are SSE instruction sets which takes advantage of so called SSE registers which may be accessed as 16x8b, 8x16b, 4x32b, 2x64b or 1x128b register. [6]

On the ARM platform, there are NEON instructions and registers, which provides great deal of operations the SSE does. NEON is, however, not as easy to use as the thread parallelization, because it requires changes to algorithm implementation. Furthermore, much higher knowledge of algorithm and its hotspots is required. To attain the best performance, it is also better to manually unroll loops and not mix NEON and ARM instruction because that causes instruction pipeline stalls. In C and C++, there are two ways to use NEON instructions. The simpler and less performance-wise is to use compiler intrinsics which are functions built in to the compiler. These functions are then transformed to NEON instruction during compilation. However, the quality of produced code is heavily dependent on compiler and generally lacks the speed of hand-written assembly even today. The second option is to use inline assembly and write algorithm (or parts of it) directly in assembly.[6,7]

2.3   Combination of Approaches

On typical multicore high-end ARM SoC, it is entirely possible to combine both approaches and gain significant speedup compared to non-optimized implementation. Each core contains its own NEON registers, which means performance gains from both methods stacks.

3   Test Environment

All measurements presented in this paper are done by using the same set of software and hardware components. The components are described in this section.

3.1   Stereo Matching Algorithms

Each process of computing disparity map can be divided into three basic parts: image preprocessing, stereo matching algorithm and image post processing. Moreover, variety of stereo matching algorithms and its implementations can be used. For this paper, the OpenCV library has been chosen. Specifically, its two algorithms - Block matching (BM) and Semi Global block matching (SGBM) were used for implementation and measurement.[8]

3.2   ARM Processor

ARM is a processor core architecture based on RISC architecture. The Toradex Apalis T30 board was chosen as a hardware platform for our research. It includes Nvidia Tegra III processor which contains four ARM Cortex A9 cores. It is supplied by 1GB RAM memory and 4GB eMMC memory for data.

As base system, OpenEmbedded/Angstrom custom linux distribution was built and installed.

3.3   Test Data and Program

The well-known test images from Middlebury Stereo Vision Page were used as test data, see fig. 3. Since performance of selected algorithms is dependent only on their parameters and not on data, single image set "Teddy" with resolution 450 x 375 pixels was used.

The test program was written in C++ with OpenCV library. Thread parallelization was done using OpenMP API. To measure SIMD performance gain, the BM algorithm was rewritten in NEON compiler intrinsics.

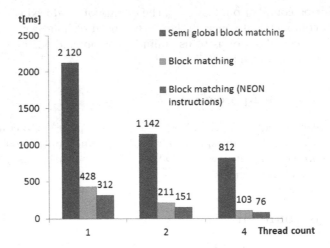

Fig. 1. Average time needed to process one stereoframe

4 Measured Results

Because these algorithms were run under Linux operating system, the measured results could be slightly affected by other tasks already running in the system. Therefore, each algorithm and each optimization was run one hundred times and the final results were computed as an average of subresults. The visualisation of these results can be seen in fig. 1 and fig. 2.

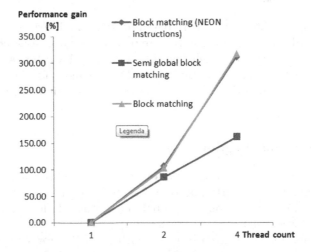

Fig. 2. Performance gain when increasing number of threads

The average time needed for computing one couple of corresponding frames (stereoframes) can be seen in fig. 1. The bar graph shows that SGBM algorithm is significantly slower then BM one; moreover, the maximal time needed for one stereoframe computing by SGBM is nearly 2.2 seconds long and the minimal one is 0.8 second long. Therefore OpenCV SGBM implementation is not suitable for online stereovision computing on ARM, even if optimizations are used. The green and blue bars represent BM algorithm with and without using NEON instructions. As can be noticed, the NEON optimizations applied on BM algorithm adds performance gain about 27 %.

The performance gain in percents can be seen in fig. 2. Different algorithms implementations are represented by each line. As can be noticed, the SGBM algorithm thread based parallelization achieves much lower performance gain then BM parallelization. At the other side, the BM performance gain is higher then 100 percent per thread. This overflow is caused by algorithm skipping edges of image as described in section 2.1.

Fig. 3. One of the source images

5   Conclusion

The BM and SGBM algorithm implementation and optimization on ARM processors was discussed in this paper. It was measured that SGBM algorithm is not suitable for realtime stereo matching computing due to its low performance. On the other side, the BM algorithm with NEON optimization and thread based parallelization is able to compute one stereo match in less then 0.1 seconds.

The twenty-seven percent performance gain was obtained by using NEON instructions together with BM algorithm. Moreover, nearly one-hundred percent per thread performance gain was achieved by using thread based optimization.

The future research should be aimed on image preprocesing and postprocesing phrases mentioned in section 3.1 and/or on better utilization of NEON units, especially on usage of directly inlined ASM instructions (as was described in section 2.2.

Acknowledgements: This paper is supported by the Internal Grant Agency at TBU in Zlin, Project No. IGA/FAI/2014/039 and by the European Regional Development Fund under the project CEBIA-Tech No. CZ.1.05/2.1.00/03.0089.

References

1. Janku, P., Dosek, R., Jasek, R.: Obstacle detection for robotic systems using combination of ultrasonic sonars and infrared sensors. In: Silhavy, R., Senkerik, R., Oplatkova, Z.K., Silhavy, P., Prokopova, Z. (eds.) Modern Trends and Techniques in Computer Science. AISC, vol. 285, pp. 321–330. Springer, Heidelberg (2014)
2. Hamza, B., Abdelhakim, K., Brahim, C.: FPGA design of a real-time obstacle detection system using stereovision. In: 2012 24th International Conference on Microelectronics (ICM), pp. 1–4 (December 2012)
3. Schmid, K., Tomic, T., Ruess, F., Hirschmuller, H., Suppa, M.: Stereo vision based indoor/outdoor navigation for flying robots. In: 2013 IEEE/RSJ International Conference on Intelligent Robots and Systems (IROS), pp. 3955–3962 (November 2013)
4. Scharstein, D., Szeliski, R., Zabih, R.: A taxonomy and evaluation of dense two-frame stereo correspondence algorithms. IEEE Computer Soc., Los Alamitos (2001) WOS:000173391800015
5. Shekhar, T.C.D., Varaganti, K.: Parallelization of face detection engine. In: 2010 39th International Conference on Parallel Processing Workshops (ICPPW), pp. 113–117 (September 2010)
6. Wan, J., Wang, R., Lv, H., Zhang, L., Wang, W., Gu, C., Zheng, Q., Gao, W.: AVS video decoding acceleration on ARM cortex-a with NEON. In: 2012 IEEE International Conference on Signal Processing, Communication and Computing (ICSPCC), pp. 290–294 (August 2012)
7. Pujara, C., Modi, A., Sandeep, G., Inamdar, S., Kolavil, D., Tholath, V.: VC-1 video decoder optimization on ARM cortex-a8 with NEON. In: 2010 National Conference on Communications (NCC), pp. 1–5 (January 2010)
8. Culjak, I., Abram, D., Pribanic, T., Dzapo, H., Cifrek, M.: A brief introduction to OpenCV. In: 2012 Proceedings of the 35th International Convention MIPRO, pp. 1725–1730 (May 2012)

Heuristic Control of the Assembly Line

Bronislav Chramcov[1], Franciszek Marecki[2], and Robert Bucki[3]

[1] Tomas Bata University in Zlín, Faculty of Applied Informatics, Zlín, The Czech Republic
b.chramcov@gmail.com
[2] Academy of Business in Dąbrowa Górnicza, Dąbrowa Górnicza, Poland
fmarecki@gmail.com
[3] Institute of Management and Information Technology, Bielsko-Biała, Poland
rbucki@wsi.net.pl

Abstract. The paper highlights the mathematical model of the assembly process for the automated line as well as sequence control algorithms for the assembled version of objects. The automatic assembly line control requires numerical simulation. Minimizing the assembly time of all objects or maximizing the number of assembled objects within the given time are treated as optimization criteria. Specification of the robot assembly line is described in detail as well as the method of controlling the order of manufactured objects. The equations of state of an automatic assembly line are presented. The simulation model includes heuristic algorithms for control determining of the assembly line. The assembly process in a line can be modeled with different assumptions.

Keywords: Modelling, Computer Simulation, Heuristic Algorithms, Control, Manufacturing.

1   Introduction

The assembly line was introduced by Ford [1]. If the same objects are installed in an assembly line, then the so-called assembly line balancing problem plays an important role. It consists of a uniform separation of assembly operations to the work stands. This is a combinatorial problem belonging to the NP class in terms of computational complexity (versions)[2,3].

Various objects are assembled in contemporary lines. Solving the problem of assembly line balancing, separately for each version, results in various sets of operations (with various times) at each station. Therefore, the optimizing sequence assembly problem of objects of different versions arises.

Assembly in contemporary lines is performed by industrial robots. These robots perform assembly operations according to established computer programs. When installing multiple versions, each robot has an assembly program for each version. In addition, before installing, the robot should recognize the assembled object (with the use of a camera).

An automatic assembly line is controlled by a computer. Robots recognize object versions and implement the selected work program. In addition, the objects are fed to

© Springer International Publishing Switzerland 2015
R. Silhavy et al. (eds.), *Intelligent Systems in Cybernetics and Automation Theory*,
Advances in Intelligent Systems and Computing 348, DOI: 10.1007/978-3-319-18503-3_19

the assembly line at the optimal sequence. Minimizing the assembly time of all objects or maximizing the number of assembled objects within the given time are treated as optimization criteria.

2   State of the Art

Modern optimization has played an important role in service-centered operations and manufacturing and as such has attracted more attention to this field [4]. Multilevel production scheduling problem is a typical optimization problem in a manufacturing system (assembly lines) which is traditionally modeled as several hierarchical sublevel problems and optimized at each level [5]. The modern methods of optimization include the use of heuristic algorithms [6]. Designed algorithms are tailored to accomplish a specialized task or goal and usually find a solution quickly and easily. Although heuristic algorithms were originally known as inaccurate, current approaches to the mathematical modeling and development of computer simulation completely changes scientific mind on these algorithms. Heuristic algorithms are thus increasingly used to optimize a range of sectors such as logistics [7], finance [8], manufacturing [9,10,11] and in a big number of other professional and scientific works. There are various methods of optimization, even today in many cases heuristic approach is very conveniently applicable. Vanderkam [12] emphasizes that heuristic algorithms are demonstrated to yield suboptimal networks in order to meet conservation goals. Although the degree of suboptimality is not known when using heuristics, some researchers have suggested that it is not significant in most cases and that heuristics are preferred since they are more flexible and can yield a solution more quickly.

Heuristic approach is widely implemented to control a manufacturing system especially to solve specific problems within the area of production planning and detailed scheduling i.e. repetitive manufacturing [13,14,15,16]. The use of heuristic algorithms successfully meets e.g. the total time minimizing criterion. Moreover, the higher the system complexity is, the more effective they prove to be in terms of quickness of finding the required solution. Their correctness is proven by the simplex method which is a numerical method for solving problems in linear programming [17,18]. Hybrid search algorithms are used for scheduling flexible manufacturing systems. These algorithms combine heuristic best-first strategy with controlled backtracking strategy as discussed in [19].

An alternative solution to cell scheduling by implementing the technique of Nagare cell is discussed by Muthukumaran and Muthu in [20]. A simulation-based three-stage optimization framework for high-quality robust solutions to the integrated scheduling problem is presented in [21]. It is considered to be a parallel machine scheduling problem with random processing/setup times and adjustable production rates.

There is a wide branch of manufacturing systems where manufacturing decisions are to be made immediately e.g. either regular parts or spare parts are manufactured in various series in automotive industry. There are many specific details which make these systems complex. Decisions are more than often made after thorough simulation of the system in order to avoid disturbances in discrete manufacturing operations.

This can be treated as searching for a satisfactory solution. Heuristics are particularly suitable for scheduling products that are to be produced on simply structured lines (such as assembly lines).

3   Formulating the Problem

Let us consider an assembly line consisting of M serial assembly station. Moreover, the initial station is distinguished $m=0$ for which chassis of assembled objects are given (such as car bodies, etc.) and the end station $m=M+1$ where assembly quality control takes place. Objects of different versions marked by w, $w=1,..., W$ are assembled in the assembly line (e.g. cars, etc). Each assembled object consists of standard and version parts. Let us assume vector of orders in the form (1), where z_w determines the number of objects to be assembled of the wth version.

$$Z = [z_w], \quad w = 1,...,W \tag{1}$$

The number N of all objects to be assembled is the sum of orders. This sum is possible to present in the form (2).

$$N = \sum_{w=1}^{W} z_w \tag{2}$$

During the assembly process carried out in the automatic line, a production pace consisting of an assembly sequence and transport of objects is distinguished. A transporter of the automated assembly line moves in cycles. The transporter is stopped during the installation of objects - at all stations. After carrying out the assembly operation by each robot, objects are moved to the next station.

During the cycle, the wth version object is passed to the initial station. At the end of the cycle, after installing the chassis, the object from the initial station is transported to the first station. Similarly, the object in the mth station, after the assembly process, is transported to the station $m+1$. An object from the last assembly station is transported to the station $M+1$ where it undergoes assembly quality control.

At each station of the automated line, there is one robot performing assembly operations. Let us assume that the robot assembly times $c_{m,w}$ are given. These times depend on the station number m, and the version w, of the object which is assembled. So, in the kth cycle ($k=1,..., K$) the times c_m^k of robot operating can be determined by means of choosing one element from each column m, of the given matrix of times $c_{m,w}$.

During the kth cycle each robot must be able to perform all the assembly operations. After the assembly process is over, the robot withdraws from the assembly space. If all robots withdraw from the assembly space, the transporter can move objects to the next station. The object transporting time can be neglected. As a consequence, the time c^k of the kth cycle is the maximal assembly time c_m^k at the stations m. The time of the kth cycle is possible to express according to the form (3).

$$c^k = \max(c_1^k, ..., c_m^k, ..., c_M^k) \tag{3}$$

The times c^k vary in the subsequent cycles k, because an object of a different version w, can be assembled at each mth station. The problem of optimal control of the automatic assembly line may consist of:

— assembling all the objects in the shortest possible time,
— assembling the maximum number of objects within a specified time.

The criterion of the shortest assembly time Q of all objects N can be written in the form (4).

$$Q = \sum_{k=1}^{N} c^k \rightarrow \min \tag{4}$$

The criterion of the maximal number of objects which are to be assembled in the given time T is the result of the condition where the number of objects K is determined according to the form (5).

$$(\sum_{k=1}^{K} c^k \le T) \wedge (\sum_{k=1}^{K+1} c^k > T) \tag{5}$$

In connection with this, the criterion can be written in the form: $K \rightarrow \max$. In a general case, version objects have different values. Therefore, the criterion value has the form of the sum of objects assembled in a predetermined time T.

4   Controlling the Order of Objects

Objects can be assembled in a different order in the automated line. Moreover, the order of assembly of objects at each station by each robot is the same because the assembly line is a series of stations without a storage buffer between them.

    The assembly time of all objects Q and the number of assembled objects within the given time T depend on the sequence of objects passed into the initial station ($m=0$). Thus, the optimization of the assembly in an automatic line is to control the order of the objects being fed into the initial station.

    Let us consider the minimization problem of the time Q of the assembly of all objects N. For this purpose, the model of an assembly process in the automatic assembly line is introduced in the form of state equations.

4.1   The State of the Assembly Line

The state of an automatic assembly line after the kth cycle is possible to express as a vector in the form (6), where s_m^k is the number of the object version which is located in the mth station after the kth cycle.

$$S^k = [s_m^k] \quad m = 0,...,M+1; \quad k = 0,1,...K \tag{6}$$

The initial state S^0 presents objects which are in the stations of the assembly line before beginning the assembly process. Objects are assembled throughout the first cycle and after the assembly process they are moved (at the end of the first cycle) to the next station. An object from the Mth station is moved to the station $M+1$ (station of the assembly quality control). At the same time after the first cycle, there is an object at the first assembly station which was placed at the initial station ($m=0$) during the first cycle. The first cycle time is calculated according to the form (7) where times c_m^1 result from objects which were at the stations of the assembly line at the initial state S^0.

$$c^1 = \max(c_1^1,...,c_m^1,...,c_M^1) \tag{7}$$

The final state S^K presents objects which are at the stations of the assembly line after the K cycles. Thus, we assume that there is a certain object left in the final state at each station. Thus, there were either $K=N$ objects assembled or K objects fulfilling the condition (5).

4.2   The Order of Assembled Objects

It is assumed that the order of assembled objects in cycle k, is possible to express as a vector in the form (8), where x_0^k is the number of the object version which was installed at the initial station in the kth cycle.

$$X = [x_0^k] \quad k = 1,...,K \tag{8}$$

This definition indicates that there is no object in the station $m=0$ after the cycle $k=0$. The assembly process in the line begins with the first cycle. Then, robots at each station m ($m=1,...,M$) assembly objects which were at these stations at the initial state. At the same time, the first object is installed in the first cycle at the initial state. This object finds its place at the first station after the first cycle. Analogously, the first object assembled in the assembly line is the one which was assembled at the last station ($m=M$) in the first cycle.

4.3   Assembly Robot Control

Assembly robots state in the cycle k is possible to express as a vector in the form (9) where u_m^k is a number of object which is assembled by the mth robot (at the mth station) in the kth cycle.

$$U^k = [u_m^k] \quad m = 1,...,M \tag{9}$$

In practice, robots recognize (by means of vision cameras) versions of objects which are at their stations and should be assembled. If the mth robot recognizes the wth version object in the kth cycle, then it starts its wth computer program. According to this program, the assembly time is known and equals c_m^k.

In the deterministic assembly model, it is assumed that $u_m^k = s_m^{k-1}$ for $m=1,...,M$. Therefore, robot control in the deterministic model consists in choosing their work mode (program) on the basis of the recognized wth version of object which is to be assembled. It is assumed in the probabilistic assembly model that:

- $u_m^k = w$ if the mth robot should assembly the recognized object in the kth cycle;
- $u_m^k = 0$ if the mth robot should not assembly the object in the kth cycle.

The object from the preceding station can be withdrawn from the assembly process (e.g. as a consequence of low quality).

4.4   The State of the Assembly Process

The equation of state of the assembly process in the assembly line takes the following form (10).

$$S^k = F(S^{k=1}, x_0^k), \quad k = 1,...K \tag{10}$$

The function F of assembly process transformation in the subsequent stations m ($m=1,...,M$) takes the form (11).

$$
\begin{aligned}
s_1^k &= x_0^k \\
s_2^k &= u_1^k \\
&.... \\
s_{m+1}^k &= u_m^k \\
&.... \\
s_{M+1}^k &= u_M^k
\end{aligned}
\tag{11}
$$

It is possible to determine, from the above equations of state, the fact that objects which are at stations m ($m=1,...,M$) of the assembly line depend on the sequence X of passing new objects into the initial station. Therefore, we can express the time c^k of the kth cycle in the form (12) and consequently we obtain equation in the form (13).

$$c^k = \max[c_1^k(x_0^k),..., c_m^k,..., c_M^k) \tag{12}$$

$$c^k = c^k(x_0^k) \tag{13}$$

Thus, the formulated criterion of minimizing the assembly time of all objects N can be written in the form (14).

$$Q = \sum_{k=1}^{N} c^k(x_0^k) \to \min \tag{14}$$

Analogously, it is possible to write the criterion of maximization which applies to numbers K of objects assembled in the determined time T in the form (15) where $K(X) \to \max$.

$$[\sum_{k=1}^{K} c^k (x_0^k) \leq T] \wedge [\sum_{k=1}^{K+1} c^k (x_0^k) > T] \tag{15}$$

5   Proposal of Heuristics Control Algorithms

The problem of determining the best sequence of N objects which are to be assembled in an automatic assembly line (by M robots/stations) is computationally difficult (belongs to the class NP in terms of computational complexity). From the theoretical point of view there exist $N!$ sequences of objects however, some of them are identical taking into account W versions of objects. Even so, with large values N (in practice over 100) it is impossible to solve this problem optimally (with the use of computer) in a limited period of time (before beginning of the so-called working shift).

On these grounds, the method of numerical simulation method [22,23] is implemented to determine the control (an assembly sequence of objects). The equations of states are basis of numerical simulation. The block diagram of the simulator consists of two loops:

— for cycles k ($k=1,...K$);
— for stations/robots m ($m=1,...M$).

States S^k and recurrently Q^k are calculated within these loops. Moreover, there are the following initial data distinguished in the simulator: M, W, N, K, T, S^0 as well as the control X i.e. the sequence of objects passed to the initial station of the line. The sequence X is determined by means of heuristic algorithms described below.

5.1   The Order Heuristic

The algorithm of the order heuristic uses the data z^w by choosing a version to be manufactured on condition there is the most/the least of it. At the same time the state of orders is distinguished before beginning of the assembly process and the current moment – during the assembly process. In case of the current state of orders z^w it is possible to consider the amount or percentage share.

5.2   The Heuristic of Assembly Times of Versions

The heuristic of assembly times of versions uses the known times $c_{m,w}$. On this basis it is possible to determine the time $C(w)$ representing the assembly process of the wth version. Objects characterized by the longest / shortest assembly time are chosen to be assembled. In this case the number of versions is to decrease throughout the assembly process – from W to 1.

5.3   The Heuristic of Work Times of Robots

The algorithm of work times of robots uses the known times $c_{m,w}$. On this basis it is possible to determine the time $C(m)$ representing the robot work time. The robot characterized by the longest work time is the so-called bottle neck of the assembly line. From this point of view it is necessary to determine a queue of assembled robots which allows this particular robot to determine the assembly cycle time.

5.4   The Heuristic of Assembly States

The algorithm of the heuristic of assembly states uses the assembly time c^k and, on this basis, chooses a version v of an object for the initial station. At the same time the assembly time $c_{v,1}^k$ should be comparable to c^k however, it should not exceed it. In this case the cycle time is not determined by the assembly time in the first station (by the first robot).

5.5   The Random Algorithm

The random algorithm is based on the principle that the version of the object which is passed to the initial station is chosen at random. The probability of drawing a version is equal or different. The different probability can be proportionate to the current orders or assembly times of versions of objects. In case of the random algorithm it is possible to generate many sequences and create a histogram of the assembly time of all objects or the number of assembled objects in the determined time T.

5.6   The Algorithm of Serializing of Versions

The combinatory algorithm is used in case when numbers of ordered objects z^w are higher than the number M of stations/robots in the line. In such a case it is possible to determine the times $\tau_{v,w}$ of intermediate processes i.e. passing from the assembly process of vth version objects only to the assembly process of the wth version objects only. It is possible to determine the optimal sequence of versions from $W!$ sequences of versions on condition that the sum of times of intermediate processes is the lowest.

6   Conclusion

Modern assembly lines are automatic, i.e. assembly is performed by industrial robots and a conveyor moves in cycles. Such lines are computer controlled. The assembly process in a line can be modeled with different assumptions, such as:

— The initial state and the final state of the assembly process equal zero (i.e. there are no objects in the line).

- The initial state of the assembly process equals zero however, the final state does not equal zero.
- The initial state of the assembly process does not equal zero and the final state does not equal zero.
- The initial state of the assembly process does not equal zero but the final state equals zero.

The problem of scheduling versions of assembled objects is computationally difficult, i.e. belongs to the class NP in terms of computational complexity. The number N of assembled objects is large (exceeds 100). Furthermore, assembly times for the line consisting of M stations/robots are not given. Therefore, the control problem of an assembly line requires the use of modeling and digital simulation.

Acknowledgements. This work was supported in part by the European Regional Development Fund within the project CEBIA-Tech No. CZ.1.05/2.1.00/03.0089.

References

1. Clarke, C.: Automotive Production Systems and Standardisation: From Ford to the Case of Mercedes-Benz. Physica-Verlag, Heidelberg (2005)
2. Lenstra, J.K., Kan, A.H.G.R., Brucker, P.: Complexity of Machine Scheduling Problems. Annals of Discrete Mathematics 1, 343–362 (1977)
3. Lenstra, J.K., Kan, A.H.G.R.: Complexity of vehicle routing and scheduling problems. Networks 11, 221–227 (1981)
4. Vasant, P.: Meta-Heuristics Optimization Algorithms in Engineering, Business, Economics, and Finance. IGI Global (2012)
5. Shi, R., Shangguan, C., Zhou, H.: Modeling and Evolutionary Optimization on Multilevel Production Scheduling: A Case Study. Applied Computational Intelligence and Soft Computing 2010, Article ID 781598, 14 pages (2010)
6. Michalewicz, Z., Fogel, D.B.: How to Solve It: Modern Heuristics. Springer (2004)
7. Niu, H., Tian, X.: An Approach to Optimize the Departure Times of Transit Vehicles with Strict Capacity Constraints. Mathematical Problems in Engineering 2013 (2013)
8. Mansini, R., Speranza, M.G.: Heuristic Algorithms for the Portfolio Selection Problem with Minimum Transaction Lots. European Journal of Operational Research 114 (1999)
9. Quan-Ke, P., Ponnuthurai, N.S., Chua, T.J., et al.: Solving manpower scheduling problem in manufacturing using mixed-integer programming with a two-stage heuristic algorithm. International Journal of Advanced Manufacturing Technology 46(9-12) (2010)
10. Georgilakis, P.S., Tsili, M.A., Souflaris, A.T.: A heuristic solution to the transformer manufacturing cost optimization problem. Journal of Materials Processing Technology 181(1-3) (2007)
11. Bensmaine, A., Dahane, M., Benyoucef, L.: A new heuristic for integrated process planning and scheduling in reconfigurable manufacturing systems. International Journal of Production Research 52(12) (2014)
12. Vanderkam, R.P.D., Wiersma, Y.F., King, D.J.: Heuristic algorithms vs. linear programs for designing efficient conservation reserve networks: valuation of solution optimality and processing time. Biological Conservation 137(3) (2007)

198   B. Chramcov, F. Marecki, and R. Bucki

13. Horn, S., Weigert, G., Beier, E.: Heuristic optimization strategies for scheduling of manufacturing processes. In: 29th International Spring Seminar on Electronics Technology, pp. 228–233. IEEE (2006)
14. Shimizu, Y., Yamazaki, Y., Wada, T.: A Hybrid Meta-heuristic Method for Logistics Optimization Associated with Production Planning. In: 18th European Symposium on Computer Aided Process Engineering, vol. 25, pp. 301–306. Elsevier (2008)
15. Bankstona, J.B., Harnett, R.M.: An Heuristic Algorithm for a Multi-product, Single Machine Capacitated Production Scheduling Problem. Computers & Operations Research 27(1) (2000)
16. Modrak, V., Semanco, P., Kulpa, W.: Performance Measurement of Selected Heuristic Algorithms for Solving Scheduling Problems. In: 11th International Symposium on Applied Machine Intelligence and Informatics (SAMI), pp. 205–209. IEEE (2013)
17. Bucki, R., Marecki, F.: Digital Simulation of Discrete Processes. Network Integrators Associates, Parkland (2006)
18. Marecki, J.: Algorithms and Applications of Artificial Intelligence. Network Integrators Associates, Parkland (2011)
19. Xiong, H., Zhou, M., Caudill, R.: A Hybrid Heuristic Search Algorithm for Scheduling Flexible Manufacturing Systems. In: International Conference on Robotics and Automation 1996, vol. 3, pp. 2793–2797. IEEE (1996)
20. Muthukumaran, M., Muthu, S.: A Heuristic Scheduling Algorithm for Minimizing Makespanand Idle Time in a Nagare Cell. Advances in Mechanical Engineering 2012 (2012)
21. Zhang, R.: A Three-Stage Optimization Algorithm for the Stochastic Parallel Machine Scheduling Problem with Adjustable Production Rates. Discrete Dynamics in Nature and Society 2013 (2013)
22. Marecki, F.: Modele matematyczneialgorytmyalokacjioperacjiizasobównaliniimontażowej. ZN PolitechnikiŚląskiej, Gliwice (1986)
23. Marecki, J., Marecki, F.: Metodysztucznejinteligencji. WSIZ, Bielsko-Biała (2012)

Matlab Adapter – Online Access to Matlab/Simulink Based on REST Web Services

Miroslav Gula and Katarína Žáková

Faculty of Electrical Engineering and Information Technology,
Slovak University of Technology,
Ilkovičova 3, 812 19 Bratislava, Slovakia
{miroslav.gula,katarina.zakova}@stuba.sk

Abstract. The paper presents a newly developed Matlab Adapter tool that enables to build Matlab/Simulink based online applications. It was built in the form of web service that is available via REST API. In this way the client application can be developed in any programming language. The server side of application was built in Java programming language. We developed several REST endpoints that enable to set parameters, to run simulations or experiments and to follow results.

Keywords: online experiments, Matlab, JMatLink, MatlabControl, Java, REST.

1 Motivation

Internet and related web technologies has firmly established their presence in our digitalized world. There is hardly any aspect of our life which is not affected by modern technology. Education is no exception. Almost every university offers some kind of support for on-line courses, on-line study materials and on-line exercises. For most of courses this support is relatively easy to implement because there is no physical equipment needed. However, in the field of automation theory and other engineering fields students are often required to work with real physical equipment in laboratories. Thus the need for remote laboratories as a part of on-line study process arises. The progress in this topic can be followed for example in [1 - 12].

Our goal was to develop a general, robust, easy to use, and easy to deploy solution, in order each real equipment using Matlab and Simulink software package could be accessed remotely.

We decided to use Matlab and Simulink as back-end control software for its widespread use in the field of automation and control theory. In addition, there already exist many real experiments that use Matlab/Simulink as control software, e.g. [13 - 15].

We used REST based web services as front-end for our solution. REST is currently very popular and widespread method for publishing internal API of the application over Internet. This is mainly because it allows high decoupling between client and server, and also because it is very simple to begin developing client application for

© Springer International Publishing Switzerland 2015
R. Silhavy et al. (eds.), *Intelligent Systems in Cybernetics and Automation Theory*,
Advances in Intelligent Systems and Computing 348, DOI: 10.1007/978-3-319-18503-3_20

REST based web services. Basically, the only thing that is needed to start to explore and test the API of REST web service is a web browser.

2   Architecture

We put a strong emphasis to develop a modular and extensible architecture using methods of software engineering. This effort led to the design of a new architecture consisting of several modules organized into several layers (Fig.1). All modules are described in more details in following sections.

The suggested architecture provides two main layers of abstraction:

- *MatlabService* – provides REST API. Since REST web services are independent of implementation language, in future the Matlab Adapter can be ported to other programming languages without breaking the functionality of existing software which will use Matlab Adapter for communication with Matlab/Simulink environment.
- *MatlabConnector* – is Java *interface*, which provides the unified API specification for communication with Matlab. Since the standard Matlab support for Java lacks of some advanced features, there were developed third party libraries to increase its functionality (e.g. JMatLink [16] or MatlabControl [17]). The proposed *MatlabConnector* provides unified API for higher layers of Matlab Adapter. It also allows introducing of new libraries and new ways of communication with Matlab into Matlab Adapter without any changes in these higher layers.

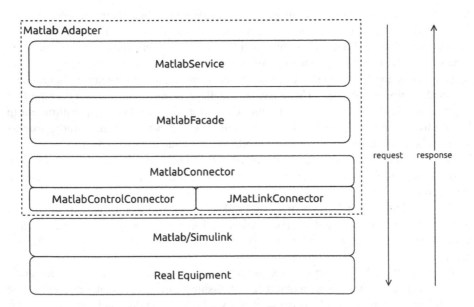

Fig. 1. The proposed architecture of the Matlab Adapter

2.1   MatlabService

This module represents entry point to the application. Its responsibility is to process REST requests, generate REST responses, and to call appropriate methods of MatlabFacade module to communicate with lower layers. It also translates Java data types and Java exceptions to and from JSON format which is main media type used by the developed REST API.

MatlabService provides several REST endpoints to allow

- start and close Matlab session,
- check whether Matlab session is active or not,
- reinitialize Matlab workspace, i.e. to clear all variables (except those used by Matlab Adapter itself),
- open and to close Simulink model for simulation or real experiment,
- check whether the particular model identified by name is opened,
- start, to stop and to pause the execution of simulation or real experiment,
- set and get one or more value(s) of specified parameter in specified model,
- return run-time data from all scopes in currently executing model of simulation or real experiment,
- upload, download or delete files into/from working directory of Matlab Adapter (this functionality is primary intended for files containing Simulink models),
- generate web view [22] of the specified model and to send it to the client as zip archive (this endpoint represents a convenient way for web based online laboratories - acting as clients of Matlab Adapter - how to obtain a web friendly and interactive representation of Simulink model that can be later embedded into the web page),
- generate list of all *DialogParameters* of all blocks of the specified model and to send it to the client in the form of JSON structure.

For implementing MatlabService module we used Jersey library [20] that represents an implementation of JAX-RS standard [21]. JAX-RS is Java standardized API for building REST web services. Using standardized API enables easier modification and extension of Matlab Adapter functionality for new developers.

2.2   MatlabFacade

MatlabFacade is helper layer. It provides simple API for MatlabService even for complex use cases. For simple use-case requiring only a single call to MatlabConnector instance on lower layer (e.g. setting or getting parameter of the experiment) it provides direct mapping between MatlabFacade method call and MatlabConnector method call. Complex use-case require multiple calls to multiple methods of MatlabConnector instance or require some other operations to be executed. MatlabFacade encapsulates this use-case into single method call on MatlabFacade instance.

This basically means that MatlabFacade method calls are mapped exactly one to one to MatlabService endpoint calls. It represents the big advantage since this design allows easy implementation of some other client-server standards (e.g. SOAP web services, web sockets, or low level TCP/IP communication) that can be considered as a replacement or an extension to MatlabService module.

2.3   MatlabConnector

As mentioned above this module is Java *interface* that abstracts communication with Matlab and provides unified API for higher layers. By implementing this interface developer can provide his/her own method of communication with Matlab.

It is important to mention that every implementation should ensure proper serialization of individual calls to Matlab session. It is important because of the fact that higher layers, mainly MatlabService, are inherently multi-threaded, but Matlab session and Matlab interpreter operate on single thread.

MatlabConnector provide several virtual methods that enable

- to start and to close Matlab session,
- to execute specified Matlab command in Matlab session,
- to get scalar value for the specified Matlab variable containing scalar,
- to get Java array with specified dimension for the specified Matlab variable containing matrix,
- to get Java string value for the specified Matlab variable containing string or equivalent data type,
- to check whether the instance of MatlabConnector is currently connected to running Matlab session.

2.4   JmatLinkConnector and MatlabControlConnector

JmatLinkConnector and MatlabControlConnector implement MatlabConnector interface and provide their own way of communication with Matlab. We decided to support two implementations of MatlabConnector to support a two redundant possibilities. It depends on the user which one he or she prefers to use.

MatlabControlConnector uses the MatlabControl library [17] for underlying implementation. MatlabControl is currently the preferred way of connecting to Matlab. It is under active development and it has relatively large user base which actively contributes to the development. The oldest version of Matlab that is supported by MatlabControl is R2007b. Previous versions of Matlab are not supported. It provides large amount of functionality, and therefore it can be a little bit more complicated to understand all features. On other hand the documentation is very well written.

MatlabControl uses JMI (Java Matlab Interface) for communication with Matlab, which is highly undocumented feature of Matlab and therefore there is always a risk of breaking compatibility between Matlab and MatlabControl in next Matlab releases.

JmatLinkConnector uses JmatLink library [16] for underlying implementation. This library runs only Matlab Engine [23] instead of whole Matlab, which allowns faster starting time of Matlab session. It is to say that the library structure is simpler than the structure of MatlabControl library. On one hand the JmatLink is easier to understand but on the other hand it provides less functionality. For example, it does not support sending Java object to Matlab (which is supported by MatlabControl). It also requires use of JNI (Java Native Interface) [24] which causes problems for multi-platform support.

However, the biggest problem is, that the library is not developed more already for longer time. The last release was done in December 2005 and Matlab R2009b was the last version where we were able to run the developed application.

As we could see, the communication with Matlab can be realized via various libraries depending on Matlab version. The developed MatlabConnector can offer unifying API providing more stable solution for higher levels of applications.

3   Implementation Details

Matlab Adapter was mainly developed using Java programming language and Java SE and Java EE technologies. There are several reasons why we chose Java. Firstly, Java is for most parts platform independent technology. Secondly, the standard Matlab kernel provides relatively good support for Java, as can be seen e.g. in [18] and [19]. Some third party Java libraries for communication with Matlab can be found in [16] and [17], too.

Matlab Adapter was implemented as standard Java web application, and for user convenience it is already packaged together with Apache Tomcat web server and Servlet container.

Matlab Adapter is standalone application. It means that the user does not have to start Matlab and then to start Matlab Adapter from the running Matlab environment. It is necessary firstly to start Matlab Adapter by starting Apache Tomcat web server. Then, the user can start and stop Matlab environment arbitrary many times whereby he or she performs desired tasks in the running Matlab session using provided REST API.

In last part we mention some Matlab functions that were used in order to support the previously described REST API.

3.1   Setting and Getting Simulink Parameters

For this purpose we used standard Matlab functions *set_param* and *get_param*. These functions provide good and reliable way how to set and get values to and from the experiment. The advantage is that the functions are also available while the Simulink simulation is running.

3.2   Obtaining Run-Time Output Data from the Experiment

To obtain run-time data from the experiment and send them to the client, we developed custom buffer component. It is implemented as a set of Matlab functions that are used directly by Matlab Adapter. They include:

- *startBufferedSimulation* – it starts execution of Simulink model; it also initialize buffer and starts buffering of values.
- *stopBufferedSimulation* – stops execution of Simulink model, and also stops buffering of values.
- *togglePauseBufferedSimulation* – toggles between pause state and running state.
- *readBuffer* – reads content of the buffer and clears it.

The basic principle of the operation is based on Simulink's method execution events [25]. The build-in function *add_exec_event_listener* enables to register listeners for execution events of an arbitrary block in the Simulink model. We call *add_exec_event_listener* function inside of *startBufferedSimulation* function in oder to register listeners on all scope blocks in particular Simulink model. In this way, each time scope block is updated it also executes the registered listener. Then, the listener can buffer actual values from the scope.

3.3   Exporting Simulink Models to Client as Web Views

To enable online visualization of Simulink block schemes we used Matlab build-in function *slwebview*. It generates Web View files into temporary directory. Then we can create a zip archive of these files and send it to the client application as response to the original request.

4   Conclusion

In this paper we described one solution for remote access to Matlab and Simulink environment. It enables execution of online experiments including experiments with real equipment. In the proposed solution the whole management of Matlab/Simulink based experiment is done remotely using developed REST API. This REST API mechanism enables integration of our application into the complete web based remote laboratory solution.

The advantage is that described solution can be used in all kinds of remote access and remote control applications that use Matlab and Simulink as control software.

Acknowledgments. This work was supported by the Slovak Grant Agency, Grant KEGA No. 032STU-4/2013 and APVV-0343-12.

References

1. Auer, M., Pester, A., Ursutiu, D., Samoila, C.: Distributed virtual and remote labs in engineering. In: IEEE International Conference on Industrial Technology, vol. 2, pp. 1208–1213 (December 2003)
2. Costa, R.J., Alves, G.R., Zenha-Rela, M.: Embedding instruments & modules into an IEEE1451-FPGA-Based weblab infrastructure. International Journal of Online Engineering (3) (2012)
3. Ozvoldova, M., Spilakova, P., Tkac, L.: Archimedes' principle - Internet accessible remote experiment. International Journal of Online Engineering 10(5), 36–42 (2014)
4. Huba, M., Šimunek, M.: Modular Approach to Teaching PID Control. IEEE Transactions on Industrial Electronics 54(6), 3112–3120 (2007) ISSN 0278-0046
5. Leva, A., Donida, F.: Multifunctional remote laboratory for education in automatic control: The CrAutoLab experience. IEEE Transactions on Industrial Electronics 55(6) (June 2008)
6. Lojka, T., Miškuf, M., Zolotová, I.: Service oriented architecture for remote machine control in ICS. In: Proceedings of SAMI 2014 - IEEE 12th International Symposium on Applied Machine Intelligence and Informatics, pp. 327–330 (2014)
7. Restivo, M.T., Mendes, J., Lopes, A.M., Silva, C.M., Chouzal, F.: A Remote Lab in Engineering Measurement. IEEE Trans. on Industrial Electronics 56(12), 4436–4843 (2009)
8. Žáková, K., Sedlák, M.: Remote Control of Experiments via Matlab. Int. Journal of Online Engineering (IJOE) 2(3) (2006)
9. Žáková, K.: WEB-Based Control Education in Matlab. In: Web-Based Control and Robotics Education, pp. 83–102. Springer, Dordrecht (2009) ISBN 978-90-481-2504-3
10. Bisták, P.: Virtual and Remote Laboratories Based on Matlab, Java and EJS. In: Fikar, M., Kvasnica, M. (eds.) Proceedings of the 17th International Conference on Process Control 2009, Štrbské Pleso, Slovakia, pp. 506–511 (2009)
11. Puerto, R., Jimenez, L.M., Reinoso, O.: Remote Control Laboratory Via Internet Using Matlab and Simulink. Computer Applications in Engineering Education 18(4), 694–702, doi:10.1002/cae.20274
12. Müller, S., Waller, H.: Efficient integration of real-time hardware and web based services into MATLAB. In: 11th European Simulation Symposium (1999)
13. Ionete, C.: LQG/LTR Controller Design for Rotary Inverted Pendulum Quanser Real-Time Experiment. In: Proceedings of the International Symposium on System Theory (SINTES 2011). Craiova, Romania, vol. 1, pp. 55–60 (2003)
14. Enikov, E.T., Polyzoev, V., Gill, J.: Low-cost take-home experiment on classical control using Matlab/Simulink Real-Time Windows Target. In: Proceedings of the 2010 American Society for Engineering Education Zone IV Conference, pp. 322–330 (2010)
15. Bolat, E.D.: Implementation of Matlab-SIMULINK Based Real Time Temperature Control for Set Point Changes. International Journal of Circuits, Systems and Signal Processing 1(1), 54–61 (2007)
16. Müller, S.: JmatLink, http://jmatlink.sourceforge.net/
17. Kaplan, J.: MatlabControl, http://code.google.com/p/matlabcontrol/
18. The MathWorks, Inc.: Overview of Java Interface, http://www.mathworks.com/help/matlab/matlab_external/product-overview.html
19. The MathWorks, Inc.: MATLAB Builder JA for Java language - Deploy MATLAB code as Java classes, http://www.mathworks.com/products/javabuilder/

20. Oracle America, Inc.: Jersey - RESTful Web Services in Java,
    `https://jersey.java.net/`
21. Java Community Process.: JSR-311: JAX-RS: The Java API for RESTful Web Services,
    `http://jcp.org/en/jsr/detail?id=311`
22. The MathWorks, Inc.: Web Views, `http://www.mathworks.com/help/`
    `rptgenext/ug/what-are-web-views.html`
23. The MathWorks, Inc.: Introducing MATLAB Engine,
    `http://www.mathworks.com/help/matlab/matlab_external/`
    `introducing-matlab-engine.html`
24. Oracle America, Inc.: Java Native Interface, `http://docs.oracle.com/javase/`
    `8/docs/technotes/guides/jni/`
25. The MathWorks, Inc.: Access Block Data During Simulation - Listen for
    Method Execution Events, `http://www.mathworks.com/help/simulink/ug/`
    `accessing-block-data-during-simulation.html#f13-92463`

FRel: A Freshness Language Model
for Optimizing Real-Time Web Search

Mariem Bambia[1] and Rim Faiz[2]

[1] LARODEC, ISG University of Tunis, Le Bardo, Tunisia
mariembanbia@yahoo.fr
[2] LARODEC, IHEC University of Carthage, Carthage Presidency, Tunisia
Rim.Faiz@ihec.rnu.tn

Abstract. An effective information retrieval system must satisfy different users search intentions expecting a variety of queries categories, comprising recency sensitive queries where fresh content is the major user's requirement. However, using temporal features of documents to measure their freshness remains a hard task since these features may not be accurately represented in recent documents. In this paper, we propose a language model which estimates the topical relevance and freshness of documents with respect to real-time sensitive queries. In order to improve recency ranking, our approach models freshness by exploiting terms extracted from recently posted tweets topically relevant to each real-time sensitive query. In our experiments, we use these fresh terms to re-rank initial search results. Then, we compare our model with two baseline approaches which integrate temporal relevance in their language models. Our results show that there is a clear advantage of using microblogs platforms, such as Twitter, to extract fresh keywords.

Keywords: Social Information Retrieval, Real-time sensitive queries, Language models, Fresh keywords.

1 Introduction

Effective Information Retrieval Systems must be able to provide relevant items for different categories of queries (e.g. regular queries, location sensitive queries, recency queries) submitted by different users with different intentions. However, there are categories of queries to which documents relevance may change over time. This is particularly significant for real-time sensitive queries, where the expected results must not only be topically appropriate to query, but also contain fresh information which is the major requirement of the user [3]. Furthermore, the different possible temporal characteristics of queries and the unknown users' intent towards the required temporality of search results explain the difficulty of understanding the query temporal nature [12]. Hence, freshness stands the fourth among the nine key reasons that motivate Internet users to prefer a given web document to another one in the retrieval results [1]. For example, a user who

---

[1] http://www.consumerwebwatch.org/pdfs/a-matter-of-trust.pdf

© Springer International Publishing Switzerland 2015
R. Silhavy et al. (eds.), *Intelligent Systems in Cybernetics and Automation Theory,*
Advances in Intelligent Systems and Computing 348, DOI: 10.1007/978-3-319-18503-3_21

submitted a query about a TV sport program would be searching for the freshest sport televisual content like the most recent sport matches and the last sport news. A sport event has been broadcasted last week and that has been topically relevant to a given query, may probably become less relevant than the freshest ones.

For this family of queries, one may offer users advanced search options to explicitly limit the search results to a specific period. This solution has two main shortcomings: 1) more inputs are required from the user. He is brought to invest the effort by means of reformulation, although his need for fresh content is immediate, 2) only a small part of the web like breaking-news can answer to some real-time sensitive queries, even if the user specified explicitly the temporal dimension, 3) Extracting temporal features from web pages remains a hard task because they may not be accurately represented in fresh documents. For instance, for the query "Presidential Debates" where relevant documents are time-sensitive to days or even hours, it could not be an obvious task to measure the true age of relevant documents for estimating their freshness. This gives rise to a major difficulty in retrieving the freshest content and estimating information freshness in response to real-time sensitive queries. The requirement of fresh content has been growing among internet users through microblogs, breaking-news and social networks. Consequently, social media seem to be interesting sources of information that can be exploited to measure and estimate freshness of web document [2], [11] and [7]. In this paper, we will argue that the freshness of a document refers to a criterion that depends on the occurrence of fresh keywords in this document. Fresh keywords are terms that recently make a buzz, popularly shared in Social Web and topically relevant to query. In our case, we assume that fresh keywords are terms occurring in recent posted resources. We consider that microblogs, breaking news and the topical trends are the best sources that may provide such keywords. The main research issue addressed in this paper is: how to exploit social information in order to estimate documents freshness in response to real-time sensitive queries?

We describe in this paper a novel language model for effective estimation of documents freshness. This paper is structured as follows: Section 2 reviews related work. Section 3 details our proposed relevance and freshness language model. Section 4 is devoted to our experimental evaluation and discusses the obtained results. The last section presents conclusion and points out possible directions for future work.

2   Related Work

There are various definitions of web documents freshness in the literature depending on the metrics and the ways of using web document features to measure this criterion. Most of the prior works on recency ranking exploit a variety of recency features when computing the rank of a document (e.g. timestamps). Such features refer to temporal signals and properties for both queries and documents used to depict documents recency to the respective queries. According to [5], the concept of document freshness introduces the idea of how old is the document. They proposed recency features to represent a page freshness (e.g. Timestamp, Linktime, Webbuzz features). The problem is that a document is labeled as very fresh, if its content is created on the same day as a query. In addition, recency features may be inaccurate and may degrade the relevance of some fresh documents because their values were not saved timely. [9] consider a document as fresh if it has not been discovered before and if it is fresh enough with respect to the user's expectations and his browsing history. They collected a set of keywords from recently browsed web pages, introduce the frequency of each term over a period of time and assign a weight to each keyword in order to extract the freshest ones. However, some terms are penalized because they appear frequently in the most pages browsed by the user over a period of time, even though they may be importantly fresh in the recent days. [15] used click-throughs in news search and content analysis technique and defined a set of temporal features (e.g. Publication age which is URL´s publication timestamp and Story age which extracts all the mentioned dates in the URL) in order to predict freshness and relevance for a given query.

Some approaches have been proposed to incorporate time into language models to handle with recency queries [4], [6] and [10]. [6] explored estimation methods to promote recent tweets. They extended time-based language models, proposed in [10], and applied time-based exponential priors to the score of the documents. The exponential prior, given by Equation1, discounts the score for each tweet according to its age in relation to the time of the query (e.g. number of days).

$$Texp(score, age) = score * \lambda \, exp[\lambda \, age] \tag{1}$$

[4] identified the time intervals of interest for each query and integrated this information in the scoring scheme (Equation2). To evaluate their framework, they rely on news articles datasets (including TREC) and on a collection of web data annotated by Mechanical Turk workers. They consider w_d which represents the lexical terms in document d, and t_d which is the timestamp for d.

$$P(d|Q) = P(w_d, t_d|q) = P(t_d|w_d, Q)P(w_d|Q) \tag{2}$$

However, extracting temporal features (e.g. dates) from web pages represents a hard task because they may not be accurately represented in fresh documents. Furthermore, fresh content resides in providing real-time information, such as breaking-news and hot topics, immediately to the user. Consequently, it is un-necessary to use some other features such as browsing history and clickthroughs information. Within this context, very little work are proposed on incorporating social information to improve real-time web search. In our approach, we assume that fresh topics can be provided by real-time sources such as microblogs and social networks. These sources are likely to potentially comprise fresh contents that can be exploited to measure documents freshness.

3 Language Model for Relevance and Freshness Estimation

The relevance is an abstract measure of how well a retrieved document meets the user's information need represented by a query. Much of works owed the concept of relevance language models viewing documents as models and queries as strings of text. However, this approximation may not be sufficient when the user expects documents which are both topically relevant as well as fresh. The needs to content recency in web search requires incorporating freshness as another significant ranking criterion [14]. In the simplest case, we incorporate fresh social information (fresh keywords) in the relevance model to form a freshness language model. We exploit the usual language model to evaluate the relevance of a document with regard to a query. This query is treated as a sample of text from a language model. In this paper, we rely on the query likelihood model [13]. Given query Q and document d, we derive a score for d with Q (Equation3).

$$RSV(Q, d) = \prod_{t \in Q} P(t|d) \tag{3}$$

In order to avoid the zero-frequency problem, that occurs when the document does not contain any query term, we use in Equation4 Dirichlet smoothing estimator [16], μ is the prior sample size on $[0, \infty[$.

$$P(t|d) = \frac{tf(t,d) + \mu P(t|c)}{|d| + \mu} \tag{4}$$

Both probabilities $P(t|d)$ and $P(t|c)$ are estimated using the Maximum Likelihood Estimator. Equation 4 estimates only the relevance of the document d to the query Q. We propose a language model that promotes documents containing fresh terms. We assume that freshness is described by a set of known fresh terms $F = \{f_1, f_2, ..f_n\}$. The language model can be seen as a translation model which evaluates how query terms are close to fresh terms.

Equation5 gives a generic way to add fresh keywords to the documents ranking model.

$$P(t|d) = \sum_{f \in F} P(t|f) * P(f|d) \tag{5}$$

We aim either to measure $P(t|f)$ that estimates the relationships strength between term t and fresh term f and to define how a term t is close to f (Equation6). This probability can be estimated using the co-occurrence or the classical definition of conditional probability [1].

We consider a simplest approach to estimate this probability.

It is estimated as a simple ratio (the number of documents) where both term t and f appear .

$$P(t|f) = \frac{NbDoc(t,f)}{NbDoc(t)+NbDoc(f)} \qquad (6)$$

With:
$NbDoc(t)$ is the number of documents containing the term t
$NbDoc(f)$ is the number of documents containing the fresh term f
$NbDoc(t,f)$ is the number of documents containing t and f

Concerning $P(f|d)$, it is estimated using Equation7.

$$P(f|d) = \frac{freq(f)}{|d|} \qquad (7)$$

With:
$freq(f)$ represents the frequency of fresh term f in the document d
$|d|$ is the document length

$P(t|C)$ is the estimated the probability of seeing the term t in the collection. It is considered as the ratio of the term frequency ft by the number of terms in all documents of the collection C (Equation8).

$$P(t|C) = \frac{tf(t)}{|C|} \qquad (8)$$

With:
$tf(t)$ represents the frequency of the term t in all document of the collection
$|C|$ is the aggregation of the frequencies of all terms in the entire collection

4 Experiments and Results

We conduct a series of experiments in order to compare the proposed freshness and relevance language model with two baseline methods: Exponential re-ranking Language Model [6] and the Temporal Relevance Language Model [4]. The general search process we used consists in re-ranking a list of documents by taking into account freshness and relevance dimensions.

4.1 Data Set and Evaluation Methodology

To build our dataset, we first collect 1000 queries from the headlines of BBC API [2]. These queries are collected during 4 days from 08/03/2014 to 11/03/2014, 250

---
[2] https://developer.bbc.co.uk/

queries per day. These queries are more likely to be sensitive to days or even hours.

Each query is submitted to Google search engine. We select the top 40 documents (Web pages). The set of Web pages is collected for all queries to form our collection (noted C). We aim to re-rank these documents by freshness and relevance. In order to form our pool of fresh terms, we use Twitter4J API to extract the more recent tweets by submitting each query to Twitter. We select the tweets starting one hour before the event related to the query until 12 hours after the event. Then, we extract the top X most frequent terms. We consider different values of X={10, 20, 40, 100} based on their frequencies in the tweets.

In practice, we proceed to the stop-words removal and stemming of tweets and Web pages thanks to Apache Lucene [3] classes including Porter Stemming Filter class. To evaluate the effectiveness of the results, we hired 50 persons from our university to assess the different results. Each assessor evaluated 20 queries. He/she were asked to assess the documents according to their freshness and relevance.

The query-document pairs are judged in the same day they were sampled from Google. We define two relevance grades according to the document-query similarity ignoring the recency dimension: relevant and not relevant. Then, we ask them to judge query-document pairs freshness: fresh or non-fresh. Table 1 shows that we mapped the separately annotated relevance and freshness into one single grade represented by a defined value. We assume promoting fresh and not relevant documents against relevant and not fresh ones.

Table 1. Aggregated Labels in sense of relevance and freshness grades

Relevance GradeFreshness Grade	Fresh	Not Fresh
Relevant	2	0.5
Not relevant	1	0

We need an evaluation metric that penalizes errors near the beginning of the ranked list and supports relevance and freshness graded by assessors. The performances are measured according to freshness and relevance dependently using the discounted cumulative relevance and freshness (DCRF):

$$DCRF_n = \sum_{i=1}^{n} \frac{RF_i}{\log_2(i+1)} \tag{9}$$

where i is the position in the re-ranked documents list, RF_i is the freshness and relevance value according to the aggregated labels graded by the editors (Table 1).

We use Mean Average Precision computed according to the RF values. As precision pertains to agreement between assessors, we resorted the Fleiss' kappa

---

[3] http://www.lucene.apache.org

coefficient. Fleiss' Kappa is envisioned to provide a quantitative measure of the agreement magnitude between assessors [8].

We compare our model with two baseline models namely Exponential Re-ranking Model (ERM) [6] and the temporal relevance language model (TRM) [4] described in Section 2. The three approaches attempt to re-rank the initial search (the top documents returned by Google). To implement both of those models, we extract the dates associated with the top documents. With these dates, we measure the maximum likelihood estimator for ERM and the timestamp t_d for TRM.

4.2   Results

Table 2 lists the *MeanDCRF* (MDCRF) and *Mean Average Precision* (MAP) of the three approaches for the four days. We observe that MAP and $MDCRF$ of baseline models results decrease as more as the event is older, whereas FRel model is consistently stronger than baselines in terms of MAP and $DCRF$. However, a document may become stale a few hours after, which affects its relevance to a recency-sensitive query. FRel MAP outperforms ERM and TWM at most 40% each day. This prevents that fresh keywords extraction from recent tweets is potentially useful to improve recency ranking for real-time sensitive queries. In order to evaluate the assessors' agreement, we used Fleiss' kappa coefficient. We obtain Fleiss' kappa of 73,60% which indicates the existence of a substantial agreement between assessors. According to Table 2, our method obtains the highest $MDCRF$ values at 72% in day 4, while the other methods are less than 50%.

From these results, we prove that our method outperforms Exponential Re-ranking Model (ERM) and Temporal Relevance Model (TRM). This is due to their assumption that recently published documents are likely to be relevant to recency queries. However, a document published at a later date may also be relevant. By reason of the different nature of the documents types analyzed (news and tweets), the temporal influence would guide retrieval in a more efficient way by using Dirichlet priors. In consequence, the models must not only rely on publication dates but also be inverted to bias the results toward older documents and analyze contents.

Table 2. Comparison with baseline models on MAP and DCRF

		ERM	TRM	FRel
Day 1	MAP	40.45	51.50	73.45
	MDCRF	53.61	49.60	68.76
Day 2	MAP	48.96	50.34	68.63
	MDCRF	42.80	46.89	86.30
Day 3	MAP	46.10	45.80	67.53
	MDCRF	46.40	41.64	64.45
Day 4	MAP	40.32	37.40	74.73
	MDCRF	48.39	38.17	72.63

Table 3. An example of recency ranking improvement comparing to Google search results. The query is "MalaysiaAirlinesFlight", and the query issue time is during from 16:31 to 20:59

rank	Document	Relevance	Freshness
(a) Ranking results by Google Search Engine			
1(4)	http://en.wikipedia.org/wiki/Malaysia_Airlines_Flight_370	relevant	non-fresh
2(5)	http://www.malaysiaairlines.com/mh370	non relevant	fresh
3(12)	http://www.independent.co.uk/news/world/australasia/mh370-airline-boss-claims-missing-flight-did-not crash-into-indian-ocean-9790455.html	non relevant	non-fresh
4(2)	http://www.bbc.com/news/world-asia-26503141	relevant	fresh
5(8)	http://www.atlantico.fr/dossier/disparition-vol-mh370-malaysia-airlines-1015555.html	non relevant	non-fresh
(b) Re-Ranking results by FRel			
1(9)	http://timesofindia.indiatimes.com/the-mystery-of-Malaysia-Airlines-flight-MH370/specialcoverage/32010069.cms	relevant	fresh
2(4)	http://www.bbc.com/news/world-asia-26503141	relevant	fresh
3(6)	http://fr.flightaware.com/live/fleet/MAS	relevant	fresh
4(1)	http://en.wikipedia.org/wiki/Malaysia_Airlines_Flight_370	relevant	non-fresh
5(2)	http://www.malaysiaairlines.com/mh370	non relevant	fresh

rank in (b)

rank in (a)

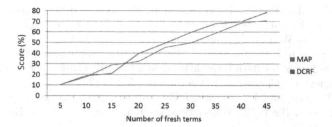

Fig. 1. Effects of fresh keywords number on scores' performance

Table 3 illustrates an example of the five top retrieved documents by Google and re-ranked by our approach for the query 'MalaysiaAirlinesFlight'. By recency judgments, we emphasize that the re-ranking result promotes relevant and fresh content to the top of the ranked list. We also evaluated the impact of the number of fresh keywords. Figure1 shows that the longer the list of the fresh keywords existing in a document is raised and varies over time, the higher DCRF and MAP obtained for the top-40 documents. This prevents that more fresh documents are promoted to the top of search results.

5   Conclusion

We propose a language model that estimates the relevance and the freshness of web documents in respect with recency sensitive queries based on topically relevant fresh keywords extracted from the recent posted tweets. The results show that our approach outperforms the baseline models. Our experiments demonstrate that there is a clear advantage of using social information to measure the freshness of documents in real-time search. There is, also, an opportunity for using Twitter as an important source to show the significance of extracting fresh terms from microblogs which constitutes essential descent of recent events. The results reveal the inevitability of a novel language model to improve recency ranking. The proposed approach gives encouraging results and opens several future directions: Our first future work is to incorporate our model in recommender systems in order to recommend the most relevant and fresh TV programs for each single user based on its watching behaviors and interests. Another line of research will be to extend the personalization through other ranking features, as the user's social behavior, and to exploit annotations to treat recency of TV programs.

References

[1] Bai, J., Nie, J.Y., Cao, G.: Context-dependent term relations for information retrieval. In: Proc. Empirical Methods in Natural Language Processing, EMNLP 2006, pp. 551–559 (2006)

[2] Ben Jabeur, L., Tamine, L., Boughanem, M.: Featured tweet search: Modeling time and social influence for microblog retrieval. In: Proceedings of International Conference on Web Intelligence, China, pp. 166–173 (2012)

[3] Dai, N., Shokouhi, M., Davison, B.: Learning to rank for freshness and relevance. In: Proceeding of the 34th International ACM SIGIR Conference on Research and Development in Information Retrieval, pp. 95–104 (2011)

[4] Dakka, W., Gravano, L., Ipeirotis, P.G.: Answering general time-sensitive queries. Proceedings of the IEEE Transactions on Knowledge and Data Engeneering 24, 220–235 (2012)

[5] Dong, A., Chang, Y., Zheng, Z., Mishne, G., Bai, J., Zhang, R., Buchner, K., Liao, C., Diaz, F.: Towards recency ranking in web search. In: Proceedings of the Third ACM International Conference on Web Search and Data Mining, WSDM 2010, pp. 11–20. ACM, New York (2010)

[6] Efron, M., Golovchinsky, G.: Estimation methods for ranking recent information. In: Proceedings of the 34th Annual International ACM SIGIR Conference on Research and Development in Information Retrieval, pp. 495–504 (2011)

[7] Huo, W., Tsotras, V.J.: Temporal top-k search in social tagging sites using multiple social networks. In: Kitagawa, H., Ishikawa, Y., Li, Q., Watanabe, C. (eds.) DASFAA 2010. LNCS, vol. 5981, pp. 498–504. Springer, Heidelberg (2010)

[8] Viera, A.J., Garrett, J.M.: Understanding interobserver agreement: The kappa statistic. Family Medecine Research Series 37(5), 360–363 (2005)

[9] Karkali, M., Plachouras, V., Vazirgiannis, M., Stefanatos, C.: Keeping keywords fresh: A bm25 variation for personalized keyword extraction. In: Proceedings of the 2nd Temporal Web Analytics Workshop, pp. 17–24 (2012)

[10] Li, X., Croft, W.B.: Time-based language models. In: ACM (ed.) Proceedings of the Twelfth International Conference on Information and Knowledge Management, New York, NY, USA (2003)

[11] Massoudi, K., Tsagkias, M., Rijke, M., Weerkamp, M.: Incorporating query expansion and quality indicators in searching microblog posts. In: Proceedings of the 33rd European Conference on IR Research, Dublin, Ireland, pp. 362–367 (2011)

[12] Moon, T., Chu, W., Lihong, L., Zheng, Z., Chang, Y.: Online learning for recency search ranking using real-time user feedback. ACM Transactions on Information Systems 30(4), 20 (2010)

[13] Ponte, J., Croft, W.: A language modeling approach to information retrieval. In: Proceedings of the 21st annual International ACM SIGIR Conference on Research and Development in Information Retrieval, pp. 275–281 (1998)

[14] Wang, H., Dong, A., Li, L., Chang, Y.: Joint relevance and freshness learning from clickthroughs for news search. In: Proceedings of the 21st International Conference on World Wide Web, pp. 579–588 (2012)

[15] Wang, H., Dong, A., Li, L., Chang, Y.: Joint relevance and freshness learning from clickthroughs for news search. In: Proceedings of the 21st International World Wide Web Conference Committee, pp. 579–588 (2012)

[16] Zhai, C., Lafferty, J.: A study of smoothing methods for language models applied to information retrieval. ACM Transactions on Information Systems 2(2), 179–214 (2004)

Simulation of the Video Feedback for Mobile Robots in Simbad Environment

Piotr Lech[1], Krzysztof Okarma[1], Konrad Derda[2], and Jarosław Fastowicz[1]

[1] West Pomeranian University of Technology, Szczecin
Faculty of Electrical Engineering
Department of Signal Processing and Multimedia Engineering
26. Kwietnia 10, 71-126 Szczecin, Poland
{piotr.lech,krzysztof.okarma,jaroslaw.fastowicz}@zut.edu.pl
[2] West Pomeranian University of Technology, Szczecin
Faculty of Electrical Engineering
ICT Students' Scientific Group "Apacz 500"
26. Kwietnia 10, 71-126 Szczecin, Poland
dk27546@zut.edu.pl

Abstract. Rapid progress in the field of machine vision applications can be especially clearly visible in robotics. For this reason many groups of image processing and analysis algorithms and solutions which have not been previously applied in automation and robotics require testing, verification and modifications before their application in any hardware solution. Such prototyping of video applications for robotics is possible using a simulation environment without the necessity of building the physical robot which can be damaged or even lost during some preliminary experiments. In order to verify and demonstrate the usefulness of such approach some experiments related to video feedback for mobile robotics have been conducted using free Java based 3D mobile robot simulator known as Simbad. Presented results confirm the great potential of such simulation environments in rapidly developing area of machine vision applications in robotics.

Keywords: video feedback, 3D simulation, mobile robots.

1 Introduction

Video feedback is currently one of the most rapidly growing fields of mobile robotics. Modern methods of designing control algorithms for mobile robots using visual feedback often utilize the simulation techniques implemented in the simulator environments. The selection of simulation environments implies the ability to verify the design assumptions before their implementation on the physical (real) platform. Additionally, such environment can be a platform for rapid prototyping of algorithms and their preliminary verification without any references to a particular physical implementation.

Therefore this paper is dedicated to the suitability of open source mobile robot simulation environment Simbad [12] for prototyping the algorithms for

© Springer International Publishing Switzerland 2015
R. Silhavy et al. (eds.), *Intelligent Systems in Cybernetics and Automation Theory*,
Advances in Intelligent Systems and Computing 348, DOI: 10.1007/978-3-319-18503-3_22

controlling the mobile robots using visual feedback. This particular simulator, namely Simbad, has been chosen not only due to its open source code but also considering the great potential associated with the Java language, including an easy portability of the code.

2   Simulation Environments

For the effectiveness of simulation a simulation environment has to ensure:

- the ability to define the world, where the mobile robot can move, as a 3D model,
- the definition of physical phenomena,
- interaction with external environment.

On the other hand, it is desirable to apply several simplifications at the stage of testing algorithms which ensure a sufficient representation of reality required for experiments. Another important issue is the possibility of addition of new elements which are not included in the basic package of the simulator.

Some exemplary popular commercial simulation packages include:

- **RoboWorks** [4] developed by Newtonium company, which is a universal tool for developing, simulation and animation of any physical system, including the issues related to the simulation of physical objects i.a. mobile robots,
- **Microsoft Robotics Developer Studio** [1], a versatile tool with high complexity, which is a powerful simulation environment for technology-related service delivery; through the connection with the service one can get the functionalities suitable for the project,
- **Webots** [6], specialized simulation environment for mobile robots, which has its own API for controlling the robots, its own physics and a large number of predefined sensors.

Apart from commercial products, there are also some open source packages available, e.g.:

- **Open Dynamics Engine (ODE)** [2], developed by Russel Smith which is high performance platform independent library with C/C++ API, integrated collision detection with friction and advanced joint types, useful for simulations of rigid body dynamics, especially vehicles, objects in virtual reality environments and virtual creatures e.g, in computer games and simulation tools.
- **ROBOOP** [3], being an object oriented toolbox in C++ for robotics simulation with an additional OpenGL interface,
- **Simbad** [5] a Java 3D robot simulatorÂ for scientific and educational purposes, especially related with artificial intelligence and machine learning for use with autonomous robots and autonomous agents.

Commercial solutions presented above are highly enhanced environments which makes a rapid implementation almost impossible especially for new developers and collaboration of group of them. In commercial packages the expansion of the environment and the use of foreign external libraries is difficult or limited. Open source solutions, due to their openness, allow to connect various components together in relatively easy way.

The Simbad simulator presented above, due to its clarity is a tool that allows developers a quick inclusion into implementation of even an advanced project. Some of the most relevant features of Simbad are related to 3D visualization and sensing, availability of various sensors, including vision sensors (color cameras), range sensors (sonars and infrared) and contact sensors (bumpers). This simulator supports both single or multi-robots simulation with Swing user interface for control which is commonly used toolkit for Java. Some relevant extensions provided separately are Neural Network library (PicoNode) and the Evolutionary Algorithms library (PicoEvo).

3   Video Feedback

The use of video feedback is currently a popular and rapidly growing field of mobile robotics [7,9,10]. Research on this topic is carried out in many directions, e.g. related to increasing the efficiency of image processing via parallelization and clustering of calculations, video based self-localization of mobile robots [14,15], cooperative work of group of robots, etc.

Considering the placement of vision sensors two types of video feedback can be distinguished:

- local, in which the sensor is mounted on the robot's body,
- global, where the sensor (or sensors) are independent elements.

4   Potential of Simbad in Video Feedback Simulations

Simbad simulator gives the opportunity to use a virtual camera which is located on the top of the robot and in the unmodified program may be rotated right and left. The element responsible for the implementation of the camera is the CameraSensor object. Its constructor used for creating the camera objects is assigned in the Robot class. The structure of code is shown below.

```
public class ImagerDemo extends Demo
{
    public class Robot extends Agent
    {
        CameraSensor camera;
        public Robot(Vector3d position, String name)
        {
            super(position, name);
```

```
            // Adding the sensors
            camera = RobotFactory.addCameraSensor;
            public void initBehavior() {}
            public void performBehavior() {}
        }
    }
}
```

Because the full code of the simulator is publicly available, the video sensor can be freely modified, e.g. by changing its resolution, transformation of the RGB camera into monochrome one or changing the direction of observation (custom defined up or down). The Simbad's ability of to use external libraries is another great advantage of the package making it even more attractive due to the possibility of using some specialized image processing libraries such as OpenCV for more advanced projects.

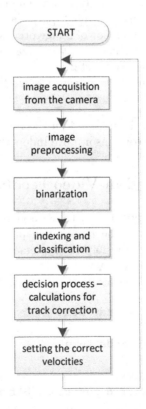

Fig. 1. Scheme of image processing and control processes used in the simulations

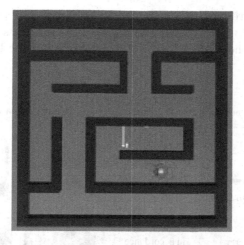

Fig. 2. Illustration of the maze implemented for experiments

5   Exemplary Simulations

In the example simulations the model of a mobile robot has been used (modeled on a real Khepera robot [13]) with differential control. The available control methods are as follows:

- `setLeftVelocity(double lv)` – setting the left wheel's linear velocity (in meters per second),
- `setRightVelocity(double rv)` – setting the right wheel's linear velocity (in meters per second),
- `setWheelsVelocity(double lv,double rv)` – setting the linear velocity of the two wheels at a time,
- `getLeftVelocity() and getRightVelocity()` – methods return the current speed of the left and right wheel respectively.

These methods are more than enough to carry out a number of issues of simulation.

5.1   Local Visual Feedback for a Single Robot

In this case a standard implementation available in the simulator as described above can be utilized. The general scheme of image processing and processes during the simulation is shown in Fig 1.

Maze
The task for the simulation is defined as follows:

- the robot is placed anywhere inside the maze,

– the task of the robot is the random exploration of maze,
– all decisions about the robot's motion utilize the classification of images obtained from the camera based on the detection of the lines.

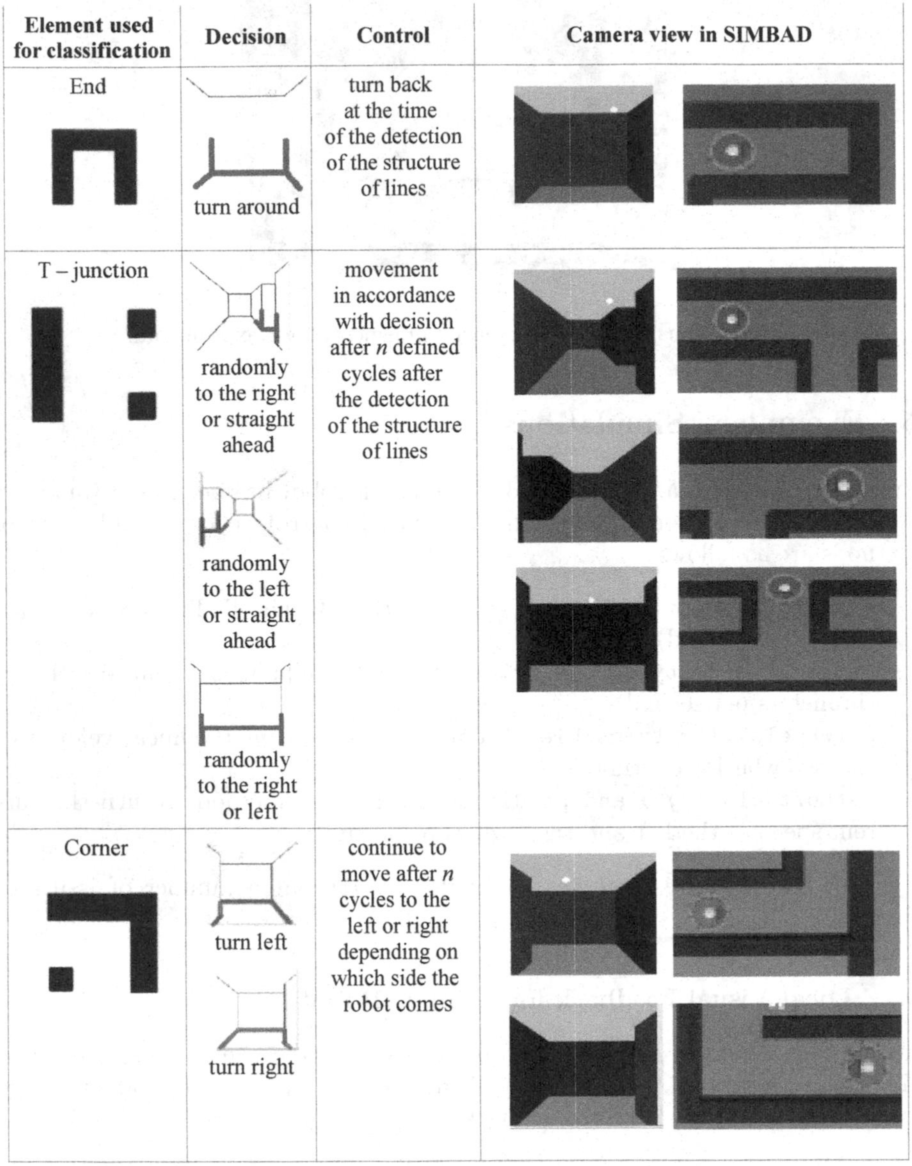

Element used for classification	Decision	Control	Camera view in SIMBAD
End	turn around	turn back at the time of the detection of the structure of lines	
T – junction	randomly to the right or straight ahead / randomly to the left or straight ahead / randomly to the right or left	movement in accordance with decision after n defined cycles after the detection of the structure of lines	
Corner	turn left / turn right	continue to move after n cycles to the left or right depending on which side the robot comes	

Fig. 3. Decisions scheme used for control of the mobile robot in the simulation of the maze

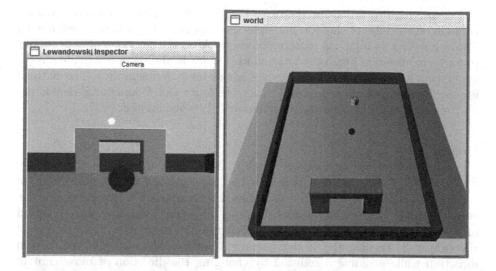

Fig. 4. Illustration of the "Goal!!!" project

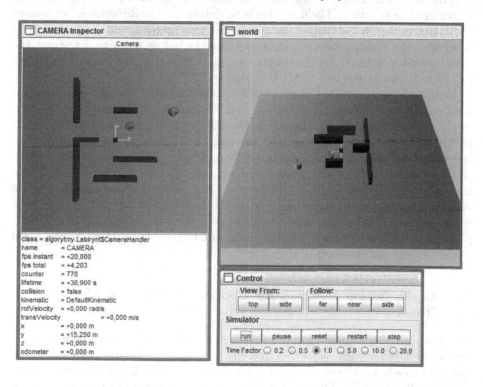

Fig. 5. Simulation with global visual feeedback

The top view of a maze is shown in Fig. 2. The decision about the rotation is made on the base of classical line detection implemented utilizing well known Canny edge detector [8] and pattern matching. After the detection of lines similar to patterns shown in Fig. 3 the rotation in the specified direction is made after n simulation cycles (the number of cycles is chosen experimentally). The pattern is represented by thick red lines. Corner elements and T-junctions, depending on the direction of movement, are symmetrical to one another.

Goal!!!

The purpose of the simulation "Goal!!!" is to place the ball in the goal by a robot. The aim to be achieved in this task is to get the correct positioning of a goal, ball and robot using the vision based techniques. For this purpose some simplifications of the control algorithm can be made. First, the start position of the robot is randomly chosen, the robot then starts moving towards any of the side walls keeping the 45 degrees angle between the wall and the motion direction while avoiding a collision by changing the direction of movement in the event of a possible collision. At the same time the robot looks for the ball and the goal. Those are the only three elements of the robot's world which form a straight line together. The first task is to determine the direction of motion of the robot to match the line of shot (determined straight line), the second is to determine the distance between the robot and the ball in order to determine the speed with which the robot should hit the ball (which translates into a necessary force for "kicking" the ball to hit the goal).

In order to determine the exact coordinates of the robot's position the method `getCoords()` is used which may correspond to obtaining a position from the GPS in the real world. The only thing that remains is to determine when the ball is on the same line with the goal and this part of the algorithm corresponds to the visual feedback. The decision to set a single line corresponds to the applied classifier which may be represented schematically as | o | which corresponds to the viewing a fragment of the ball between the goalposts. The classifier is not very accurate and the ball does not always hit the center of the goal, but the most commonly is placed properly between the goalposts, depending on the starting point in which the robot was located at the start of the simulation. A screenshot of the project is illustrated in Fig. 4.

5.2   Global Visual Feedback for Cooperative Work of Multiple Robots

The purpose of the simulation is to replicate a hall in which the robots move, avoiding collisions with each other and the walls. For simplicity a random movements of robots have been assumed. The camera used for implementation of the visual feedback is placed just under the ceiling and covers the entire space of the hall in side its field of view. Unfortunately, in the Simbad 3D simulator's world it is impossible to place the camera itself as the object from which the images can be acquired. However, it is possible to modify the code responsible for plugging the camera and placing the stationary robot equipped with the camera anywhere in space. For this purpose it is required to modify the `CameraHandler` class in order to observe the ground as shown below:

```
public class CameraHandler extends Agent
{
    public CameraHandler(int height)
    {
        super(new Vector3d(0, height, 0), "CAMERA");
        // Add sensors
        camera = RobotFactory.addCameraSensor(this);
        camera.rotateX(-Math.PI / 2);
    }
}
```

The command `super(new Vector3D (0, height, 0), "CAMERA");` is responsible for "hanging" the camera over the scene. Rotation of the camera towards the ground is made using the command `camera.rotateX(-Math.PI/2);`. The image from the camera located above the scene is available for any of the robots. The screenshot of the project is shown in Fig. 5 where the "CAMERA Inspector" window illustrates the global video feedback camera's view and the "world" window illustrates the simulated scene. Zero values of `rotVelocity` and `transVelocity` are related to the stationary camera placed over the scene together with a stationary robot. An additional verification of observed images may be conducted using the image based floor detection algorithms presented e.g. in the paper [11].

The simulation is based on the placing the robots in space with obstacles blocking the movements whereas the movements of robots are random and change in the case of possible collision with an obstacle or another robot. Collision detection is based on determining the specified minimum allowable distance from the robot to the obstacle and in the case when it is achieved a random change of direction occurs.

6   Conclusions

Simbad simulator is a universal environment useful for simulations related to the control of mobile robots in a 3D environment. Full access to its source code allows easy expansion of the program with many new elements which do not exist in the system. The simulator is also suitable for sophisticated tasks which is the robots' control with the use of video feedback. Some exemplary algorithms associated with local and global video feedback implemented by the authors and presented in the paper fully confirm its the usefulness for rapid prototyping of machine vision applications for mobile robotics.

References

1. Microsoft Robotics Developer Studio,
   http://www.microsoft.com/en-us/download/details.aspx?id=29081
2. Open Dynamics Engine, www.ode.oeg
3. ROBOOP, http://roboop.sourceforge.net
4. RoboWorks software by Newtonium, http://www.newtonium.com
5. Simbad, http://simbad.sourceforge.net
6. Webots by Cyberbotics, http://www.cyberbotics.com
7. Bonon-Font, F., Ortiz, A., Oliver, G.: Visual navigation for mobile robots: A survey. Journal of Intelligent and Robotic Systems 53(3), 263–296 (2008)
8. Canny, J.: A computational approach to edge detection. IEEE Transactions on Pattern Analysis and Machine Intelligence 8(6), 679–698 (1986)
9. Chatterjee, A., Rakshit, A., Singh, N.N.: Vision Based Autonomous Robot Navigation – Algorithms and Implementations. SCI, vol. 455. Springer, Heidelberg (2013)
10. Desouza, G.N., Kak, A.C.: Vision for mobile robot navigation: a survey. IEEE Transactions on Pattern Analysis and Machine Intelligence 24(2), 237–267 (2002)
11. Fazl-Ersi, E., Tsotsos, J.K.: Region classification for robust floor detection in indoor environments. In: Kamel, M., Campilho, A. (eds.) ICIAR 2009. LNCS, vol. 5627, pp. 717–726. Springer, Heidelberg (2009)
12. Hugues, L., Bredeche, N.: Simbad: An autonomous robot simulation package for education and research. In: Nolfi, S., Baldassarre, G., Calabretta, R., Hallam, J.C.T., Marocco, D., Meyer, J.-A., Miglino, O., Parisi, D. (eds.) SAB 2006. LNCS (LNAI), vol. 4095, pp. 831–842. Springer, Heidelberg (2006)
13. Mondada, F., Franzi, E., Guignard, A.: The Development of Khepera. In: Experiments with the Mini-Robot Khepera, Proceedings of the First International Khepera Workshop, pp. 7–14. HNI-Verlagsschriftenreihe, Heinz Nixdorf Institut (1999)
14. Okarma, K., Lech, P.: A fast image analysis technique for the line tracking robots. In: Rutkowski, L., Scherer, R., Tadeusiewicz, R., Zadeh, L.A., Zurada, J.M. (eds.) ICAISC 2010, Part II. LNCS, vol. 6114, pp. 329–336. Springer, Heidelberg (2010)
15. Okarma, K., Tecław, M., Lech, P.: Application of super-resolution algorithms for the navigation of autonomous mobile robots. In: Choraś, R.S. (ed.) Image Processing & Communications Challenges 6. AISC, vol. 313, pp. 147–154. Springer, Heidelberg (2015)

EgoTR: Personalized Tweets Recommendation Approach

Slim Benzarti[1] and Rim Faiz[2]

[1] LARODEC, ISG University of Tunis, Le Bardo, Tunisia
Slim_benzarti@yahoo.fr
[2] LARODEC, IHEC University of Carthage, Carthage Presidency, Tunisia
Rim.Faiz@ihec.rnu.tn

Abstract. Twitter and LinkedIn are two popular networks each in its territory. Nowadays, people use both of them in order to update their social (Twitter) and professional (LinkedIn) life. However, an information overload problem, caused by the data provided from these two networks separately, troubled many users. Indeed, the main goal of this work is to provide personalized recommendations that satisfy the user's expectations by exploiting the user generated content on Twitter and LinkedIn. We propose a method of recommending personalized tweet based on user's information from twitter and LinkedIn simultaneously. Our Final method considers two main elements: keywords extracted from Twitter and LinkedIn. Those extracted from Twitter are filtered by criteria such as hashtags, URL expansion and Tweets similarity. In order to evaluate our framework performance, we applied our system on a set of data collected from Twitter and LinkedIn. The experiments show that the proposed categorization of the elements is successfully important and our method outperforms several baseline methods.

Keywords: Twitter, LinkedIn, Tweet Recommendation, Content based, Personalization, Skills, and Interests.

1   Introduction

In recent years, Twitter has rapidly become a popular social information network. It is a microblogging platform where short posts are shared between several of users. Recent statistics show that more than 500M users generate more than 300M tweets every day. In other hand, LinkedIn is also a popular social network specialized in the professional user's life that significantly ahead of its competitors and its membership grows by approximately two new members every second. People use social networks not only to maintain social links with other people, but also for several other purposes, as well as sending messages, chatting and gathering news URLs. However, Information Retrieval Systems (IRS) faces new challenges due to the growth and diversity of available data. This huge quantity of information can be exploited and a lot of relevant data can be inferred to answer user's information needs in both social and professional life. In Twitter, the problem of overlapped content is observed when, some relevant tweets are flooded by other ones that might not interest the user at all and which oblige him to seek for the needed information by doing his own scan.

© Springer International Publishing Switzerland 2015

R. Silhavy et al. (eds.), *Intelligent Systems in Cybernetics and Automation Theory*,
Advances in Intelligent Systems and Computing 348, DOI: 10.1007/978-3-319-18503-3_23

227

In addition, professional interests must not be forgotten, because it is also an important factor to decide whether a tweet is useful or not. Profiling user's personal interests this way may be inaccurate and may not reflect the right user's future intensions. Besides the user's own tweet and retweet history used by many researchers, there are many other kinds of available important information, such as "favorite statues" which is utilized for the very first time to improve the user modeling. In another direction, the user's professional life and his relations with his colleagues can also greatly influence his behavior. By adding LinkedIn as another source of information, recommendations will perfectly satisfy the user expectations and facilitate the task of updating the news concerning his professional interests. In order to fully utilize such information, we propose an egocentric user-based approach exploiting the user generated content extracted from two separate networks to capture personal interests. This method is a promising technology for recommender systems. The recommendation will be based on the observed user posts so that the unobserved user preferences can be inferred from it. In personalized tweet recommendation, tweets are regarded as a set of words, and the user's preferences are obtained by analyzing his interactions.

Our approach exploit first, Twitter features such as tweets, retweet, favorite content, and LinkedIn skills mentioned explicitly by the user to capture his interests in order to create a combined profile which greatly represents his expectations. Then, we will exploit other features to create a tweet profile, composed by a bag of keywords, by combining several criteria such as hashtags, URL expansion and tweets similarity. The use of such features makes our model fully utilize the information mentioned on these two on line networks and do better personalize recommendations Indeed, the recommendation depends on measuring the similarity between the user profile and the tweet profile using the cosine formula. This paper is structured as follows: In Section 2, we survey the related works on personalized tweet recommendation. Section 3 introduces the proposed model. We evaluate our framework in Section 4. Finally, we discuss our work and summarize conclusions and future works.

2   Related Work

2.1   User Personalization

At first, user profile construction is either done in a static way, by gathering information that hardly changes like name, age and so on, or in a dynamic way, by gathering data that frequently changes. User's data are obtained explicitly by the user himself or implicitly by observing his behavior and interactions during his session (history, clicks, pages visited, etc.). According to [18], the user profile contains information such as: 1) Basic information which refers to the name, age, address, etc. 2) Knowledge of the user which is extracted generally from his web page navigation. 3) Interests which are well-defined through a set of keywords. 4) Feedback which design collected information from user's activity and could be deduced from number of clicks and time allowed in consulting resource, etc. 5) Preferences which are characteristics of user describing preferences in specific links or nodes, they could indicated preference in page style presentation, color, etc. [12] defined tags as the means by which users utilize for many purposes like: contributing and sharing, making an opinion, making attention, etc. In [13], the authors discuss the tags usefulness and they conclude that, on the one hand it's used for guiding other users to have information, and on the other hand to receive information about a user due to the history of tagging. [15] analyze the semantics of hashtags in more detail and reveal that tagging in Twitter rather used to join public discussions than organizing content for future retrieval. [17] have defined metrics to characterize hashtags respecting four dimensions: frequency, specificity, consistency, and stability over time. [20] explored the retrieval of hashtags for recommendation and introduced a method which takes in consideration user interests to find hashtags that are often applied to posts related to this topic. [7] Presents an in-depth comparison of three measures of influence, in-degree, re-tweets, and mentions, to identify and rank influential users. Based on these measures, they also investigate the dynamics of user influence across topics and time. Amalthaea, by [21], is one of many systems which create keyword profiles by extracting keywords from Web pages. In fact, the profile is built using one keyword vector for each user and then it is compared to document's vectors using cosine similarity. WebMate by [16] represent user profiles using one keyword vector per user's interest. Twopics introduced by [19], which is an entity-based profiling approach, aims to discover the topics of interest for Twitter users by examining the entities they mention in their Tweets. Other researchers were interested in exploring in depth URL mentioned by the user. [1] Evaluated and showed that the URL expansion strategy achieve 70-80% accuracy. Given the links between tweets and external sources, entities and topics extracted from articles, can be propagated to the corresponding tweets to further contextualize and enhance the semantics of Twitter activities. In this approaches, the authors used only tweets as input to extract the user's topic profile. We suggest including more information such as Tweets, retweets and favorites, which will be taken into account in the detection of user interest. Generally, the main purpose of constructing user's profile in information systems is to adapt the user generated content to answer his expectations.

To Insure a Relevant Profile Construction, We Must Cope with this Problem. The User Profile Evolution Consists Mainly of Apprehending Interest's Change, and Propagating these Changes in the Profile's Representation

2.2   Recommender Systems

2.2.1   Content-Based Recommendation Systems

[24] developed recommender systems that match user preferences (discovered from users profiles), with features extracted from locations (such as tags and categories), to make recommendations. Further recommendation systems provide a user with possible friends based on user's interactions in several social networks [5] and [10]. Other authors use the user location histories that reveal preferences in the friend's recommendation process. Consequently, users with similar past locality have similar preferences and are more likely to become friends. A study on MySpace [2] reveals that user's social connections are related to their geographical belonging. [3] analyze the rating data from Movie-Lens and finds that people at different places have diverse preferences. For instance, people who live in Minnesota are more interested in crime and war movies, whereas users from Florida are more concerned by fantasy and animation movies. To cope with this observation, [4] and [14] have proposed several

algorithms to increase the relevance of the search results. Content recommendations in Twitter aim at evaluating the importance of information for a given user and directing the user's attention to certain items. [8] focus on recommending URLs posted in Twitter messages and propose to structure the problem of content recommendations into three separate dimensions: discovering the source of content, modeling the interests of the users to rank content and exploiting the social network structure to adjust the ranking according to the general popularity of the items.

2.2.2   Collaborative Filtering Recommendation Systems

In a real-time system for on-line web content using a collaborative filtering method was proposed to perform more varied and personalized recommendations within a geographical area. However [8] didn't investigate user modeling in detail, but represent users and their tweets by means of a bag of words, from which they remove stop-words. [6] assumes that most available music recommender systems are based on collaborative filtering methods; they recommend music to user by considering some other user's ratings for the same music pieces. This technique is quite widely utilized, including music shopping services like Amazon or iTunes. However, this recommendation method suffers from the cold start problem. In the proposed approach, three major elements on Twitter are considered: tweet topic level factors, user social relation factors and explicit features such as authority of the publisher, and quality of the tweet. Collaborative filtering approaches exploit information about users who like similar items a long time ago. As discussed in [8], the only issue is that this kind of method obliges each news post to receive instantaneous feedback from numerous users before being recommended to other users. Some other systems rely on textual description of item that could be recommended, for instance, profiles that describe the user's interests regarding the items in the system. Finally, such systems require a means of measuring the compatibility between users and items in order to know which items to recommend to which users. Their results show that using learning to rank over three types of features helped to incorporate real-time web content while further improving the relevance ranking.

2.3   Profiles Merging

Several methods have been developed to construct the user's interests. Three approaches are commonly used:

- **The direct approach**: is to directly ask users what they like, for example, by listing all categories of interests and asking them to make selections.
- **The semi-direct approach**: is to ask users to assign notes to items (e.g. products they have purchased) they have manipulated.
- **The indirect approach**: is to get the user preferences from his friend's votes, by collecting their past ratings concerning a given topic.

The accuracy of a user profile of a recommender system can be increased by integrating data from other recommender systems. As users often have accounts on different Web 2.0 platforms, the combination of user profiles from different platforms might increase the quality of a user profile as well. [1] combined form-based and tag-based

Web 2.0 profiles. The tag-based profiles were extracted from Flickr, Delicious and StumbleUpon and could not only successfully overcome the cold-start problem but also improve the quality of comprehensive single-platform-based profiles. A significant observation is that the authors detected a very small overlap of the tags that a user used in different systems during the combination of the profiles. This leads to the assumption that users use different social networks for diverse goals: while users use LinkedIn to connect to their business partners, and those who use Twitter might tweet about both private and business related topics.

All the previous presented representations do not consider the time feature which is an important dimension in the user's context. To address this problem, we will attribute a weight which expresses the importance of the keyword and classified them in order to take into account the user interests evolution.

2.4   The Proposed Personalized Tweet Recommendation

The focus of our work is how to extract user profiles from Social Web content, then how to recommend relevant tweets. The problem here is the unstructured nature of the Social Networks data as well as the fact that users do not only interact on Twitter or mention explicitly their skills on LinkedIn related to their work and professional knowledge, but that a huge quantity of tags is often about their social life and other topics.

2.4.1   User Profile Construction

EgoTR is a system for constructing a rich user profile and recommending interesting tweets according to the user interests. Given generated data, EgoTR aims to infer and output relevant recommendations in order to satisfy user expectations.

The profile construction consists in extracting a set of terms that reflects the preferences of the user. First, in order to have a concise representation of the user from LinkedIn, we extract user's information from his profile to process textual data collections into a shape that facilitates discovering knowledge from user mentioned information. The goal is not only to rank or to select user's information, but also to extract salient interests from the user profile in order to build more meaningful and rich representation of their future expectations.

Then, we have to remove from the original user data all information presenting troubles. We proceed in two steps: We filter stop words by eliminating special characters which are very common. Second, we reduce a term to its morphologic root, in order to distinguish the variations of the word itself. Finally, we compute the Term Frequency (TF) to each keyword, consequently a weight will be associated to each one and it best reflects the user's interests. The interests that are returned will be ranked, not only by their frequency, but also by taking into account the number of user's friends who voted for this term. The more the word has votes, the more it is considered important in representing the interests of the user.

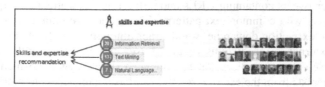

Fig. 1. Skills and expertise

The numbers enclosed by the red circles represent the user's friend's ratings or votes (Fig. 1). For the first item "Information Retrieval", 21 persons confirmed that the user is competent in this domain, 13 voted for the second one "Text Mining" and 8 persons confirmed that he is competent in "Natural Language". In our proposed user profiling method, these keywords will have different weights depending on the number of reviews. It should be noted that all used criteria are normalized between 0 and 1.

$$CI_L(i) = W(U; i) * V(i) \tag{1}$$

Where $CI_L(i)$ is the weight of the selected user's interest (Obtained after calculating the TF), U represents the user and i is the extracted interest. It should be noted that we attributed more importance to this category of keywords, while they are explicitly mentioned by the user and approved by their friends.

EgoTR coordinates input from various sources to enrich user profiles. In order to enhance personalization, we will exploit user generated content on Twitter by analyzing his tweets, retweets and favorites. In which follows, we will detail the followed steps to enrich the user profile. This process involved the use of certain parameters, namely: keywords, hashtags and URLs. We aim to enrich Twitter posts by adding semantic to facilitate browsing the interests behind the user's posts. Furthermore, we retrieved the tweets that were mentioned explicitly by the user as favorite. Twitter users do not only generate contents by posting tweets, but also by sharing links to external resources that can refer to news articles, blogs, company Web pages, other social networks, etc.

To find the words that best represent the semantic content of a tweet, we will use the TF.IDF function. It should be noted that we will attribute different weights to the chosen criteria, since they express different user interests. After performing several combinations using all the selected criteria, we give the formula 2 below which presents the linear combination of the most relevant criteria that we used from Twitter i.e. tweets, retweets, favorites that give the highest value.

$$CI_{TW}(i) = \alpha CI_T(i) + \beta CI_{RT}(i) + \gamma CI_{FAV}(i) \tag{2}$$

The parameters α, β, γ represent the correlation coefficients and their values will be further detailed in the next section: It should be noted that all used criteria are normalized between 0 and 1. After performing the weighting step, a classification according to weight is performed. The user profile that is delivered by EgoTR describes users' characteristics like names, locations or date of birth as well as interests.

2.4.2   Tweet Profile Construction

Our solution is divided into two major steps:

First, we filter tweets containing URLs and extracting keywords that require identification of tweets with common text patterns. Then, once we obtain the URLs and the outputs of the extraction demarche, we abstract uniquely identifiable profile information on the external service like: username or user id. Later, we link the user's Twitter profile to these external services. As a result, we extract keywords which will be considered as interests from the external sources, and as a gain access to more information about the user. Finally, we forward data embedded in URLs to the Twitter profiles.

URL Expansion: To further enrich the semantic of tweets, we implemented a strategy that allows obtaining accurate information about the content of the URLs mentioned by the user. Twitter users do not only generate contents by posting tweets, but also by sharing links to external resources that can refer to news articles, blogs, company Web pages, other social networks, etc. We go in depth and examine the tags that contain URLs in order to infer information about the topic of a post and we assume that those topics are indicative to user's interests.

Tweets Similarity: For a given tweet, we recover the tweets that are similar, and then we classify the results according to their relevance (by calculating the tweets similarity between the obtained tweets and the initial one) to a specific interval of time. It allows taking into account the profile evolution over the time. The similarity between tweets is calculated using the function provided by Lucene. If we take two tweets, and we want to calculate their similarity we rely on sim(t,t').

We denote the current tweet modeled as a vector t:

$$F(\vec{t}, t'') = \frac{\sum_{\vec{t}' \neq \vec{t}} Sim(\vec{t}, \vec{t}')}{|T_t''| - 1} \tag{3}$$

The Hashtags Use: In this step we integrate the use of the same hashtags in order to enrich the tweet's profile. We propose to find tweets that contain similar hashtags and expand the content by detecting the keywords that may be relevant.

$$F(t) \begin{cases} 1 \ if \ it \ contains \ an \ URL \\ 0 \quad otherwise \end{cases}$$

2.4.3 Merging User Profiles

In this step, we are going to merge the interests previously extracted from both of Twitter and LinkedIn in a single new profile. According to the interests express the level of user attention for a given item by analyzing past interactions. Twitter is known for the vast amount of information exchanged between users, which give us an accurate idea about the user. LinkedIn is a professional social network, where the user mentions relevant information related to his skills and related to their skills and professional profile. In addition to these interests, we will take the votes of the user's friends into consideration. Seen the credibility of LinkedIn and the interest that gives the user who mentions expressly his skills, we will assign more weight to keywords that are part of this network than those extracted from Twitter. Consequently, we will fix $\alpha = 0, 4$ and $\beta = 0, 6$:

$$W_F(i) = \alpha \, CI_{TW}(i) + \beta \, CI_L(i) \tag{4}$$

W_F expresses the final weight obtained after combining CI_{TW} which denotes the interest extracted from Twitter. CI_L represents those retrieved from LinkedIn.

3   Tweets Recommendation Based on User Profile

In our work, content-based recommendation is adopted while we rely on a textual description of the tweets and user's profiles that contains the user's interests. Thus, we will measure the compatibility between user interests, that are extracted from their Twitter interactions and interests mentioned explicitly on LinkedIn, and tweets in order to know which ones are relevant and able to be recommended. To measure the compatibility, we use the cosine function which measures the user-tweet similarity.

The cosine coefficient measures the cosine of the angle between the user profile interests and the tweet profile keywords. It can be computed by normalizing the dot product of the two profiles with respect to their norms where Fti;u denotes the weight of a term in the user profile and Vti;d is its weight in the tweet profile.

$$Cos(\vec{u}, \vec{d}) = \frac{\sum_i (F_{t_i,u} . V_{t_i,d})}{\sqrt{\sum_i, (F_{t_i,u})^2} * \sqrt{\sum_i (v_{t_i,d})^2}}$$ (5)

Content based Content based recommendations aim at evaluating the importance of information for a given user and directing the user's attention to certain items where tweets are regarded as items, and the preferences of users on the tweets are the correlation between users and items. To recommend specific tweet i who is supposed to be the most attractive to the user, the system must find the relative position of the interesting items within the total order of items for a specific user u. To this end, for each user, we aggregate his rankings in the test set by accumulating the weight of the item in order to produce a single total list. The items are again sorted in descending order of their accumulated frequencies. The main goal is to help users to discover new items of interest, therefore we add an additional restriction that the item to be recommended has to be novel for the user, and we remove from the suggestion list all occurrences of the pair (u; i). Finally, we generate a Top-10 recommendation list by selecting the 10 items with the highest score.

3.1   Experiments and Evaluations

Dataset description: We systematically collected tweets through the Twitter API. We retrieved concretely the activity of 7236 users; these people published 31450 tweets, over a two-month period: 1 February, 2013-31 March, 2013 (inclusive). More specifically, we have extracted 20392 tweets, 6389 retweeted statues and 4669 favorites. More details of our dataset are presented in Table 1. We have also retrieved the name, surname, date of birth, occupation, skills and expertise, the votes of the user's friends relative to the 7236 users from LinkedIn through LinkedIn Profile API.

In the following section, we describe the experiment run in the study to evaluate the performance of our approach. To improve the performance of the linear combination, that express the use of different criteria to construct the user profile based on Twitter interactions, we vary the parameters of the proposed equation that allows finding the values of the correlation coefficients α; β; γ that gives the best result.

The formula is

$$CI_{TW}(i) = \alpha \, CI_T \, (i) + \beta \, CI_{RT} \, (i) + \gamma \, CI_{FAV} \, (i)$$ (6)

and the best correlation coefficients that we will use are:

$\alpha = 0.5$

$\beta = 0.2$

$\gamma = 0.3$.

We can see that the overall performance scores of our proposed features combination of tweets, retweets and favorites gets 68.53 % precision, 64.25 % recall. The obtained results confirm the hypotheses that the combination of tweets, retweets and favorites expresses significantly the user interest.

We observed that the (tweets, retweets and favorites)-based user modeling strategy, performs better than the (tweets, retweets) based method by 4%. However, there is no significant difference in performance between (tweets, favorites)-based with 64.42% and (tweets, Hashtags)-based user modeling strategy 63.55%. We thus find the first evidence for our hypothesis that confirms that the use of favorite's statues criteria, in addition to tweets and retweets, enhances the user modeling better than any other baseline strategy. The result of this combination (68.53%) exceeds the most elevated value of all combinations deemed successful by nearly 5 %. The enriched obtained profile presents a rate of 71.88 % that beyond the profile based only on Twitter interactions by 3.35 %. The evaluation's results confirm that the merge of information from different sources improves the user personalization.

The performance of our system in the recommendation phase depends on the number of interests in the user profile and the number of keywords in the tweet profile. It is noted that the system either provides useful recommendations for the user and bad ones. We notice that when the user profile is poor in terms of interests, erroneous recommendations dominate. As the system acquires more keywords, recommendations become more accurate by answering the exact needs of the user.

From the obtained results, we can announce that the recommendation evolution is based mainly on the number of existing keywords in both profiles (User profile and tweet profile).

3.2   Baseline Approaches Comparison

We used Mean Average Precision (MAP), a popular rank evaluation method to evaluate the proposed approach EgoTR. For a single user, average precision is defined as the average of the P@n values for his tweets, retweets and favorites:

$$AP = \frac{\sum_{n=1}^{N} P@n * RF(n)}{|T|}. \tag{7}$$

where n is the number of tweets, |T| is the total number of retweets and favorites criteria for the targeted user, RF(n) is a binary function that describes whether the user has retweeted or checked as favorite the nth tweet in the result list. Finally, MAP can be obtained by averaging the AP values of all the users. We have compared our proposed model to several others baseline approaches.

Retweeted Times: is an objective estimation of the popularity of a tweet. This ranking strategy ignores personalization and assumes that the user's interests are the same as general publics. Profiling: This ranking strategy calculates the similarity between a tweet and the user's profile and shows the tweets sorted by similarity score. In comparison, the EgoTR model assimilates content based models by describing the information as weighted keywords, and gives 0, 6852 MAP. Also, it takes advantages of retrieving interests from Twitter tags and explicit interests from LinkedIn.

According to the previous results, we conclude that our proposed method made a great improvement to tweets recommendation performance. The result can be explained by the fact that the model includes more parameters to describe the personal interests and it is also explained by the terms contained in the profile of the tweet, which help detecting in detail the user's preferences.

4   Discussion

In this section we will discuss the results of P@n (n=1,3,5,10) and MAP on the test set. Chronological strategy gets 0,2287 MAP because, retweeting a status depends on user personal interests more than on the time of the post. The method of ranking by the number of retweeted times, performs poorly with 0,2865 MAP. This means that there is still a wide gap between personal interests and the focus of public attention, which indicates that personalization is very important on Twitter. The profiling method is a classic content based method and gives much better performance with 0,4538 MAP.

Our personalized recommending tweets approach, takes advantage of content filtering based recommendation by extracting contextual information from several online social networks (Twitter and LinkedIn) and incorporating them in our system, our experiments prove that it is helpful for detecting personal interests. The evaluation of our experiments shows the effectiveness of our system regarding problems previously addresses in the first section. The outcome of the described experiments clearly shows the benefits of our EgoTR. Our goal is to prove the importance of our contribution based on extracting useful information from several user profiles. The experiments proved that we got advantage of the term frequency because it reflects how often the user used a term. Second, we noticed that it is important to combine explicit and implicit information mentioned by the user.

The chosen feature combination (tweet, retweet and favorites) allows us proving the effectiveness of our system regarding problems addresses in this proposition. Another point we analyzed during our experiments was the user's activities. In fact, the results show that the more the user interacts with the social systems (Twitter), the more terms our system collects and the more relevant items the profile contains. Consequently, the system allows constructing a rich profile that helps performing a more accurate and targeted personalized tweet recommendation. Thus, we deduced that the use of heterogeneous social annotations, from several sources, provides accurate information for modeling the user profile. Finally, by comparing our system with three baseline systems (Chronological strategy, Retweet strategy and the Profiling strategy), the results reveal the efficiency of our EgoTR approach. We deduce that, according to evaluation metrics Precision and the Mean Average Precision (MAP), our system performs better results.

5   Conclusion and Future Work

In this paper, we have introduced EgoTR, a framework for modeling, enriching, and recommending useful tweets exploiting data available on social networks. Currently, we investigate the extraction of user generated content from both Twitter and LinkedIn that model the users' interests and evaluate them in the context of recommending relevant tweets. We have conduct experiments in the extraction of user data that model his interests. In this work, we rely specifically on Twitter interactions i.e tweets, retweets and favorite statues, then on LinkedIn by analyzing the user mentioned skills. The future directions of this work will focus on gathering data from more than two sources by exploiting other social networks reflecting the user interests and expertise such as CiteULike and dbpedia. We can also provide an opportunity for the user to interact with our system by asking questions that may reflect its interests and their evolutions.

References

1. Abel, F., Gao, Q., Houben, G.-J., Tao, K.: Semantic enrichment of twitter posts for user profile construction on the social web. In: Antoniou, G., Grobelnik, M., Simperl, E., Parsia, B., Plexousakis, D., De Leenheer, P., Pan, J. (eds.) ESWC 2011, Part II. LNCS, vol. 6644, pp. 375–389. Springer, Heidelberg (2011)
2. Adomavicius, G., Tuzhilin, A.: Toward the next generation of recommender systems: A survey of the state-of-the-art and possible extensions. IEEE Trans. on Knowl. and Data Eng., IEEE Educational Activities Department 17(6), 734–749 (2005)
3. Bianne Bernard, A.L., Menasri, F., Al-Hajj Mohamad, R., Kermorvant, C., Mokbel, C., Likforman-Sulem, L.: Dynamic and contextual information in hmm modeling for handwritten word recognition. IEEE Trans. Pattern Anal. Mach. Intell., IEEE Educational Activities Department 33(10), 2066–2080 (2011)
4. Arase, Y., Xie, X., Duan, M., Hara, T., Nishio, S.: A game based approach to assign geographical relevance to web images. In: WWW, pp. 811–820. ACM (2009)
5. Backstrom, L., Leskovec, J.: Supervised random walks: predicting and recommending links in social networks. In: Proceedings of the Fourth ACM International Conference on Web Search and Data Mining, WSDM 2011, pp. 635–644. ACM, New York (2011)
6. Bischo, K., Firan, C.S., Nejdl, W., Paiu, R.: Can all tags be used for search? In: Proceedings of the 17th ACM Conference on Information and Knowledge Management, CIKM 2008, pp. 193–202. ACM, New York (2008)
7. Cha, M., Mislove, A., Gummadi, K.P.: A measurement-driven analysis of information propagation in the Flickr social network. In: Proceedings of the 18th International Conference on World Wide Web, WWW 2009, pp. 721–730. ACM, New York (2009)
8. Chen, J., Nairn, R., Nelson, L., Bernstein, M., Chi, E.: Short and tweet: experiments on recommending content from information streams. In: CHI 2010: Proceedings of the 28th International Conference on Human Factors in Computing Systems, pp. 1185–1194. ACM, New York (2010)
9. Gilbert, E., Karahalios, K.: Predicting tie strength with social media. In: Proceedings of the SIGCHI Conference on Human Factors in Computing Systems, CHI 2009, pp. 211–220. ACM, New York (2009)

10. Gupta, M., Li, R., Yin, Z., Han, J.: Survey on social tagging techniques. SIGKDD Explor. Newsl. 12(1), 58–72 (2010)
11. Helic, D., Trattner, C., Strohmaier, M., Andrews, K.: on the navigability of social tagging systems. In: Proceedings of the 2010 IEEE Second International Conference on Social Computing, SOCIALCOM 2010, pp. 161–168. IEEE Computer Society, Washington, DC (2010)
12. Hidetoshi, K., Keiji, Y.: GeoVisualRank: a ranking method of geotagged imagesconsidering visual similarity and geo-location proximity. In: Srinivasan, S., Ramamritham, K., Kumar, A., Ravindra, M.P., Bertino, E., Kumar, R. (eds.), pp. 69–70. ACM (2011)
13. Huang, J., Chen, J., Cai, H., Friedland, R.P., Koubeissi, M.Z., Laidlaw, D.H., Auchus, A.P.: In Diffusion Tensor MRI Tractography reveals altered brainstem fiber connections accompanying agenesis of the corpus callosum (2011)
14. Jennings, N.R., Sycara, K., Wooldridge, K.: A roadmap of agent research and development. Autonomous Agents and Multi-Agent Systems 1(1), 7–38 (1998)
15. Laniado, D., Mika, P.: Making sense of twitter. In: Patel-Schneider, P.F., Pan, Y., Hitzler, P., Mika, P., Zhang, L., Pan, J.Z., Horrocks, I., Glimm, B. (eds.) ISWC 2010, Part I. LNCS, vol. 6496, pp. 470–485. Springer, Heidelberg (2010)
16. Mezghani, M., Zayani, C.A., Amous, I.: and. Gargouri F A user profile modelling using social annotations: a survey. In: Proceedings of the 21st International Conference Companion on World Wide Web, WWW 2012 Companion, pp. 969–976. ACM, New York (2012)
17. Michelson, M., Macskassy, S.: A Discovering users' topics of interest on twitter: a first look. In: Proceedings of the Fourth Workshop on Analytics for Noisy Unstructured Text Data, AND 2010, pp. 73–80. ACM, New York (2010)
18. Mokrane, B., Dimitre, K.: Personnalisation de l'information: aperçu de l'état de l'art et définition d'un modèle flexible de profils. In: CORIA, pp. 201–218 (2005)
19. Moukas, A., Moukas, R., Maes, P.: Amalthaea: An evolving multi-agent information filtering and discovery system for the www (1998)
20. Ramaswamy, S., Rastogi, R., Shim, K.: Efficient algorithms for mining outliers from large data sets. In: Proceedings of the 2000 ACM SIGMOD International Conference on Management of Data, SIGMOD 2000, pp. 427–438. ACM, New York (2000)
21. Rashid, A.M., Karypis, G., Riedl, J.: Learning preferences of new users in recommender systems: an information theoretic approach. SIGKDD Explor. Newsl. 10(2), 90–100 (2012)
22. Sandholm, T., Ung, H.: Real-time, location-aware collaborative filtering of web content. In: Proceedings of the 2011 Workshop on Context-Awareness in Retrieval and Recommendation, CaRR 2011, pp. 14–18. ACM, New York (2011)
23. Schubert, P.: and. Koch M. The power of personalization: Customer collaboration and virtual communities. In: Proc. Americas Conf. on Information Systems, AMCIS 2002, Dallas, TX, pp. 1953–1965 (2002)
24. Zhou, L.T.J., Li, M.: User-level sentiment analysis incorporating social networks. In: Proceedings of the 17th ACM SIGKDD International Conference on Knowledge Discovery and Data Mining, KDD 2011, pp. 1397–1405. ACM, New York (2011)

Simulation of Hybrid Fuzzy Adaptive Control
of Pneumatic Muscle Actuator

Mária Tóthová, Ján Piteľ, and Alexander Hošovský

Technical University of Košice, Faculty of Manufacturing Technologies,
Department of Mathematics, Informatics and Cybernetics, Prešov, Slovakia
{maria.tothova,jan.pitel,alexander.hosovsky}@tuke.sk

Abstract. The pneumatic muscle actuator is highly nonlinear system and it is difficult to control it using only a linear controller with fixed gains. The hybrid fuzzy adaptive control scheme with reference model was designed to control such actuator. It uses a multiplicative signal adaptation with a linear controller in the feedforward and a fuzzy controller in the adaptive feedback loop. In the paper there are presented some simulation results of this control. The nonlinear dynamic model of one-DOF actuator based on the advanced geometric muscle model was used in simulation.

Keywords: pneumatic artificial muscle, dynamic simulation, hybrid control, fuzzy control.

1 Introduction

Pneumatic muscle actuator (PMA) offers several appealing characteristics such as high power-to-weight ratio, natural compliance, flexibility in contact, sensitivity to touch which are of interest for biomechanical and biorobotic applications [1,2,3], as well as industrial robotic applications [4,5,6]. But its disadvantage is hysteresis, non-linear character and other issues associated with pneumatic systems in general [7,8]. That is why it is difficult to control it using only a linear PID controller with fixed gains which is not able to meet the adequate performance for the whole range of actuator movement, as well as for a varying inertia moment [9]. The accurate control of such actuator has remained a challenging task to apply various advanced control techniques [10]. In [11] an accurate positioning of pneumatic artificial muscle at different temperatures using a sliding mode controller was tested. Different nonlinear control strategies (backstepping vs. sliding mode) were tested in [12] to achieve the robustness of applied control. In [13] several control loops were used in order to overcome the negative effects of the muscle elasticity and compressibility of air. Sophisticated control methods based on computational intelligence can be used too. In [14] the switching algorithm using learning vector quantization neural network was proposed to identify the external load of the actuator. Fuzzy logic control of pneumatic muscle system using a linearizing control scheme was introduced in [15]. Here a hybrid fuzzy adaptive control with reference model is used to control of the antagonistic PMA.

© Springer International Publishing Switzerland 2015   239
R. Silhavy et al. (eds.), *Intelligent Systems in Cybernetics and Automation Theory*,
Advances in Intelligent Systems and Computing 348, DOI: 10.1007/978-3-319-18503-3_24

2   Materials and Methods

2.1   Pneumatic Muscle Actuator

An antagonistic PMA consists of two pneumatic artificial muscles which act against each other with their tensile forces (F_1, F_2) through the gear mechanism. It is designed in order to generate a retraction force or rotary movement [16]. A pair of ON/OFF twin-solenoid valves (for filling and for discharging) is needed to operate one muscle. The compressed air flows into the muscles in the form of pressure pulses through the inlet solenoid valves and thereby the resulting pressures (P_1, P_2) in the muscles are manipulated variables. The outlet solenoid valves are used to discharge muscles and the throttle valves are used to adjust the actuator dynamics. The resulting position of the actuator arm is given by an angle of rotation arm (φ) and it is determined by a balance of both tensile forces depending on the air pressure in each muscles and load (m) of the actuator arm [17].

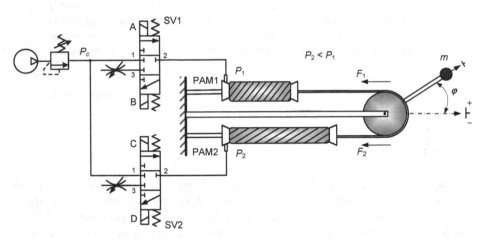

Fig. 1. Principle diagram of the antagonistic PMA

2.2   Hybrid Fuzzy Adaptive Control of PMA

With the aim to design the control system of the antagonistic PMA for lower cost industrial applications a simplification of control algorithm in relation to the actuator stiffness control was taken into account. The control philosophy based on pressure change only in one muscle was designed [18, 19]. The second muscle (passive) has a constant maximal pressure and subserves as nonlinear spring toward to discharged and controlled first muscle (active) [20]. Thus stiffness of the PMA will be self-aligning and the stiffness control loop can be avoided. Another task was to design the control system which would be capable of handling the PMA's nonlinear properties and also could be able to compensate the changes of the inertia moment due to the variation in external load of the PMA.

A hybrid adaptive control using multiplicative signal adaptation with a linear controller in the feedforward branch and a fuzzy controller in the adaptive feedback loop was used (Fig. 2). The actual position φ of the actuator arm is compared with the desired position φ_d. The control deviation e is fed to the linear controller with constant gains and the control error from the reference model (error in the dynamics is represented by $e_M = \varphi_M - \varphi$) is fed to the parallel branch with fuzzy controller which compensation signal in a form of adaptation gain K_{AM} is multiplied with the control signal u_{PD} of the linear controller [21]. Decoder determines which solenoid valve of the PMA will be operated by PWM manipulated signals A, B, C, D.

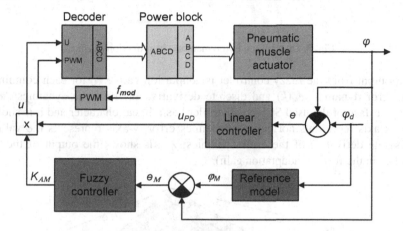

Fig. 2. Hybrid fuzzy adaptive control scheme with reference model

Three steps in the design of this control system were important:
- the reference model derivation – a model of second order was chosen and it is more described in [22],
- the linear controller gains – PD controller was used and its basic gains were designed using the PID tuning tool of Simulink Control Design,
- the fuzzy controller design – a zero-order Sugeno fuzzy controller was used due to low computational demands.

2.3   Sugeno Fuzzy Controller

A zero-order Sugeno fuzzy controller was type of DISO (double input, single output) with triangular membership functions (*trimf*). The error dynamics e_M and the discrete derivative of the error dynamics Δe_M were the input variables to the fuzzy controller and the adaptation gain K_{AM} was the output variable. Seven membership functions distributed over the relevant universe (work interval) were used: NB - negative big, NM - negative medium, NS - negative small, Z - zero, PS - positive small, PM - positive medium and PB - positive big (Fig. 3). These membership functions are the same for both input signals $e_M(k)$ and $\Delta e_M(k)$ to the fuzzy controller. The input and output variables were normalized into interval <-1;1>.

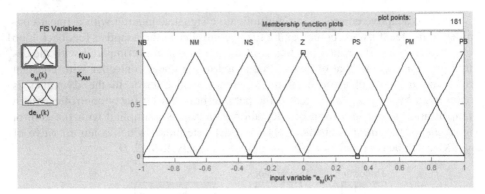

Fig. 3. Distribution of the membership functions (*trimf*)

The output from the fuzzy controller is adaptation gain K_{AM} for each combination of the error dynamics $e_M(k)$ and discrete derivative of the error dynamics $\Delta e_M(k)$. A fuzzy surface of the used Sugeno controller has a linear character and it is shown in Fig. 4 (x-axis represents normalized dynamics error, y-axis represents the values of the discrete derivative of the error dynamics, z-axis shows the output of the fuzzy controller in the form of adaptation gain).

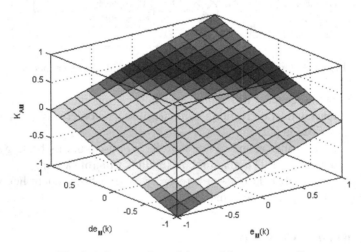

Fig. 4. A fuzzy surface of the used Sugeno controller

3   Results

Based on Fig. 2 the simulation model of hybrid fuzzy adaptive control of the antagonistic PMA with linear PD controller in the feedforward branch and Sugeno fuzzy controller in the adaptive feedback loop was designed in Matlab/Simulink environment (Fig. 5). The colors of the blocks in the block diagram in Fig. 2 correspond to the colors of the corresponding blocks in the simulation model in Fig. 5.

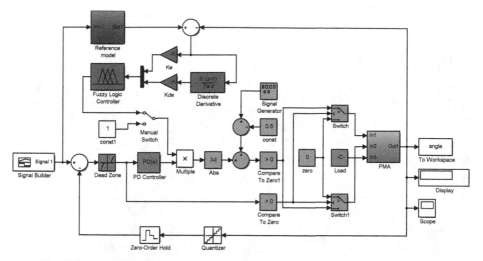

Fig. 5. Simulation model of hybrid fuzzy adaptive control of the antagonistic PMA

An important part of the designed simulation model in Fig. 5 is a dynamic model of the actuator represented by PMA subsystem. This subsystem has been created on the basis of Fig. 1 using an advanced geometric muscle model and it is more described in [23, 24, 25]. The power block in the block diagram in Fig. 2 represents the power module needed for solenoid valves control and it contains power transistors and a power supply. Because the effect of this module on the relevant dynamic characteristics of the actuator is negligible, it was not included in the simulation model. The control part consists of modeled linear PD controller which output signal is multiplied by the output signal from the fuzzy controller of the adaptation subsystem with reference model.

The simulation model of hybrid fuzzy adaptive control (Fig. 5) contains the block named "*Fuzzy Logic*" in the adaptive feedback loop into which the designed Sugeno fuzzy controller was inserted from the working environment (*workspace*). The scaling gains of the error dynamics and its discrete derivative were set to $K_e = 0.0057$ and $K_{\Delta e} = 0.027$. These values were experimentally determined [26]. The gains of the linear PD controller were set to $K_P = 0.25$ and $K_D = 0.001$. The reference (setpoint) signal has changed in the range from -28 deg to +28 deg. The time period of simulation was set to 20 s and the actuator arm has achieved 15 desired positions. The step changes were chosen randomly and the responses were simulated without load and with load 3.34 kg of the actuator.

In Fig. 6 and 7 the responses of the simulated system with hybrid fuzzy adaptive controller and the only linear controller to several step changes in arm position are shown. As it can be seen in Fig. 6 there are no significant differences between the only linear control and hybrid fuzzy adaptive control (after several steps of adaptation) when the actuator is no load. But with increased load to 3.34 kg the responses of the only linear control are exceedingly oscillatory. On the other hand, hybrid fuzzy adaptive controller was capable to follow the desired position without any significant oscillations.

244 M. Tóthová, J. Piteľ, and A. Hošovský

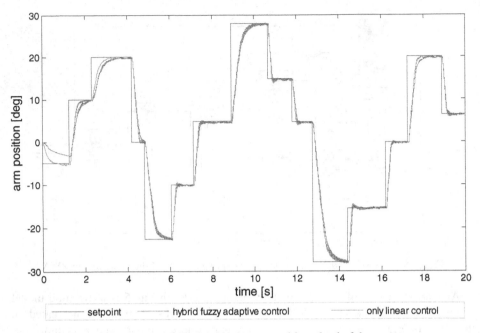

Fig. 6. Responses of the simulated system without load of the actuator

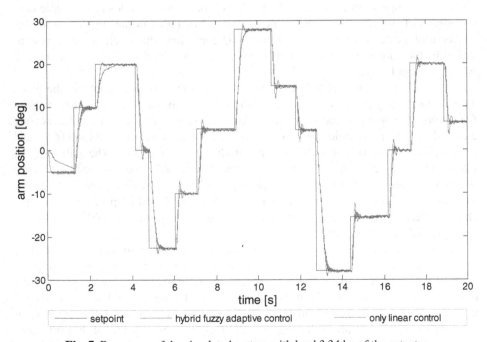

Fig. 7. Responses of the simulated system with load 3.34 kg of the actuator

4   Conclusion

In this work it was proved by simulation of hybrid fuzzy adaptive control of the antagonistic PMA:

- The dynamic model of the antagonistic PMA based on advanced geometric muscle model designed in author's previous works can be used for simulation of control of such actuator with different external loads of the actuator arm.
- A relatively simple Sugeno fuzzy controller with triangular membership functions used in the adaptive feedback loop of the hybrid control system with reference model improves control of the antagonistic PMA under conditions of varying inertia moment.

In future work a Mamdani fuzzy controller will be tested in the adaptive feedback loop of the hybrid control system and also another type of membership functions will be used. An attention will be also paid to optimize the scaling gains of the fuzzy controller using different optimization methods offered by Matlab/Simulink toolboxes.

Acknowledgments. The research work is supported by the Project of the Structural Funds of the EU, title of the project: Research and development of the intelligent nonconventional actuators based on artificial muscles, ITMS code: 26220220103.

References

1. Líška, O., More, M., Janáčová, D., Charvátová, H.: Design of Rehabilitation Robot Based on Pneumatic Artificial Muscles. In: Mathematical Methods and Optimization Techniques in Engineering, pp. 151–154. Europment, Antalya (2013)
2. Židek, K., Piteľ, J., Galajdová, A., Fodor, M.: Rehabilitation Device Construction Based on Artificial Muscle Actuators. In: Proceedings of the 9th IASTED International Conference: Biomedical Engineering, BioMed 2012, pp. 855–861. Iasted Press, Innsbruck (2012)
3. Židek, K., Šeminský, J.: Automated Rehabilitation Device Based on Artificial Muscles. In: Annals of DAAAM for 2011 & Proceedings of the 22nd International DAAAM Symposium "Intelligent Manufacturing & Automation: Power of Knowledge and Creativity", pp. 1113–1114. DAAAM, Vienna (2011)
4. Hošovský, A., Havran, M.: Dynamic Modeling of One Degree of Freedom Pneumatic Muscle-Based Actuator for Industrial Applications. Tehnički Vjesnik 3/19, 673–681 (2012)
5. Tóthová, M., Piteľ, J., Mižáková, J.: Electro-Pneumatic Robot Actuator with Artificial Muscles and State Feedback. Applied Mechanics and Materials 460, 23–31 (2014)
6. Straka, Ľ.: Operational Reliability of Mechatronic Equipment Based on Pneumatic Artificial Muscle. Applied Mechanics and Materials 460, 41–48 (2014)
7. Boržíková, J., Balara, M.: Mathematical Model of Contraction Characteristics of the Artificial Muscle. Manufacturing Engineering 6(2), 26–29 (2007)
8. Sárosi, J.: Study on Viscoelastic Behaviour of Pneumatic Muscle Actuator. Annals of Faculty Engineering Hunedoara - International Journal of Engineering XII, 83–86 (2014)
9. Balara, M.: The Upgrade Methods of the Pneumatic Actuator Operation Ability. Applied Mechanics and Materials 308, 63–68 (2013)

10. Kinsner, W., et al.: Challenges in Engineering Education of Cognitive Dynamic Systems. In: Proc. 2012 Canadian Engineering Education Association (CEEA 2012), pp. 1–12. University of Manitoba, Winnipeg (2012)
11. Sárosi, J.: Accurate Positioning of Pneumatic Artificial Muscle at Different Temperatures Using LabVIEW Based Sliding Mode Controller. In: Proceedings of the 9th IEEE International Symposium on Applied Computational Intelligence and Informatics (SACI 2014), pp. 85–89. IEEE Press, Timisoara (2014)
12. Carbonell, P., Jiang, Z.P., Repperger, D.: Nonlinear Control of a Pneumatic Muscle Actuator: Backstepping vs. Sliding Mode. In: IEEE Conference on Control Applications, pp. 167–172. IEEE Press, Saint-Petersburg (2001)
13. Wang, Y., et al.: Study of Smooth and Accurate Position Controls of Pneumatic Artificial Muscle Actuators for Robotic Arms. Advanced Materials Research 317-319, 799–806 (2011)
14. Ahn, K.K., Thanh, D.C., Ahn, Y.K.: Intelligent Switching Control of a Pneumatic Artificial Muscle Manipulator. JCME International Journal 48(4), 657–667 (2005)
15. Balasaubramanian, K., Rattan, K.S.: Fuzzy Logic Control of a Pneumatic Muscle System Using a Linearizing Control Scheme. In: 22nd International Conference of North American Fuzzy Information Processing Society, Chicago, pp. 432–436 (2003)
16. Piteľ, J., Tóthová, M.: Operating Characteristics of Antagonistic Actuator with Pneumatic Artificial Muscles. Applied Mechanics and Materials 616, 101–109 (2014)
17. Tóthová, M., Piteľ, J., Boržíková, J.: Operating Modes of Pneumatic Artificial Muscle Actuator. Applied Mechanics and Materials 308, 39–44 (2013)
18. Piteľ, J., Balara, M., Boržíková, J.: Control of the Actuator with Pneumatic Artificial Muscles in Antagonistic Connection. Transactions of the VŠB – Technical University of Ostrava, LIII (2), 101–106 (2007)
19. Piteľ, J., Tóthová, M.: Design of Hybrid Adaptive Control of Antagonistic Pneumatic Muscle Actuator. In: 34th IASTED International Conference on Modeling, Innsbruck (in press, 2015)
20. Vagaská, A.: Mathematical Description and Static Characteristics of the Spring Actuator with Pneumatic Artificial Muscle. Applied Mechanics and Materials 460, 65–72 (2014)
21. Tóthová, M., Piteľ, J.: Reference Model for Hybrid Adaptive Control of Pneumatic Muscle Actuator. In: 9th IEEE International Symposium on Applied Computational Intelligence and Informatics (SACI 2014), pp. 105–109. IEEE Press, Timisoara (2014)
22. Hošovský, A., Michal, P., Tóthová, M., Biroš, O.: Fuzzy Adaptive Control for Pneumatic Muscle Actuator with Simulated Annealing Tuning. In: 12th IEEE International Symposium on Applied Machine Intelligence and Informatics (SAMI 2014), pp. 205–209. IEEE Press, Herľany (2014)
23. Tóthová, M., Piteľ, J.: Dynamic Model of Pneumatic Actuator Based on Advanced Geometric Muscle Model. In: 9th International Conference on Computational Cybernetics (ICCC 2013), pp. 83–87. IEEE Press, Tihany (2013)
24. Tóthová, M., Piteľ, J.: Simulation of Actuator Dynamics Based on Geometric Model of Pneumatic Artificial Muscle. In: 11th International Symposium on Intelligent Systems and Informatics (SISY 2013), pp. 233–237. IEEE Press, Subotica (2013)
25. Tóthová, M., Piteľ, J.: Dynamic Simulation of Pneumatic Muscle Actuator in Matlab/Simulink Environment. In: 12th IEEE International Symposium on Intelligent Systems and Informatics (SISY 2014), pp. 209–213. IEEE Press, Subotica (2014)
26. Havran, M.: Computer Aided Control of Non-Conventional Actuator of Manipulators. Dissertation work, 136 p. Prešov (2012)

Case Study of Learning Entropy for Adaptive Novelty Detection in Solid-Fuel Combustion Control

Ivo Bukovsky and Cyril Oswald

Department of Instrumentation and Control Engineering
Czech Technical University in Prague
{Cyril.Oswald,Ivo.Bukovsky}@fs.cvut.cz

Abstract. This paper deals with the case study of usability of the Learning Entropy approach for the adaptive novelty detection in MIMO dynamical systems. The novelty detection is studied for typical parameters of linear systems including time delay. The solid-fuel combustion process is selected as a representative of typical non-linear dynamic MIMO system. The complex mathematical model of a biomass-fired 100kW boiler is used for verification of the potentials of the proposed method, and the motivation for novelty detection in solid-fuel combustion processes is discussed in this paper.

Keywords: adaptive systems, novelty detection, learning entropy, combustion process, multiple input, control, incremental learning, quadratic neural unit.

1   Introduction

A solid-fuel combustion is one of main sources of heat energy in both large (power plants and heat plants) and small-scale (local heat sources) devices. It is also a huge source of pollution. In recent decades, the aim of developed countries is to reduce the overall pollution produced by the modern society. The solid-fuel fired heat energy sources are logic target of continuously tightened emission regulations. The goal of the new heat source design, or the modernization of existing equipment, is not just to reach demanded power output with minimal cost, but to reach the emission limits, too. Nowadays, it is increasingly difficult and more expansive to reach the emission limits with the technical measures only and the modern control algorithms help to meet the regulations. The controlled solid-fuel combustion in heat-water boiler of any size is very complex nonlinear MIMO process from a control systems point of view. The immediate energy output and emissions in flu gas is given not only by a set of inputs values (fuel feed, combustion air feed) and properties (fuel quality, combustion air temperature, etc.) but also by the current internal dynamics of the combustion process. Therefore, it is difficult to estimate which input has a main impact to the observed output (the energy output, the flu gas composition) and if the changes in the combustion process are caused by the control intervention or by the uncontrolled changes in the combustion process. We understand Adaptive Novelty Detection (AND) to be an approach for detection of:

© Springer International Publishing Switzerland 2015
R. Silhavy et al. (eds.), *Intelligent Systems in Cybernetics and Automation Theory*,
Advances in Intelligent Systems and Computing 348, DOI: 10.1007/978-3-319-18503-3_25

— unusual system transitions to a new system state that is not caused by the observed system inputs, or
— parameter perturbations, or
— sensor faults,

where the detection is based on adaptation (learning) of predictive models (adaptive filters, neural networks) to data in a real time. Typically for control engineering are, we can consider perturbations of the gain, time constant, or input-delay parameters. In the solid-fuel combustion, AND can be employed for detection of unusual changes in the inner process dynamic and thereby contribute to make the right control decision (postpone the reaction, change the inner model, etc.).

For Novelty Detection (ND) in computational intelligence area, we shall first distinguish between statistical approaches [1] and neural–network approaches [2]. Regarding the statistical novelty detection approaches in time series, Sample Entropy of Pincus [3] and its extensions [4] have to be mentioned here. In this paper, we focus on the utilization of learning–system novelty detection approaches that clearly include neural networks in general [5]. Learning–systems based novelty detection can include fault detection approaches [6] [7], non-stationary changes [8], recently encode-decode detection approach [9], or concept drifts detection utilizing the incremental learning techniques [10]. In this study, the AND is based on recently introduced algorithm of Learning Entropy (LE) that was recently introduced for time series [11] with recent case studies for possible enhancement of time series prediction [12] and also for learning algorithms of adaptive filters with discussion on its possible connotations to detection of system perturbations or sensor perturbations in [13]. The basic principles of the AND and the presentation of the method based on the LE method will be mentioned in section 2 of this article. In section 3, there will be presented an example of a new states detection on a simulated mathematical model of a linear MISO system. The application of the LE applied to AND for a fundamental linear MIMO plant followed by the complex model of solid-fuel combustion data will be showed in section 3. Section 4 discusses connotation for real use and for further research. Main notation terms and symbols are defined In Nomenclature, and x denotes a scalar, $\mathbf{x}$ is for a vector, $\mathbf{X}$ stands for a matrix, constant sampling discrete-time adaptive models are considered, the discrete-time index k is introduced when necessary, and further notation is explained at place.

2   Applied Approaches

2.1   Multiple-Input Polynomial Models HONU

This subsection introduces the implemented principles of Higher-Order Neural Units (HONU) for MISO dynamical systems as they are later in this paper introduced for adaptive novelty detection with learning entropy. In general, a linear or polynomial-type, i.e., HONU, adaptive model for a MISO system can be defined as follows

$$\tilde{y}(k) = f_r(\mathbf{y}, \mathbf{U}, \mathbf{w}), \tag{1}$$

where $y(k)$ denotes the adaptive model output, f_r denotes mapping function HONU of polynomial order r, $\mathbf{y}$ denotes vector of recent history of real measured values as follows

$$\mathbf{y} = \begin{bmatrix} y(k - n_{ya}) \\ y(k - n_{ya} - 1) \\ \vdots \\ y(k - n_{yb}) \end{bmatrix}, \tag{2}$$

where n_{ya} and n_{yb} are configuration parameters of vector $\mathbf{y}$, $\mathbf{U}$ is a multidimensional array of m rows that are vectors of external inputs to the model configured as

$$\mathbf{U} = \begin{bmatrix} \mathbf{u_1} \\ \mathbf{u_2} \\ \vdots \\ \mathbf{u_m} \end{bmatrix} = \begin{bmatrix} u_1(k - n_{u1a}) & u_1(k - n_{u1a} - 1) & \dots & u_1(k - n_{u1b}) \\ u_2(k - n_{u2a}) & u_2(k - n_{u2a} - 1) & \dots & u_2(k - n_{u2b}) \\ \vdots & \vdots & \ddots & \vdots \\ u_m(k - n_{uma}) & u_m(k - n_{uma} - 1) & \dots & u_m(k - n_{umb}) \end{bmatrix} \tag{3}$$

where parameters n_{uia} and n_{uib} define optional setups of used history of external variables $\mathbf{u_i}$ as model inputs, and finally $\mathbf{w}$ in (1) represents all adaptable parameters (neural weights) as the $n \times (n+1)$ vector

$$\mathbf{w} = [w_0 \quad w_1 \quad \dots \quad w_n], \tag{4}$$

where w_0 is a weight that represents an absolute term in the model, i.e. the neural bias from the perspective of neural networks.

For a linear mapping function, i.e. for $r=1$, model (2) can be expressed in a long-vector multiplication form as follows

$$\tilde{y}(k) = f_{r=1}(\mathbf{y}, \mathbf{U}, \mathbf{w}) = \mathbf{w} \cdot \mathbf{x}, \tag{5}$$

where the vector $\mathbf{x}$ is a long-vector representation of $[\mathbf{y}, \mathbf{U}]$ that is augmented with a unit as

$$\mathbf{x} = [x_0 \quad x_1 \quad \dots \quad x_n] = [1 \quad \mathbf{y}^{\mathbf{T}} \quad \mathbf{u_1} \quad \mathbf{u_2} \quad \dots \quad \mathbf{u_m} \quad]^{\mathbf{T}}. \tag{6}$$

To preserve the vector multiplication form of HONU also for a nonlinear mapping function $f_r>1$ in (2), the augmented vector x can yield a long-column-vector of polynomial terms denoted further as **colx**; for the case of second polynomial order $r=2$, i.e. for QNU, $\mathbf{x}$ yields a column vector **colx** whose elements are defined as follows

$$\mathbf{colx} = [\{colx_{i,j} = x_i \cdot x_j; i = 0 \dots n, j = i \dots n\}]. \tag{7}$$

Then the adaptive model for QNU yields

$$\tilde{y}(k) = f_{r=2}(\mathbf{y}, \mathbf{U}, \mathbf{w}) = \mathbf{w} \cdot \mathbf{colx}, \tag{8}$$

where **colx** is given through (2)(3)(6)(7); and accordingly, the indexing of the row-vector **w** for QNU then yields

$$\mathbf{w} = [\{w_{i,j}; i = 0 \ldots n, j = i \ldots n\}], \tag{9}$$

In this study LNU and QNU (HONU with r=1, 2) are considered.

In the next subsection, the adaptive algorithm of Learning Entropy is introduced for multiple-input dynamical systems.

2.2   Learning Entropy Novelty Detection for Incremental Learning

The principle of LE is extendable to incrementally learning dynamical systems in general as the LE algorithm evaluates weight increments of an adaptive model regardless its mathematical structure. For a general incremental learning rule as follows,

$$\mathbf{w}(k+1) = \mathbf{w}(k) + \Delta\mathbf{w}, \tag{10}$$

where $\Delta\mathbf{w}$ is a vector of learning increments of all adaptable parameters. The Learning Entropy can be practically calculated as the Approximate Individual Sample Learning Entropy (AISLE) [11] that is the ratio of learning increments of unusually large magnitude in respect to the recent learning history. AISLE for the model (1) is given as follows

$$E_A(k) = \frac{1}{n_\alpha \cdot n_w} \sum_{\alpha_{min}}^{\alpha_{max}} \sum_{i=0}^{n_w} h_\alpha \left(|\Delta w_i| > |\overline{\Delta w_i}|\right) \text{ where } \begin{matrix} h_\alpha(\text{TRUE}) = 1 \\ h_\alpha(\text{FALSE}) = 0 \end{matrix}, \tag{11}$$

where $\boldsymbol{\alpha}=[\alpha_{min},\ldots,\alpha_{max}]$ is $1 \times n_\alpha$ vector of detection sensitivities to overcome the otherwise single-scale nature of the detection [6] [11] and the recently average magnitude of each learning increment at time k is here calculated as

$$|\overline{\Delta w_i}| = \frac{1}{M} \sum_{\kappa=k-M}^{k-1} |\Delta w_i(\kappa)|, \tag{12}$$

where M (i.e. m in [11]) is the number of samples of considered recent learning history for LE. Contrary to [11] [14], it shall be highlighted that formula (12) applies to adaptive models with time indexing according to (1) in this paper.

3   Experimental Analysis of LE for a Multiple Input System

3.1   Novelty Detection in a 2-Input Linear System with Delays

To demonstrate clearly the potentials of the proposed approach, its potentials can be demonstrated on artificial data sampled from a continuous-time linear dynamical system that is externally excited by two inputs u_1 and u_2 as follows

$$Y(s) = G_1(s) \cdot U_1(s) + G_2(s) \cdot U_2(s), \tag{13}$$

where U_1, U_2 are Laplace transforms of input variables, Y is the Laplace transform of a plant output y, s stands for the Laplace operator, and the individual transfer functions are defined as follows

$$G_i(s) = \frac{K_i}{T_i \cdot s + 1} e^{-Td_i \cdot s} \text{ where } i = 1, 2. \tag{14}$$

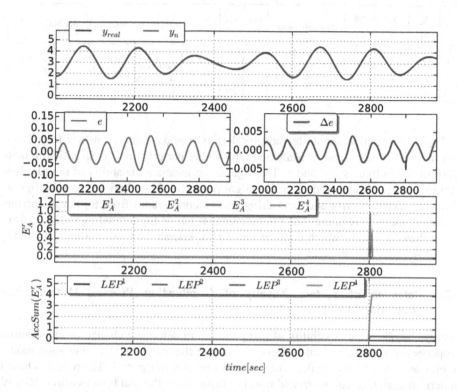

Fig. 1. Detection of time delay perturbation at t_3=2800 [sec] (17) by LNU (r=1) , where the error is defined as $e = y - \overline{y}$, and LEP denotes the accumulative sums of LE (Learning Entropy Profiles [5]).

The novelties to be detected are introduced as three suddenly appearing perturbations as follows

$$K_1(t \geq t_1) = 1.1 \cdot K_1, \tag{15}$$

$$T_2(t \geq t_2) = 1.1 \cdot T_2, \tag{16}$$

$$Td_2(t > t_3) = 1.1 \cdot Td_2. \tag{17}$$

where t is a continuous index of time.

Experiments were carried out for the sinusoidaly forced system (13) with setup given in table 1, with two sinus inputs u_1 and u_2 of different frequencies and for perturbations (15-17).

Table 1. Plant (13-14), perturbations (15-17), model (LNU, r=1) and detection setups

	Plant Setup				Adaptive Model and Detection Setups					
i	K_i [/]	T_i [sec]	T_{di} [sec]	Perturbations time:	n_{uia} [/]	n_{uib} [/]	α=[100, 50, 30, 10, 7] , M=500			
1	1	3	3	t_1=2 400 [sec] t_2=2 600 [sec] t_3=2 800 [sec]	3	10	n_{ya}=1 n_{yb}=5 [/]	sampling: Δt = 2 sec	μ=1E-3 μ_{LE}=1E-5	pre-trainnig: 0...1000 sec epochs=5
2	2	5	5		3	10				

Typical simulation results for novelty detection in data of continually forced system induced as gain, time constant and input delay perturbations as specified in table 1 are shown in figures 1, 4 and 5. The Figures demonstrate the ability of the LE algorithm to detect novelty in data due to slight parameter perturbations, and the figures also show detail of error and error difference (second graphs from top) at the perturbation moment. It is interesting to observe, that LE of higher orders r=3,4 perform better than for r=1,2 , which conforms to our experiences with time series as in [11] [12] [13].

3.2   Detection of System Novelty of Combustion Process Mathematical Simulation Model

Finally, we show the possibility of using the LE to detect the change of unmeasured properties of the combustion process. We utilize the mathematical simulation model of biomass-fired 100kW boiler schematically depicted in figure 2. The model is based in part on analytical equations and tuned to behavior of the real biomass-fired 100kW boiler in the operating point. In this simulation experiment, the primary air and the secondary air supply is periodically and steeply changed. The task is to detect the step change of the unmeasured fuel humidity.

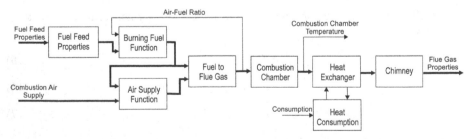

Fig. 2. The diagram of the mathematical simulation model of the biomass combustion in the 100kW heat-water boiler

Courses of the primary and the secondary air supply and the combustion chamber temperature, which are used as the inputs and the output of the polynomial MISO model, are in figure 6 in the appendix depicted. The results are depicted in figure 3. The LE algorithm successfully detects the change in the fuel humidity parameter carried on at time 8000 seconds. The lag in detection is caused by the used delay in the fuel impact to the system which is close to 600 seconds in the used mathematical simulation model.

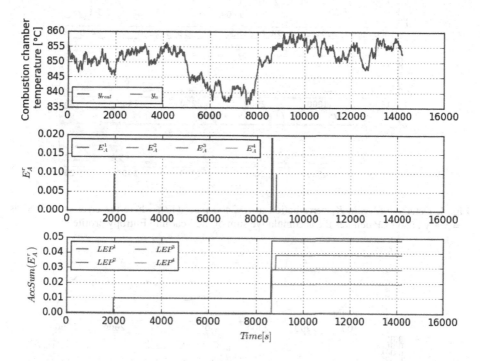

Fig. 3. The LE based novelty detection algorithm found the strong novelty at time of the fuel humidity change. The small peak at time 2000 seconds indicates a small reaction of the LE algorithm to the change on the primary air supply.

4   Discussion

The results shown in figures 1, 4 & 5 demonstrates the potentials of the algorithm to detect novelty in a system via real-time evaluation of real-time data. For a linear 2-input system, novelty caused by perturbations of a single gain, a time constant, as well as an input-time delay were detected by increased Learning Entropy of a linear adaptive model (HONU, r=1) that was adapted by a fundamental gradient descent incremental learning. Similarly, a quadratic neural unit (HONU, r=2) was also used and the results were very similar for the linear MISO system. Except the sample-by-sample gradient descent learning on the linear MISO system, we also recorded promising capability of LE when using batch learning of Levenberg-Marquardt; reporting

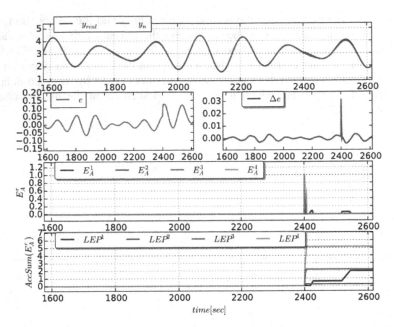

Fig. 4. Detection of gain K_1 perturbation at t_1=2400 [sec] (17), where the error is defined as $e = y - \bar{y}$, and LEP denotes the accumulative sum of the Learning Entropy profile [5]

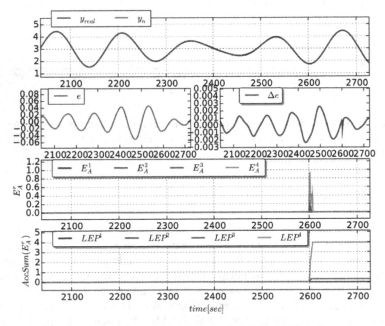

Fig. 5. Detection of time constant T_2 perturbation at t_2=2600 [sec] (17), where the error is defined as $e = y - \bar{y}$, and LEP denotes the accumulative sum of the Learning Entropy profile [5]

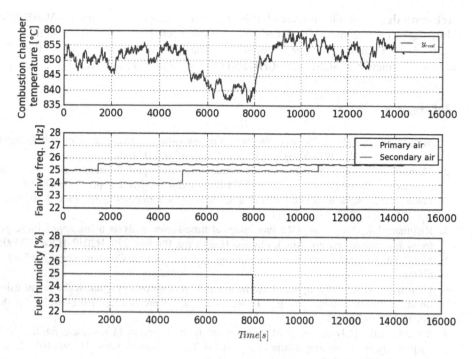

Fig. 6. Courses of the combustion chamber temperature (the output), the combustion air supplies (the controlled inputs) and the fuel humidity (the unmeasured input variable) during the simulation experiment. The output is burdened by model internal noise.

this however exceeds the scope of this paper. The results were achieved for sinusoidaly, i.e. continuously, exciting inputs that, of course, represents only a limited scope of control engineering problems. For step inputs to a system, another research for adaptive novelty detection via Learning Entropy shall be carried out as it is more difficult task that requires further research.

5   Conclusions

The paper originally introduced adaptive novelty detection via Learning Entropy algorithm for multiple-input systems, and its potentials for adaptive novelty detection in combustion process data were experimentally studied on a complex nonlinear model of a combustion process. The Learning Entropy algorithm reliably detected novelty in a linear dynamical MISO system with the use of linear neural unit as well as with a quadratic neural unit (HONU, $r=2$). For the nonlinear complex MIMO model of a combustion process, the algorithm indicates potentials to detect novelty due to unmeasurable variation of fuel quality that is important consideration in control and optimization of combustion processes.

Acknowledgement. This research has been partially supported by grant TA02020836 "Intelligent control methods of residual-biomass fired boiler ecological control" of Technology Agency of the Czech Republic, partially by the Grant Agency of the CTU in Prague, grants No. SGSI21177/0HK2/3TI12 and SGS14/179/OHK2/3T/12.

References

1. Markou, M., Singh, S.: Novelty detection: a review—part 1: statistical approaches. Signal Process. 83, 2481–2497 (2003), doi:10.1016/j.sigpro.2003.07.018
2. Markou, M., Singh, S.: Novelty detection: a review—part 2: neural network based approaches. Signal Process. 83, 2499–2521 (2003), doi:10.1016/j.sigpro.2003.07.019
3. Pincus, S.M.: Approximate entropy as a measure of system complexity. Proc. Natl. Acad. Sci. U S A 88, 2297–2301 (1991)
4. Richman, J.S., Moorman, J.R.: Physiological time-series analysis using approximate entropy and sample entropy. Am. J. Physiol. Heart Circ. Physiol. 278, H2039–H2049 (2000)
5. Marsland, S.: Novelty detection in learning systems. Neural Comput. Surv. 3, 157–195 (2003)
6. Demetriou, M.A., Polycarpou, M.M.: Incipient fault diagnosis of dynamical systems using online approximators. IEEE Trans. Autom. Control 43, 1612–1617 (1998), doi:10.1109/9.728881
7. Trunov, A.B., Polycarpou, M.M.: Automated fault diagnosis in nonlinear multivariable systems using a learning methodology. IEEE Trans. Neural Netw. 11, 91–101 (2000), doi:10.1109/72.822513
8. Alippi, C., Roveri, M.: Just-in-Time Adaptive Classifiers #x2014;Part I: Detecting Nonstationary Changes. IEEE Trans. Neural Netw. 19, 1145–1153 (2008), doi:10.1109/TNN.2008.2000082
9. Alippi, C., Bu, L., Zhao, D.: A prior-free encode-decode change detection test to inspect datastreams for concept drift. In: 2013 Int. Jt. Conf. Neural Netw., IJCNN, pp. 1–6 (2013)
10. Elwell, R., Polikar, R.: Incremental Learning of Concept Drift in Nonstationary Environments. IEEE Trans. Neural Netw. 22, 1517–1531 (2011), doi:10.1109/TNN.2011.2160459
11. Bukovsky, I.: Learning Entropy: Multiscale Measure for Incremental Learning. Entropy 15, 4159–4187 (2013), doi:10.3390/e15104159
12. Bukovsky, I., Homma, N., Cejnek, M., Ichiji, K.: Study of Learning Entropy for Novelty Detection in lung tumor motion prediction for target tracking radiation therapy. In: 2014 Int. Jt. Conf. Neural Netw., IJCNN, pp. 3124–3129 (2014)
13. Bukovsky, I., Oswald, C., Cejnek, M., Benes, P.M.: Learning entropy for novelty detection a cognitive approach for adaptive filters. Sens. Signal Process. Def. SSPD 2014, pp. 1–5 (2014)
14. Bukovsky, I., Kinsner, W., Bila, J.: Multiscale analysis approach for novelty detection in adaptation plot. Institution of Engineering and Technology, pp. 27–27 (2012)

Appendix

Nomenclature

α ... bias of detection sensitivity

e ... error between a model and real

E_A ... Approximated Learning Entropy

r ... polynomial order, also order of E_A

f_r ... polynomial mapping HONU

h_α ... detection function

MIMO ... multiple-input multiple-output

MISO ... multiple-input single-output

n_α ... length of vector $\boldsymbol{\alpha}$

n_w ... number of adaptable parameters

n_x ... length of vector $\mathbf{x}$

n_{yia}, n_{uib} ... configuration parameters of model

T ... time constant of a linear system

T_d ... input time delay

$\boldsymbol{u}$ ... vector of control (or external) inputs

$\boldsymbol{w}$ ... vector of all adaptable parameters

$\mathbf{x}$ ... augmented vector of model variables

y ... real output (measured)

$\bar{y}$ ... model output (predicted value)

One Approach to Adaptive Control
of a Nonlinear Distributed Parameters Process

Petr Dostal, Jiri Vojtesek, and Vladimir Bobal

Tomas Bata University in Zlín, Faculty of Applied Informatics,
Nam T.G. Masaryka 5555, 760 01 Zlin, Czech Republic
{dostalp,vojtesek,bobal}@fai.utb.cz

Abstract . The paper provides a procedure for the design of adaptive control of
a nonlinear distributed parameter process represented by a tubular chemical
reactor. The presented method is based on approximation of a nonlinear model
of the process by its external linear model with a structure obtained from simulated dynamic characteristics. The parameters of the external linear model
are estimated using corresponding delta model. To derive of controllers, the
polynomial approach is used. The procedure is tested on the nonlinear model of
the process.

1   Introduction

Nonlinear distributed parameter processes often constitute production units in
chemical, biochemical or power industry. Their mathematical models are described
by sets of nonlinear partial differential equations (NPDRs). The methods of modelling
and simulation of such processes are described e.g. in [1], [2] or [3].

   The control of nonlinear distributed parameter processes often represents very
complex problem. The control problems can be due to the process nonlinearity, its
distributed nature, high sensitivity of the state and output variables to input changes,
and some others. Evidently, the process with such properties is hardly controllable by
conventional control methods, and, its effective control requires application some of
advanced methods.

   Typical processes suitable for the verification of various advanced control methods
are tubular chemical reactors (TCRs). Detailed static and dynamic analysis of   the
TCR with countercurrent cooling is presented e.g. in [4].

   One of the methods to some extent eliminate the unfavorable properties of nonlinear distributed parameter systems uses adaptive strategies based on an appropriate
choice of a continuous-time external linear model (CT ELM) with recursively estimated parameters. These parameters are consequently used for parallel updating of
controller's parameters. Some results obtained in this field were presented by authors
of this paper e.g. in [5] and [6].

   For the CT ELM parameter estimation, either the direct method, see, e.g. [7],
or application of an external delta model with the same structure as the CT model
can be used. The basics of delta models have been described e.g. in [8] and [9].

© Springer International Publishing Switzerland 2015   259
R. Silhavy et al. (eds.), *Intelligent Systems in Cybernetics and Automation Theory*,
Advances in Intelligent Systems and Computing 348, DOI: 10.1007/978-3-319-18503-3_26

Although delta models belong into discrete models, they do not have such disadvantageous properties connected with shortening of a sampling period as discrete z-models. In addition, parameters of delta models can directly be estimated from sampled signals. Moreover, it can be easily proved that these parameters converge to parameters of CT models for a sufficiently small sampling period (compared to the dynamics of the controlled process), as shown in [10].

This paper deals with continuous-time adaptive control of a tubular chemical reactor with a countercurrent cooling as a nonlinear single input – single output process. With respect to practical possibilities of a measurement and control, the mean reactant temperature and the output reactant temperature are chosen as the controlled outputs, and, the coolant flow rate as the control input. The nonlinear model of the reactor is approximated by a CT external linear model with a structure chosen on the basis of computed controlled outputs step responses. The parameters of the CT ELM then are estimated via corresponding delta model. The control structure with two feedback controllers is considered, e.g. [11]. The resulting controllers are derived using the polynomial approach, e.g. [12] and the pole assignment method, see, e.g. [13]. The method is tested on a mathematical model of a tubular chemical reactor.

2   Model of the Reactor

An ideal plug-flow tubular chemical reactor with a simple exothermic consecutive reaction $A \rightarrow B \rightarrow C$ in the liquid phase and with the countercurrent cooling is considered. Heat losses and heat conduction along the metal walls of tubes are assumed to be negligible, but dynamics of the metal walls of tubes are significant. All densities, heat capacities, and heat transfer coefficients are assumed to be constant. Under above assumptions, the reactor model can be described by five PDRs in the form

$$\frac{\partial c_A}{\partial t} + v_r \frac{\partial c_A}{\partial z} = -k_1 c_A \tag{1}$$

$$\frac{\partial c_B}{\partial t} + v_r \frac{\partial c_B}{\partial z} = k_1 c_A - k_2 c_B \tag{2}$$

$$\frac{\partial T_r}{\partial t} + v_r \frac{\partial T_r}{\partial z} = \frac{Q_r}{(\rho c_p)_r} - \frac{4U_1}{d_1(\rho c_p)_r}(T_r - T_w) \tag{3}$$

$$\frac{\partial T_w}{\partial t} = \frac{4}{(d_2^2 - d_1^2)(\rho c_p)_w}[d_1 U_1(T_r - T_w) + d_2 U_2(T_c - T_w] \tag{4}$$

$$\frac{\partial T_c}{\partial t} - v_c \frac{\partial T_c}{\partial z} = \frac{4 n_1 d_2 U_2}{(d_3^2 - n_1 d_2^2)(\rho c_p)_c}(T_w - T_c) \tag{5}$$

with initial conditions

$$c_A(z,0) = c_A^s(z), \ c_B(z,0) = c_B^s(z), \ T_r(z,0) = T_r^s(z), \ T_w(z,0) = T_w^s(z), \ T_c(z,0) = T_c^s(z)$$

and boundary conditions

$$c_A(0,t) = c_{A0}(t), \ c_B(0,t) = c_{B0}(t), \ T_r(0,t) = T_{r0}(t), \ T_c(L,t) = T_{cL}(t).$$

Here, t is the time, z is the axial space variable, c are concentrations, T are temperatures, v are fluid velocities, d are diameters, ρ are densities, c_p are specific heat capacities, U are heat transfer coefficients, n_1 is the number of tubes and L is the length of tubes. The subscript $(\cdot)_r$ stands for the reactant mixture, $(\cdot)_w$ for the metal walls of tubes, $(\cdot)_c$ for the coolant, and the superscript $(\cdot)^s$ for steady-state values.

The reaction rates and heat of reactions are nonlinear functions expressed as

$$k_j = k_{j0} \exp\left(\frac{-E_j}{RT_r}\right), \ j = 1, 2 \tag{6}$$

$$Q_r = (-\Delta H_{r1}) k_1 c_A + (-\Delta H_{r2}) k_2 c_B . \tag{7}$$

The fluid velocities are calculated via the reactant and coolant flow rates as

$$v_r = \frac{4 q_r}{\pi n_1 d_1^2} \ , \ v_c = \frac{4 q_c}{\pi (d_3^2 - n_1 d_2^2)} \tag{8}$$

The parameter values with correspondent units used for simulations are given in Table 1.

From the system engineering point of view, $c_A(L,t) = c_{Aout}$, $c_B(L,t) = c_{Bout}$, $T_r(L,t) = T_{rout}$ and $T_c(0,t) = T_{cout}$ are the output variables, and, $q_r(t)$, $q_c(t)$, $c_{A0}(t)$, $T_{r0}(t)$ and $T_{cL}(t)$ are the input variables.

Table 1. Used parameter values

$L = 8$ m	$U_1 = 2.8$ kJ/m²s K	$U_2 = 2.56$ kJ/m²s K	
$d_1 = 0.02$ m	$c_{pr} = 4.05$ kJ/kg K	$c_{pw} = 0.71$ kJ/kg K	$c_{pc} = 4.18$ kJ/kg K
$d_2 = 0.024$ m	$\rho_r = 985$ kg/m³	$\rho_w = 7800$ kg/m³	$\rho_c = 998$ kg/m³
$d_3 = 1$ m	$E_1/R = 13477$ K	$E_2/R = 15290$ K	$(-\Delta H_{r1}) = 5.8 \cdot 10^4$ kJ/kmol
$n_1 = 1200$	$k_{10} = 5.61 \cdot 10^{16}$ 1/s	$k_{20} = 1.128 \cdot 10^{18}$ 1/s	$(-\Delta H_{r2}) = 1.8 \cdot 10^4$ kJ/kmol

Among them, for the control purposes, mostly the coolant flow rate can be taken into account as the control variable, whereas other inputs entering into the process can be accepted as disturbances. In this paper, the mean reactant temperature given by

$$T_m(t) = \frac{1}{L} \int_0^L T_r(z,t) \, dz \tag{9}$$

and the reactant output temperature $T_{rout}(t)$ are considered as the controlled outputs.

3 Computation Models

For computation of both steady-state and dynamic characteristics, the finite differences method is employed. The procedure is based on substitution of the space interval $z \in \ <0, L>$ by a set of discrete node points $\{z_i\}$ for $i = 1, \dots, n$, and, subsequently, by approximation of derivatives with respect to the space variable in each node point by finite differences. Two types of finite differences are applied, either the backward finite difference

$$\left.\frac{\partial y(z,t)}{\partial z}\right|_{z=z_i} \approx \frac{y(z_i,t)-y(z_{i-1},t)}{h} = \frac{y(i,t)-y(i-1,t)}{h} \tag{10}$$

or the forward finite difference

$$\left.\frac{\partial y(z,t)}{\partial z}\right|_{z=z_i} \approx \frac{y(z_{i+1},t)-y(z_i,t)}{h} = \frac{y(i+1,t)-y(i,t)}{h}. \tag{11}$$

Here, a function $y(z,t)$ is continuously differentiable in the interval $<0,L>$, and, $h = L/n$ is the discretization step.

3.1   Dynamic Model

Applying the substitutions (10), (11) in (1) – (5), and, omitting the argument t in parenthesis, PDRs (1) – (5) are approximated by a set of ODRs in the form

$$\frac{dc_A(i)}{dt} = -\left[b_0 + k_1(i)\right]c_A(i) + b_0\, c_A(i-1) \tag{12}$$

$$\frac{dc_B(i)}{dt} = k_1(i)c_A(i) - \left[b_0 + k_2(i)\right]c_B(i) + b_0\, c_B(i-1) \tag{13}$$

$$\frac{dT_r(i)}{dt} = b_1 Q_r(i) - (b_0 + b_2)T_r(i) + b_0\, T_r(i-1) + b_2\, T_w(i) \tag{14}$$

$$\frac{dT_w(i)}{dt} = b_3\left[T_r(i) - T_w(i)\right] + b_4\left[T_c(i) - T_w(i)\right] \tag{15}$$

$$\frac{dT_c(m)}{dt} = -(b_5 + b_6)T_c(m) + b_5\, T_c(m+1) + b_6\, T_w(m) \tag{16}$$

for $i = 1, \ldots, n$ and $m = n - i + 1$, and with initial conditions $c_A(i,0) = c_A{}^s(i)$, $c_B(i,0) = c_B{}^s(i)$, $T_r(i,0) = T_r{}^s(i)$, $T_w(i,0) = T_w{}^s(i)$ and $T_c(m,0) = T_c{}^s(m)$ for $i = 1, \ldots, n$ and $m = n - i + 1$.

The boundary conditions enter into Eqs. (12) – (14) and (16) for $i = 1$.

Now, nonlinear functions in Eqs. (12) – (16) take the discrete form

$$k_j(i) = k_{j0}\exp\left(\frac{-E_j}{RT_r(i)}\right),\ j = 1, 2 \tag{17}$$

$$Q_r(i) = (-\Delta H_{r1})k_1(i)c_A(i) + (-\Delta H_{r2})k_2(i)c_B(i) \tag{18}$$

for $i = 1, \ldots, n$.

The parameters b in Eqs. (12) – (16) are calculated from formulas

$$\begin{gathered} b_0 = \frac{v_r}{h}, b_1 = \frac{1}{(\rho c_p)_r}, b_2 = \frac{4U_1}{d_1(\rho c_p)_r}, b_3 = \frac{4d_1 U_1}{(d_2^2 - d_1^2)(\rho c_p)_w}, \\ b_4 = \frac{4d_2 U_2}{(d_2^2 - d_1^2)(\rho c_p)_w}, b_5 = \frac{v_c}{h}, b_6 = \frac{4n_1 d_2 U_2}{(d_3^2 - n_1 d_2^2)(\rho c_p)_c} \end{gathered} \tag{19}$$

Here, the formulas for computation of T_m and $T_{r\,\mathrm{out}}$ take discrete forms

$$T_m(t) = \frac{1}{n} \sum_{i=1}^{n} T_r(z_i, t), \quad T_{r\,\mathrm{out}}(t) = T_r(z_n, t) \tag{20}$$

3.2   Steady-State Model

Computation of the steady-state characteristics is necessary not only for a steady-state analysis but the steady state values $y^s(i)$ also constitute initial conditions in ODRs (12) – (16) (here, y presents some of the variable in the set (12) – (16)).

The steady-state model can simply be derived equating the time derivatives in (12) – (16) to zero. Then, after some algebraic manipulations, the steady-state model takes the form of difference equations

$$c_A^s(i) = \frac{b_0}{b_0 + k_1^s(i)} c_A^s(i-1) \tag{21}$$

$$c_B^s(i) = \frac{1}{b_0 + k_2^s(i)} \left[k_1^s(i) c_A^s(i) + b_0 c_B^s(i-1) \right] \tag{22}$$

$$T_r^s(i) = \frac{1}{b_0 + b_2} \left[b_1 Q_r^s(i) + b_0 T_r^s(i-1) + b_2 T_w^s(i) \right] \tag{23}$$

$$T_w^s(i) = \frac{1}{b_3 + b_4} \left[b_3 T_r^s(i) + b_4 T_c^s(i) \right] \tag{24}$$

$$T_c^s(m) = \frac{1}{b_5 + b_6} \left[b_5 T_c^s(m+1) + b_6 T_w^s(m) \right] \tag{25}$$

for $i = 1, \ldots, n$ and $m = n - i + 1$.

Nonlinear functions accordant with a steady-state are

$$k_j^s(i) = k_{j0} \exp\left(\frac{-E_j}{R T_r^s(i)} \right), \quad j = 1, 2 \tag{26}$$

$$Q_r^s(i) = (-\Delta H_{r1}) k_1^s(i) c_A^s(i) + (-\Delta H_{r2}) k_2^s(i) c_B^s(i) \tag{27}$$

Now, the formulas for computation T_m and $T_{r\,\mathrm{out}}$ have forms

$$T_m^s = \frac{1}{n} \sum_{i=1}^{n} T_r^s(z_i), \quad T_{r\,\mathrm{out}}^s = T_r^s(z_n) \tag{28}$$

3.3   Steady-State and Dynamic Characteristics

Typical reactant temperature profiles along the reactor tubes computed for $c_{A0}^s = 2.85$, $c_{B0}^s = 0$, $T_{r0}^s = 323$, $T_{c0}^s = 293$ and $q_r^s = 0.15$ for various coolant flow rates are shown in Fig. 1(left graph). A presence of a maximum on the reactant temperature profiles is a common property of many tubular reactors with exothermic reactions.

A dependences of the reactant mean temperature and the reactant output temperature on the coolant flow rate is shown in Fig. 1(right graph). The form of both curves documents a nonlinear relation between supposed controlled outputs and the coolant flow rate which is considered as the control input.

Dynamic characteristics were computed in the neighborhood of the chosen operating point $q_c^s = 0.27\ m^3/s$, $T_m^s = 334.44\ K$, $T_{rout}^s = 326.10\ K$. For the dynamic analysis and subsequent control purposes, the controlled outputs are defined as deviations from steady values

$$y_1(t) = \Delta T_m(t) = T_m(t) - T_m^s$$
$$y_2(t) = \Delta T_{rout}(t) = T_{rout}(t) - T_{rout}^s$$

(29)

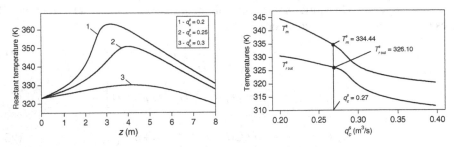

Fig. 1. Reactant temperature profiles for various coolant flow rates(on the left) and dependences of the reactant mean and output temperatures on the coolant flow rates(on the right)

Such form is frequently used in the control. The deviation of the coolant flow rate is denoted as

$$\Delta q_c = q_c(t) - q_c^s .$$

(30)

The responses of both outputs to the coolant flow rate step changes are shown in Fig. 2.

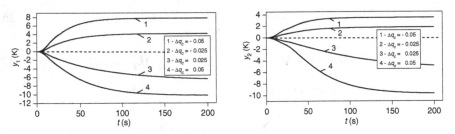

Fig. 2. Reactant mean temperature step responses(left graph) and reactant output temperature step responses(right graph)

The above shown responses demonstrate more expressive nonlinear properties of the reactant output temperature to input changes than the reactant mean temperature. This fact is evident also from the gain values computed as

$$g_s = \lim_{t \to \infty} \frac{y(t)}{\Delta q_c}$$

(31)

and presented in Tab. 2.

Table 2. Gains for various input changes

Δq_c	- 0.025	- 0.05	0.025	0.05
Main reactant temperature				
g_s	-155.4	-166.2	-263.5	-205.1
Output reactant temperature				
g_s	-69.6	-72.0	-194.3	-193.5

Moreover, the dynamics of the reactant output temperature is slower in comparison with the dynamics of the reactant mean temperature.

4   CT and Delta ELM

For the control purposes, the control input variable are considered in the form

$$u(t) = 10 \; \frac{q_c(t) - q_c^s}{q_c^s} \tag{32}$$

This expression enables to obtain control input and controlled output variables of approximately the same magnitude.

A choice of the CT ELM structure does not stem from known structure of the model (1) – (5) but from a character of simulated step responses. It is well known that in adaptive control a controlled process of a higher order can be approximated by a linear model of a lower order with variable parameters. Taking into account profiles of curves in Fig. 2 with zero derivatives in $t = 0$, the second order CT ELM has been chosen for both controlled outputs in the form of the second order linear differential equation

$$\ddot{y}(t) + a_1 \dot{y}(t) + a_0 y(t) = b_0 u(t) \tag{33}$$

where $y = y_1$ or $y = y_2$, and, in the complex domain, as the transfer function

$$G(s) = \frac{b_0}{s^2 + a_1 s + a_0}. \tag{34}$$

Establishing the δ operator

$$\delta = \frac{q-1}{T_0} \tag{35}$$

where q is the forward shift operator and T_0 is the sampling period, the delta ELM corresponding to (33) takes the form

$$\delta^2 y(t') + a_1' \delta y(t') + a_0' y(t') = b_0' u(t') \tag{36}$$

where t' is the discrete time.

When the sampling period is shortened, the delta operator approaches the derivative operator, and, the estimated parameters a', b' reach the parameters a, b of the CT model (33).

5   Delta Model Parameter Estimation

Substituting $t' = k - 2$, equation (36) can be rewritten to the form

$$\delta^2 y(k-2) + a_1' \delta y(k-2) + a_0' y(k-2) = b_0' u(k-2) \tag{37}$$

In the paper, the recursive identification method with exponential and directional forgetting was used.

Establishing the regression vector

$$\Phi_\delta^T(k-1) = \left(-\delta y(k-2) \quad -y(k-2) \quad u(k-2) \right) \tag{38}$$

where

$$\delta y(k-2) = \frac{y(k-1) - y(k-2)}{T_0} \tag{39}$$

the vector of delta model parameters

$$\Theta_\delta^T(k) = \left(a_1' \ a_0' \ b_0' \right) \tag{40}$$

is recursively estimated from the ARX model

$$\delta^2 y(k-2) = \Theta_\delta^T(k) \Phi_\delta(k-1) + \varepsilon(k) \tag{41}$$

where

$$\delta^2 y(k-2) = \frac{y(k) - 2y(k-1) + y(k-2)}{T_0^2}. \tag{42}$$

6   Controller Design

The control system with two feedback controllers is depicted in Fig. 3.

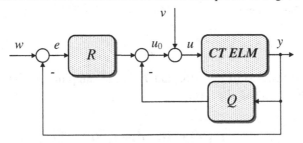

Fig. 3. Control system with two feedback controllers

In the scheme, w is the reference signal, v denotes the load disturbance, e the tracking error, u_0 output of controllers, u the control input and y the controlled output. The transfer function $G(s)$ of the CT ELM is given by (34).

The reference w and the disturbance v are considered as step functions with transforms

$$W(s) = \frac{w_0}{s}, \quad V(s) = \frac{v_0}{s} \tag{43}$$

The transfer functions of both controllers are in forms

$$R(s) = \frac{r(s)}{\tilde{p}(s)}, \quad Q(s) = \frac{\tilde{q}(s)}{\tilde{p}(s)} \tag{44}$$

where $\tilde{q}$, r and $\tilde{p}$ are coprime polynomials in s fulfilling the condition of properness $\deg r \leq \deg \tilde{p}$ and $\deg q \leq \deg \tilde{p}$.

The controller design described in this section appears from the polynomial approach. The general requirements on the control system are formulated as its internal properness and strong stability (in addition to the control system stability, also the controller stability is required), asymptotic tracking of the reference and load disturbance attenuation. The procedure to derive admissible controllers can be performed as follows:

Transforms of the basic signals in the closed-loop system take following forms (for simplification, the argument s is in some equations omitted)

$$Y(s) = \frac{b}{d} \left[r W(s) + \tilde{p} V(s) \right] \tag{45}$$

$$E(s) = \frac{1}{d} \left[(a\tilde{p} + b\tilde{q}) W(s) - b\tilde{p} V(s) \right] \tag{46}$$

$$U(s) = \frac{a}{d} \left[r W(s) + \tilde{p} V(s) \right]. \tag{47}$$

Here,

$$d(s) = a(s)\,\tilde{p}(s) + b(s)\left[r(s) + \tilde{q}(s) \right] \tag{48}$$

is the characteristic polynomial with roots as poles of the closed-loop.

Establishing the polynomial t as

$$t(s) = r(s) + \tilde{q}(s) \tag{49}$$

and substituting (49) into (48), the condition of the control system stability is ensured when polynomials $\tilde{p}$ and t are given by a solution of the polynomial Diophantine equation

$$a(s)\tilde{p}(s) + b(s)t(s) = d(s) \tag{50}$$

with a stable polynomial d on the right side.

With regard to the transforms (43), the asymptotic tracking and load disturbance attenuation are provided by divisibility of both terms $a\tilde{p} + b\tilde{q}$ and $\tilde{p}$ in (46) by s. This condition is fulfilled for polynomials $\tilde{p}$ and $\tilde{q}$ having forms

$$\tilde{p}(s) = s\,p(s), \quad \tilde{q}(s) = s\,q(s). \tag{51}$$

Subsequently, the transfer functions (44) take forms

$$Q(s) = \frac{q(s)}{p(s)}, \quad R(s) = \frac{r(s)}{s\, p(s)} \tag{52}$$

and, a stable polynomial $p(s)$ in their denominators ensures the stability of controllers.

The control system satisfies the condition of internal properness when the transfer functions of all its components are proper. Consequently, the degrees of polynomials q and r must fulfill inequalities

$$\deg q \le \deg p, \quad \deg r \le \deg p + 1. \tag{53}$$

Now, the polynomial t can be rewritten to the form

$$t(s) = r(s) + s\, q(s). \tag{54}$$

Taking into account the solvability of (50) and conditions (53), the degrees of polynomials in (50) and (52) can be easily derived as

$$\deg t = \deg r = \deg a, \deg q = \deg a - 1$$
$$\deg p \ge \deg a - 1, \deg d \ge 2 \deg a \tag{55}$$

Denoting $\deg a = n$, polynomials t, r and q have forms

$$t(s) = \sum_{i=0}^{n} t_i s^i, \quad r(s) = \sum_{i=0}^{n} r_i s^i, \quad q(s) = \sum_{i=1}^{n} q_i s^{i-1} \tag{56}$$

and, relations among their coefficients are

$$r_0 = t_0, \quad r_i + q_i = t_i \text{ for } i = 1, \ldots, n. \tag{57}$$

Since by a solution of the polynomial equation (50) provides calculation of coefficients t_i, unknown coefficients r_i and q_i can be obtained by a choice of selectable coefficients $\beta_i \in \langle 0,1 \rangle$ such that

$$r_i = \beta_i t_i, \quad q_i = (1 - \beta_i) t_i \text{ for } i = 1, \ldots, n. \tag{58}$$

The coefficients β_i distribute a weight between numerators of transfer functions Q and R.

Remark: If $\beta_i = 1$ for all i, the control system in Fig. 5 reduces to the 1DOF control configuration ($Q = 0$). If $\beta_i = 0$ for all i, and, both reference and load disturbance are step functions, the control system corresponds to the 2DOF control configuration.

For the second order model (34) with $\deg a = 2$, the controller's transfer functions take specific forms

$$Q(s) = \frac{q(s)}{p(s)} = \frac{q_2 s + q_1}{s + p_0}$$
$$R(s) = \frac{r(s)}{s\, p(s)} = \frac{r_2 s^2 + r_1 s + r_0}{s(s + p_0)} \tag{59}$$

where

$$r_0 = t_0, \; r_1 = \beta_1 t_1, \; r_2 = \beta_2 t_2$$
$$q_1 = (1-\beta_1)t_1, \; q_2 = (1-\beta_2)t_2 \qquad (60)$$

The controller parameters then result from a solution of the polynomial equation (50) and depend upon coefficients of the polynomial d. The next problem here is to find a stable polynomial d that enables to obtain acceptable stabilizing controllers.

In this paper, the polynomial d with roots determining the closed-loop poles is chosen as

$$d(s) = n(s)(s+\alpha)^2 \qquad (61)$$

where n is a stable polynomial obtained by spectral factorization

$$a^*(s)a(s) = n^*(s)n(s) \qquad (62)$$

and α is the selectable parameter.

Note that a choice of d in the form (61) provides the control of a good quality for aperiodic controlled processes.

The coefficients of n then are expressed as

$$n_0 = \sqrt{a_0^2} \;, \quad n_1 = \sqrt{a_1^2 + 2n_0 - 2a_0} \qquad (63)$$

and, the controller parameters p_0 and t can be obtained from solution of the matrix equation

$$\begin{pmatrix} 1 & 0 & 0 & 0 \\ a_1 & b_0 & 0 & 0 \\ a_0 & 0 & b_0 & 0 \\ 0 & 0 & 0 & b_0 \end{pmatrix} \times \begin{pmatrix} p_0 \\ t_2 \\ t_1 \\ t_0 \end{pmatrix} = \begin{pmatrix} d_3 - a_1 \\ d_2 - a_0 \\ d_1 \\ d_0 \end{pmatrix} \qquad (64)$$

where

$$d_3 = n_1 + 2\alpha, \; d_2 = 2\alpha n_1 + n_0 + \alpha^2$$
$$d_1 = 2\alpha n_0 + \alpha^2 n_1, \; d_0 = \alpha^2 n_0 \qquad (65)$$

Now, it follows from the above introduced procedure that tuning of controllers can be performed by a suitable choice of selectable parameters β and α.

The controller parameters r and q can then be obtained from (60).

The adaptive control system is shown in Fig. 4.

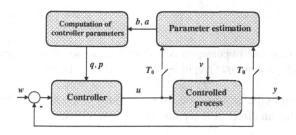

Fig. 4. Adaptive control scheme

7   Control Simulation

Also the control simulations were performed close to the operating point $q_c^s = 0.27 \ m^3/s$, $T_m^s = 334.44 \ K$, $T_{rout}^s = 326.10 \ K$. For the start (the adaptation phase), the P controller with a small gain was used in all simulations.

With respect to more expressive nonlinearity and slower dynamics of the reactant output temperature in comparison with the reactant mean temperature, the changes of references as well as the control running time intervals were chosen different for both outputs.

The effect of the pole α on the controlled responses is transparent from Fig. 5. For both outputs, two values of α were selected. The control simulations show sensitivity of controlled outputs to α. The higher values of this parameter speed the control, however, they provide greater overshoots (undershoots). Other here not mentioned simulations showed that a careless selection of the parameter α can lead to controlled output responses of a poor quality, to oscillations or even to the control instability.

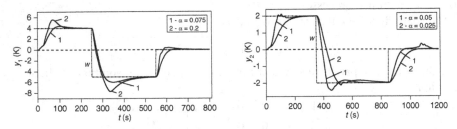

Fig. 5. Effect of α ($\beta_1 = 1$, $\beta_2 = 0.5$) on controlled output y_1 (left graph) and y_2 (right graph)

Moreover, an increasing α leads to higher values and changes of the control input as shown in Fig. 6. This fact can be important in control of real technological processes.

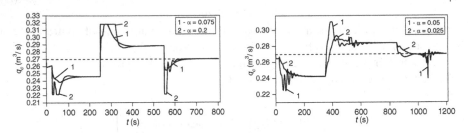

Fig. 6. Coolant flow rate responses in control of reactant mean temperature(left graph) and reactant output temperature(right graph) – effect of α ($\beta_1 = 1$, $\beta_2 = 0.5$)

The controlled output y_1 response for two values β_2 is shown in Fig. 7 (left graph). It can be seen that an effect of this parameter is insignificant.

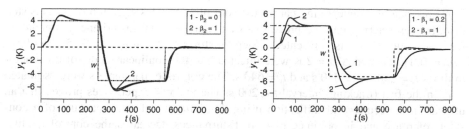

Fig. 7. Controlled output responses: effect of β_2(left graph), ($\alpha = 0.1$, $\beta_1 = 1$) and β_1(right graph), ($\alpha = 0.15$, $\beta_2 = 0$)

The controlled output responses documenting an effect of the parameter β_1 are in Fig. 7(right graph) and Fig. 8 (left graph). In both cases, a higher value of β_1 results in greater overshoots (undershoots) whereas its influence on the speed of control is inexpressive.

Corresponding control input responses can be seen in Fig. 8(right graph) and Fig. 9(left graph). There, an increasing β_1 leads to greater values of inputs, however, it can reduce occurred oscillations, as shown in Fig. 9 – left graph.

Of interest, the evolution of estimated CT ELM parameters in control of the reactant mean temperature is shown in Fig. 9 – right graph.

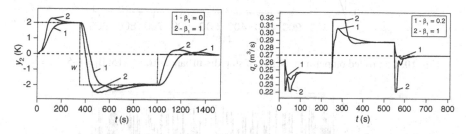

Fig. 8. Controlled output responses: effect of β_1(left graph), ($\alpha = 0.04$, $\beta_2 = 0$) and coolant flow rate responses in control of reactant mean temperature – effect of β_1 (right graph), ($\alpha = 0.15$, $\beta_2 = 0$)

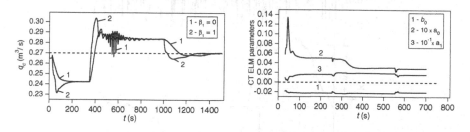

Fig. 9. Coolant flow rate responses in control of reactant output temperature – effect of β_1 ($\alpha = 0.15$, $\beta_2 = 0$) – left graph and CT ELM parameter evolution ($\alpha = 0.15$, $\beta_1 = 1$, $\beta_2 = 0$) – right graph

A presence of an integrating part in the controller enables rejection of various step disturbances entering into the process. As an example, step disturbances attenuation for the output y_1 is presented. Step disturbances $\Delta c_{A0} = 0.15$ kmol/m^3, $\Delta q_r = -0.03$ m^3/s and $\Delta T_{r0} = 2$ K were injected into the nonlinear model of the reactor in times $t_v = 220$ s, $t_v = 440$ s and $t_v = 640$ s. The controller parameters were estimated only in the first (tracking) interval $t < 200$ s. The authors' experiences proved that an utilization of recursive identification using the delta model after reaching of a constant reference and in presence of step disturbances decreases the control quality. From this reason, during interval $t \geq 200$ s, fixed parameters were used. The controlled output responses y_1 are shown in Fig. 10.

To illustrate an effect of an additive random disturbance, the result of the controlled output y_1 simulation in a presence of the random signal $v(t) = c_{A0}(t) - c_A^s$ is shown in Fig. 11.

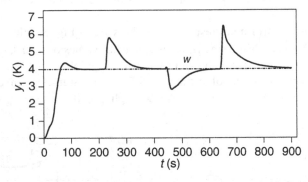

Fig. 10. Controlled output in presence of step disturbances ($\alpha = 0.15$, $\beta_1 = 0.5$, $\beta_2 = 0$)

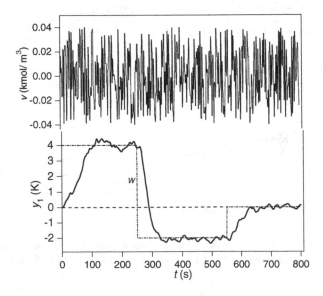

Fig. 11. Controlled output in the presence of random disturbance in c_{A0} ($\alpha = 0.15$)

8 Conclusions

In this paper, one approach to continuous-time adaptive control of nonlinear distributed parameter processes was proposed. As a typical representative of this class of processes, the tubular chemical reactor was considered. The control strategy is based on the preliminary steady-state and dynamic analysis of the process and on the assumption of the temperature measurement along the reactor. The proposed algorithm employs an alternative continuous-time external linear model with parameters obtained through recursive parameter estimation of a corresponding delta model. The control system structure with two feedback controllers is considered. Resulting continuous-time controllers are derived using the polynomial approach and given by a solution of the polynomial equation. Tuning of their parameters is possible via closed-loop pole assignment. The presented method has been tested by computer simulation on the nonlinear model of the tubular chemical reactor with a consecutive exothermic reaction. The simulation results demonstrate an applicability of the presented control strategy.

References

1. Luyben, W.: Process modelling, simulation and control for chemical engineers. McGraw-Hill, New York (1989)
2. Ingham, J., Dunn, I.J., Heinzle, E., Přenosil, J.E.: Chemical Engineering Dynamic: Modelling with PC Simulation. VCH Verlagsgesellschaft, Weinheim (1994)
3. Corriou, J.P.: Process Control. Theory and Applications. Springer, London (2004)
4. Dostál, P., Bobál, V., Vojtěšek, V.: Simulation of steady-state and dynamic behaviour of a tubular chemical reactor. In: Proc. 22nd European Conference on Model-ling and Simulation, Nicosia, Cyprus, pp. 487–492 (2008)
5. Dostál, P., Bobál, V., Blaha, M.: One approach to adaptive control of non-linear processes. In: Proc. IFAC Workshop on Adaptation and Learning in Control and Signal Processing, ALCOSP 2001, Cernobbio-Como, Italy, pp. 407–412 (2001)
6. Dostál, P., Bobál, V., Gazdoš, F.: Adaptive control of nonlinear processes: Continuous-time versus delta model parameter estimation. In: IFAC Workshop on Adap-tation and Learning in Control and Signal Processing, ALCOSP 2004, Yokohama, Japan, pp. 273–278 (2004)
7. Rao, G.P., Unbehauen, H.: Identification of continuous-time systems. IEE Proc.-Control Theory Appl. 152, 185–220 (2005)
8. Middleton, R.H., Goodwin, G.C.: Digital Control and Estimation - A Unified Approach. Prentice Hall, Englewood Cliffs (1990)
9. Mukhopadhyay, S., Patra, A.G., Rao, G.P.: New class of discrete-time mod-els for conti-nuos-time systems. International Journal of Control 55, 1161–1187 (1992)

10. Stericker, D.L., Sinha, N.K.: Identification of continuous-time systems from samples of input-output data using the d-operator. Control-Theory and Advanced Technology 9, 113–125 (1993)

11. Dostál, P., Gazdoš, F., Bobál, V., Vojtěšek, J.: Adaptive control of a contin-uous stirred tank reactor by two feedback controllers. In: Proc. 9th IFAC Workshop Adaptation and Learning in Control and Signal Processing, ALCOSP 2007, Saint Petersburg, Russia, pp. P5-1–P5-6 (2007)

12. Kučera, V.: Diophantine equations in control – A survey. Automatica 29, 1361–1375 (1993)

13. Bobál, V., Böhm, J., Fessl, J., Macháček, M.: Digital Self-tuning Controllers. Springer, Berlin (2005)

Laboratory Systems Control
with Adaptively Tuned Higher Order Neural Units

Ivo Bukovsky, Peter Benes, and Matous Slama

Department of Instrumentation and Control Engineering
Czech Technical University in Prague, Prague, Czech Republic
{ivo.bukovsky;petermark.benes;matous.slama}@fs.cvut.cz

Abstract. This paper summarizes the design theory of linear and second order polynomial adaptive state-feedback controllers for SISO systems using the Batch-propagation Through Time (BPTT) learning algorithm. Deeper focus is given towards real time implementation on various laboratory experiments, with an accompaniment of corresponding theoretical simulations, to demonstrate the feasibility of use of polynomial adaptive state-feedback controllers for real time control. Raspberry Pi and open-source scripting language Python are also exhibited as a suitable implementation platform, for both testing and rapid prototyping as well as for teaching of adaptive identification and control.

Keywords: Adaptive Control, Higher-Order Neural Units (HONUs), Gradient Descent (GD), Batch-Propagation Through Time (BPTT), Raspberry Pi.

1   Introduction

Non-linear neural units as a class of HONUs (Higher-Order Non-linear Neural Units) are polynomial neural models which belong within the scope of polynomial neural networks [1]-[7]. Amongst their advantages in comparison to conventional neural networks, belongs the adjustable quality of non-linearization whilst at the same time being linear in parameter structure, furthermore having the consequence of good convergence during learning relatively simple algorithms as such that of the incremental gradient decent (GD) algorithm and Levenberg-Marquardt (L-M), or their modification as a version for recurrent adaptive models [8]-[10]. A study of utilisation of HONUs, first, through to third order i.e; linear, quadratic and cubic neural units (LNU, QNU and CNU) respectively may be found in the works [8] & [12]-[13]. Recently foundations on stability of the GD algorithm for static and dynamic HONUs for dynamic system identification, where stability of the adaptive algorithm in every step is evaluated via correspondence with a calculated spectral radius of the matrix of dynamics of weights of the adaptive system can be found in the work [23]. Till now studies of HONUs as a means for online control of Single-Input-Single-Output (SISO) engineering systems, have been rather theoretically focussed, in exception we may note the practical implementation of a statically tuned system model as well as

© Springer International Publishing Switzerland 2015   275
R. Silhavy et al. (eds.), *Intelligent Systems in Cybernetics and Automation Theory*,
Advances in Intelligent Systems and Computing 348, DOI: 10.1007/978-3-319-18503-3_27

online implemented HONU feedback controller, for control of a laboratory Bathyscaphe system [13]. Thus, this paper aims to present the usage of HONUs, as recurrent models for dynamic system identification and furthermore, as a controller. A key contribution being to exhibit the use of offline adaptively trained dynamic neural models for system identification (contrary to [13]), followed by static application of an extended HONU state feedback controller, as a model with constant parameters, for control. A further added contribution of this paper is also to show case the usability of the open-source programming language Python along with the low cost utilisation of Raspberry Pi, as a reliable experimental platform, for prototyping and testing real time dynamical system control, as well as being a suitable hardware medium for tertiary based educational purposes.

2   Applied Methods

In this section polynomial function based HONUs in the sense of adaptive models for dynamical system identification and control, will be recalled. We will direct the focus of this paper to SISO systems, which may be identified and further adaptively controlled via use of the famous GD, and L-M algorithm in the sense of a batch form of neural weight training.

The mathematical structure of the HONU with further non-linear degree of order r may be written in the following long vector representation.

$$y = \sum \left\{ w_{i,j,...,r} \cdot x_i \cdot x_j \cdots x_r \right\} \quad \text{where} \quad i=0...n, \quad j=i...n, \quad ..., \quad \text{a} \quad x_0 =1 \tag{1}$$

Which may be written in abbreviation of the above long vector representation, as follows,

$$y = \mathbf{w} \cdot \mathbf{colx} \tag{2}$$

where y is the output of the HONU model and the long vector representation of the weights is,

$$\mathbf{w} = \left[\left\{ w_{i,j,...,r} \right\} \right] \quad \text{where} \quad i = 0...n, j = i...n, \quad ..., \tag{3}$$

Furthermore, the long column vector of HONU inputs, is given by the following form

$$\mathbf{colx} = \left[\left\{ x_i \cdot x_j \cdots x_r \right\} \right] \quad \text{where} \quad i = 0...n, j = i...n, ..., \tag{4}$$

Given the input vector of previous measured samples of the engineering process, represented via the following vector $\mathbf{x}$

$$\mathbf{x} = \left[x_0 = 1 \ x_1 \ x_2 \ ..x_n \right]^T = \left[1 \ y(k) \ y(k-1) \ y(k-n) \ u(k) \ ..u(k-n) \right]^T \tag{5}$$

In the sense of static neural units, the input vector **x** contains only external inputs, (in other words, only stepwise delayed values of the desired value or setpoint d, input value u or measured values of the real system yr). In the case of dynamic neural units, the input vector **x,** contains besides these stepwise unit delayed measured values, also the previous values of y, being the output of the neural model itself. As for the adaptive identification of the system (generally SISO systems), it is necessary to employ the incremental GD weight update rule, with as a further extension, the incorporation of a normalized learning rate. However, as the focus of this paper is driven through implementation of a batch form of neural weight training, being that of the BPTT method. Eq. (6) recalls the structure behind this method, as follows,

$$\Delta \mathbf{w} = (\mathbf{J}^T.\mathbf{J} + \frac{1}{\mu}.\mathbf{I})^{-1}.\mathbf{J}^T.\mathbf{e} \tag{6}$$

where the neural weights are updated via the following form $\mathbf{w}=\mathbf{w}+\Delta\mathbf{w}$. Here J is the Jacobian matrix of partial derivatives with μ, representing the learning rate of the weight update rule. In extension, the employed learning rate may be normalised for stability during the learning process of identification for the neural weights. Stability of the adaptive algorithm without normalization is analogically derived in the work [23] and the resulting positive influence of normalization on the stability of the learning algorithm may be recalled in the work [6]. A further property behind the use of BPTT is the suppression of noise during HONU identification and further control [7]-[8]. This is due to the algorithm focussing primarily on the main governing law behind the experimental data rather than contemporary dynamics of the system. After, and furthermore during identification of the dynamical system at hand, via such HONU, that means via models as per (1)-(5), we may extend a further HONU as an adaptive feedback controller. For this, we offer the principle denoted in the following scheme, Fig. 1.

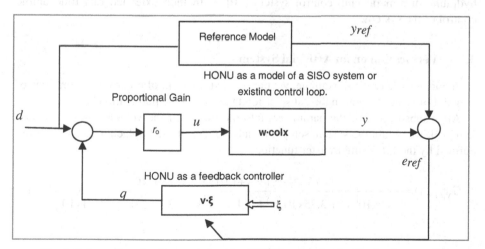

Fig. 1. Scheme for adaptive control with a HONU as an adaptive feedback controller

As per Fig. 1, inputs to the feedback controller ξ must be designed in accordance with a concrete case. Analogically to Eq. (5), ξ may be constructed via several delayed steps of the controlled value yr, and furthermore, its deviation with the employed reference model, denoted as e_{ref} (as per Fig. 1).

From this, it is then possible to test whether an adaptive feedback controller model via implementation of the GD algorithm or a modification with the L-M algorithm as a form of batch training (BPTT) of the adaptive models neural weights, is more suitable (for quick testing of such HONU configurations, it is possible to utilize the Python application presented in the work [14]). The incremental adaptive update rule of the HONU as a feedback controller is dictated via the following extension of the GD rule, with Eq. (8), being an adaptive gain for processes with non-constant gain across its range of desired values.

$$v_{i,j,\ldots r}(k+1) = v_{i,j,\ldots r}(k) + \mu_v \cdot e_{ref}(k) \cdot \frac{\partial y(k)}{\partial v_{i,j,\ldots r}} \tag{7}$$

$$r_0(k+1) = r_0(k) + \mu_{r_o} \cdot e_{ref}(k) \cdot \frac{\partial y(k)}{\partial r_o} \tag{8}$$

where r_o is the proportional gain parameter also adapted incrementally via GD or via a batch form of training with extension of the L-M algorithm, analogically to Eq (6).

3   Experimental Analysis

In this section we aim to demonstrate the functionality of the algorithm primarily through two online laboratory control tasks [20] & [22]. In addition, we aim to draw the parallels of HONU properties via a comparison of a theoretical simulation, via an active car automobile model in publication [19] as well as offline implementation of a hydraulic-pneumatic tank control system [20], with then extended real time online control of the system.

3.1   Verification on an Artificial System

As a concrete theoretical example, we present a simulation of control for vibrations of a model of an active automobile suspension taken from the publication [19].

After substitution of the parameters into the model of the vehicle suspension system [19], we obtain the system setup after adjustment of the Laplacian transformation, defined by the following transfer function.

$$G_{y_r,u}(s) = \frac{(1.15\,s^2 + 0.146s + 1.01\mathrm{x}10^3) \cdot 3.33\mathrm{x}10^{-2}}{3.45\mathrm{x}10^1\,s^4 + 3.75\mathrm{x}10^2 s^3 + 4.48\mathrm{x}10^4 s^2 + 3.71\mathrm{x}10^4 s + 1.44\mathrm{x}10^6} \tag{9}$$

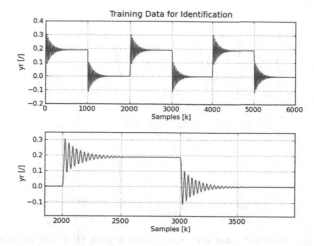

Fig. 2. Training data and identification of the system in (9) dynamic LNU trained by BPTT (HONU, r=1, green line), vehicle suspension system (blue). Superimposition in the lower graph implies accuracy in identification

Where the controlled value yr simulates the distance of the vehicle tyre from the body of the vehicle itself and u is representing the action interference, into the rigidity of the stiffness of the suspension. Regarding the scheme of control for the system (9), a cascade PI controller is implemented for further steady state error minimisation. Following successful identification of the vehicle suspension system (Fig. 2), via an LNU trained with a BPTT training method. Fig. 3 depicts the substantial enhancement on the control performance of the initial cascade PI controller setup, with extension of a state feedback LNU with constant parameters, identified via BPTT training. In this simulation the introduction of several perturbations, interfering with the dynamical output of the system, were introduced to test the capability of the LNU control algorithm. In spite this complex linear system, the LNU (HONU, r=1) as a state feedback controller of constant parameters is able to compensate these perturbations and return to close vicinity of the desired behaviour, even with a perturbation of larger magnitude as at the 3000th sample, thus validating the performance of such HONU controller.

3.2   Verification on Laboratory Systems

Following adequate functionality of the previous theoretical simulation, we extend the application of HONUs to two real time experimental laboratory tasks. The first of which being the optimisation to the already adequate control circuit of the water levitation laboratory task [21], with a PI controller to which a QNU with constant parameters, identified by the BPTT algorithm on the training data (desired and controlled value) of the original PI control loop, is extended.

Following online application as in Fig. 4, the QNU controller not only adheres to the desired behaviour of the water levitation experiment more closely, but further to

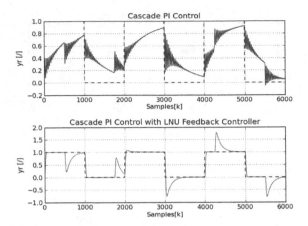

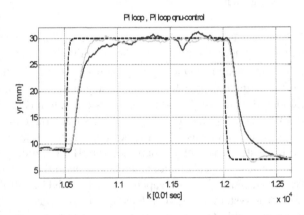

Fig. 3. Control of a simulated automotive suspension system [19] with introduced perturbations. Above: Control loop with cascade PI control only. Below: Cascade PI controller with extension of an LNU with constant parameters identified via the BPTT algorithm

Fig. 4. Further optimization of the laboratory task [21], using a PI controller with extension of a QNU, identified via BPTT algorithm. Blue = PI controlled system only, Green= PI controlled system with extension of a QNU feedback controller

this, achieves a faster response to the required steady state of the experiment. This property is key to the previously tested theoretical contributions of the work [12], where both the quadratic and cubic HONU, respectively exhibits faster performance in achieving a settled desired output, in comparison to a HONU of first order (r=1). What may also be noted is that the QNU implemented as a constant parameter feedback controller, being a model linear in its parameters structure is computationally as expensive as extending another linear based controller to this control setup up, yet still exhibits substantially improved performance in comparison to the conventional PI implementation. A further area of study to this, would be comparison of online adaptation of the neural weights for the feedback HONU controller, here according to theory, in prolonged application, such setup should further minimise the error between the desired behaviour and real system output, and increasingly do so over continuous repetition of the desired values for operation of the engineering system. This however, deserves a further publication with implementation of constant parameter state feedback HONUs, itself being a wide enough area of focus for this publication.

A similar notion may be exhibited through the implementation of HONU adaptive control on the laboratory task of a hydraulic-pneumatic system [20] & [22]. In addition to the justification of HONU adaptive control performance for SISO laboratory systems, this experimental setup, has been implemented via a rather new technological setup, being that of the a cost effective mini computer Raspberry Pi (RPi) [15] with an operating system based on Linux and an open source programming language Python [16] with libraries *Numpy* [17] and *Matplotlib* [18]. For implementation of this method, the type B model of RPi was used. This model features an ARM1176JZF-S (700 MHz) processor, with graphic processing card from VideoCore IV and 512 MB of RAM.

Here we utilise a software implemented PI controller and an extended adaptive state feedback controller QNU (HONU, r=2). RPi is thus used for the physical implementation of control, where the actual programmed algorithm is implemented via the programming language Python. The Input value u into the system set up, is realised through the revolutions of the pump from which the storage tank is pumping fluid into the upper tank. The real output y which is the height of the level in the lower tank from the system, is measured by a differential sensor TMDG 338 Z3H, where its height level is a function of the pressure. For connection of RPi to the system, we utilise the GertBoard development board from which two moduli are used, explicitly a 10 Bit AD converter for the voltage signal of the pressure sensor and the PWM module for control of the pump.

Fig. 5 & Fig. 6, depicts the adaptive control of both offline and online implementation of adaptive control on the hydraulic-pneumatic system [20], via HONUs. Prior to real time online application, Fig. 5 depicts one of the most significant offline tuned HONUs as applied to the measured data of the hydraulic-pneumatic system. Here following adequate identification via a dynamic QNU (DQNU) with RTRL training, a QNU featuring one previous sample of the identified plant output and 3 previous samples of the deviation between the desired behaviour of the PI controlled plant and identified plant output, serves as an input vector of the adaptive QNU controller neural model. The training of which, was achieved via the BPTT algorithm over 5 epochs or runs of the algorithm (via use of the software application presented in [14]). Whilst the offline tuned QNU in Fig. 5 adheres almost identically with the required behaviour of the system, this controller is indeed irrespective of the natural factors of actuator limitation and furthermore, non-linearity that may be present within the sensors used for the control loop. This result nevertheless, serves as a promising indicator for the most optimum HONU adaptive controller setup to be applied in real time for the given experimental setup.

Fig. 6 thus, shows the resulting QNU applied as an online state feedback controller in extension to the previously tuned PI controller setup. Here the QNU feedback

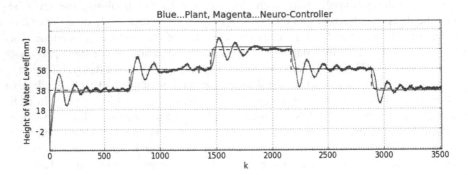

Fig. 5. Offline QNU adaptive control of the hydraulic-pneumatic system [20] system (Magenta = QNU with BPTT training, Blue=Real PI controlled plant data), using Raspberry Pi and Python programming language.

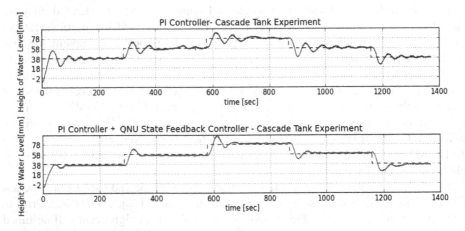

Fig. 6. Implementation of real time, online adaptive control of the hydraulic-pneumatic system [20], via an online QNU with constant parameters (magenta) as an extension with a PI controller setup (blue), using Raspberry Pi and Python programming language.

controller, applied online with constant parameters, tuned via the offline results in Fig. 5, featuring one sample of the previously identified plant output and 3 previous samples of the error between the desired behaviour and identified model output (the sampling time being 0.4 seconds), is employed. Fig. 6 depicts substantially closer adhesion to the desired behaviour (black dashed line) as compared to the previously implemented PI controller, although not as ideal as the results in Fig. 5, the QNU feedback controller in Fig. 6 indeed performs substantially better than the previous PI controller in adhering to the desired steady state. Recalling Fig. 3 & Fig. 4 we may observe that the HONU controllers implemented, achieve not only close adhesion to the desired behaviour, but also faster response to the desired behaviour of the controlled system. It is thus an interesting point of study as to the rather significant overshooting which still takes place on the hydraulic-pneumatic system after extension of the QNU feedback controller. In particular on higher water levels, the extended QNU controller exhibits slightly different performance, delivering slightly larger degree of overshooting as compared with that of control in the vicinity of lower levels, which may be thus due to the non-linear nature of such non-linear control loop as implemented via this configuration and is thus another remark for future study. In spite this, the HONU feedback controller derived via offline tuning, through use of a dynamic QNU model for system identification, proved to be a substantially better behaving controller in real implementation in comparison to the conventional, linear PI controller of which was previously employed.

4   Conclusion

This paper exhibited through use of both a theoretical simulation and two real experimental tasks, the versatility and usability of HONUs as state feedback controllers, particularly in the sense of dynamically offline tuned HONUs as a model for the applied system, with extension of an online HONU controller, of constant parameters in feedback. From the two real experimental tasks presented, application of the offline tuned HONU neural weights with a real time implemented HONU as a controller of constant parameters, deemed to indeed improve the performance of the previously applied conventional PI controller setups. As in both the offline and online experimental results, the employed HONUs not only adhered more closely to the required system output but also improved its speed of convergence towards the settled desired behaviour. As a further remark, this paper also proves the usability of a low cost hardware platform with an open source development environment as a suitable means for not only testing control based algorithms on real experimental setups, but furthermore also as a suitable means for tertiary based educational purposes.

Acknowledgements. The authors of this paper would like to acknowledge the following study grant for its support SGS12/177/OHK2/3T/12

References

1. Ivakhnenko, A.G.: Polynomial Theory of Complex Systems. IEEE Tran. on Systems. Man, and Cybernetics 1(4), 364–378 (1971)
2. Taylor, J.G., Coombes, S.: Commbes, Learning higher order correlations. Neural Networks 6, 423–428 (1993)
3. Kosmatopoulos, E., Polycarpou, M., Christodoulou, M., Ioannou, P.: High-Order Neural Network Structures for Identification of Dynamical Systems. IEEE Trans. on Neural Networks 6(2), 422–431 (1995)
4. Nikolaev, N.Y., Iba, H.: Learning Polynomial Feed-forward Neural Network by Genetic Programming and Back-propagation. IEEE Trans. on Neural Networks 14(2), 337–350 (2003)

5. Nikolaev, N.Y., Iba, H.: Adaptive Learning of Polynomial Networks: Genetic Programming, Back-propagation and Bayesian Methods. In: Genetic and Evolutionary Computation, vol. XIV, 316 p. Springer, New York (2006) ISBN: 0-387-31239-0
6. Gupta, M.M., Liang, J., Homma, N.: Static and Dynamic Neural Networks: From Fundamentals to Advanced Theory. IEEE Press and Wiley-Interscience, John Wiley & Sons, Inc. (2003)
7. Gupta, M.M., Bukovsky, I., Homma, N., Solo, M.G.A., Hou, Z.-G.: Fundamentals of Higher Order Neural Networks for Modelling and Simulation. In: Zhang, M. (ed.) Artificial Higher Order Neural Networks for Modelling and Simulation. IGI Global (2012)
8. Bukovsky, I., Homma, N., Smetana, L., Rodriguez, R., Mironovova, M., Vrana, S.: Quadratic Neural Unit is a Good Compromise between Linear Models and Neural Networks for Industrial Applications. In: The 9th IEEE International Conference on Cognitive Informatics, ICCI 2010, July 7-9. Tsinghua University, Beijing (2010)
9. Williams, R.J., Zipser, D.: A learning algorithm for continually running fully recurrent neural networks. Neural Comput. 1, 270–280 (1989)
10. Williams, R.J., Peng, J.: Gradient-based learning algorithms for recurrent neural networks and their computational complexity. In: Chauvin, Y., Rumelhart, D.E. (eds.) Back-Propagation: Theory, Architectures, and Applications, pp. 433–486. Lawrence Erlbaum, Hillsdale (1992)
11. Mandic, D.P.: Generalised normalised gradient descent algorithm. IEEE Signal Process. Lett. 11, 115–118 (2004)
12. Bukovsky, I., Redlapalli, S., Gupta, M.M.: Quadratic and Cubic Neural Units for Identification and Fast State Feedback Control of Unknown Non-Linear Dynamic Systems. In: Fourth Int. Symp. on Uncertainty Modelling and Analysis, ISUMA 2003, pp. 330–334. IEEE Computer Society, Maryland (2003) ISBN 0-7695-1997-0
13. Smetana, L.: Nonlinear Neuro-Controller for Automatic Control Laboratory System. Master's Thesis, Czech Tech. Univ. in Prague (2008) (in Czech)
14. Benes, P.M.: Software Application for Adaptive Identification and Controller Tuning Student's Conference STC, Faculty of Mechanical Engineering, CTU in Prague (2013)
15. Raspberry Pi Foundation, http://www.raspberrypi.org/about (visited January 2014)
16. Van Rossum, G., de Boer, J.: Linking a stub generator (AIL) to a prototyping language (Python). In: Proceedings of the Spring 1991 Europen Conference, Troms, Norway, May 20-24 (1991)
17. NumPy, package for scientific computing with Python, http://www.numpy.org/ (visited January 2014)
18. Hunter, J.D.: Matplotlib: A 2D Graphics Environment. Computing in Science and Engineering 9(3), 90–95 (2007), doi:10.1109/MCSE.2007.55
19. Li, H., Liu, H., Gao, H., Shi, P.: Reliable Fuzzy Control for Active Suspension Systems with Actuator Delay and Fault. IEEE Transactions on Fuzzy Systems 20(2) (April 2012)
20. Hydro-pneumatic laboratory system, Automatic Control Laboratory (111), Czech Technical University in Prague,
http://www1.fs.cvut.cz/cz/u12110/ar/index_c.htm
(visited January 2014)
21. Water levitation, Automatic Control Laboratory (111), Czech Technical University in Prague, http://www1.fs.cvut.cz/cz/u12110/ar/ulohy/
levitace.htm (visited January 2014)
22. Sláma, M.: Using Minicomputer Raspberry Pi for Adaptive Identification and Control of Hydro-pneumatic Laboratory System with Python. Diploma Thesis, FS, ČVUT in Prague, defended (January 2014) (in Czech)
23. Bukovský, I., Rodriguez, R., Bíla, J., Homma, N.: Prospects of Gradient Methods for Nonlinear Control. Strojárstvo Extra 5(5), art. no. 36, 1–5 (2012) ISSN 1335-2938

Using Simulink in Simulation of Dynamic Behaviour of Nonlinear Process

Jiri Vojtesek and Petr Dostal

Tomas Bata University in Zlin, Faculty of Applied Informatics
Nam. T.G.Masaryka 5555, 760 01 Zlin, Czech Republic
vojtesek@fai.utb.cz
http://www.utb.cz/fai

Abstract. The contribution shows benefits of using mathematical software Matlab and its add-on for graphical programming Simulink. There are shown two ways of using Simulink for simulating of the dynamic behaviour of the nonlinear system. The first method is used for those who have already done simulating program in Matlab and need to implement it inside Simulink scheme. The second option is to use special S-function used for simulink. Proposed methods are tested on the simulation of the dynamic behaviour of the mathematical model of the continuous stirred-tank reactor as a typical nonlinear equipment from the industry.

Keywords: Modelling, Simulation, Matlab, Simulink, Dynamic Behaviour, CSTR, Runge-Kutta's method.

1   Introduction

It is known that the role of the computer simulation grows very rapidly with the increasing options and computation power of the computation technology not only in the sphere of the personal computers but also in the industrial computers and microcontrollers [1].

The simulation is also connected with the modelling as a predecessor of the simulation [2]. The essential task of the modelling is to find as accurate model as possible but the maximal reduction of the complexity of the system. Simulation play role in the modelling as a verifier it shows behaviour of the proposed model and the results are then compared with the results from the measurements on the real process and if results are comparable, we can say that the designed mathematical model is good abstract representation of the real process [3]. If characteristics are different, we must come back to the modelling part and find the new model.

There can be used several mathematical software for simulations. The most famous are for example Matlab [4] or Wolfram's Mathematica [5]. The advantage of the Matlab is that the programming code is very similar to the C language which means that the created and verified M-file could be then transformed to the C language and compiled to the executable program. This program then should run on industrial computer or microcontroller without the use of Matlab

© Springer International Publishing Switzerland 2015
R. Silhavy et al. (eds.), *Intelligent Systems in Cybernetics and Automation Theory*,
Advances in Intelligent Systems and Computing 348, DOI: 10.1007/978-3-319-18503-3_28

himself. The second way is the use of Matlab compiler which can also create executable program directly from the M-file.

We can imagine the simulation of the behaviour of the mathematical model as a mathematical solution of one or the set of algebraic or differential linear or nonlinear equations depending on the type of model we have [2]. There are several methods for numerical solution of the ordinary differential equations (ODE) like Euler's method, Runge-Kutta's methods [6] etc. The big advantage of the Runge-Kutta's methods is that they can be found as a build-in function in Matlab [7] and also Mathematica which means that the user do not need to know functionality of these methods, only how to input the mathematical model and set the initial conditions and parameters of the simulation.

Matlab shifts simulation even on higher level from education point of view with the use of special tool called Simulink where programming is represented by the connections of the graphical blocks, sources, sinks etc. into one scheme where each block could be set individually [8].

Matlab and Simulink also provides tools for creating of the controller which is usually the next step after the simulation and verification of the mathematical model. Moreover, they can be used for controlling of the real process with the use of Real-Time toolbox.

This contribution shows features of the Matlab and Simulink in the field of simulation of the dynamic behaviour of the nonlinear lumped-parameters system represented by the Continuous Stirred-Tank Reactor (CSTR) with the cooling in the jacket [9]. The mathematical model of such process consist of the set of four Ordinary Differential Equations (ODE) which are solved numerically with the use of Runge-Kutta's fourth order method.

All simulations performed in this paper are done in the mathematical software Matlab, version 7.0.1.

2   Modeling and Simulation Methods

Modelling and simulation are usually the first step in the designing of the suitable controller for the system. It is common, that the behaviour of the industrial processes is unknown and we expect them as a *black boxes* which means that we do not know internal structure of the system, but only input-output measurements. Mathematical modelling and computer simulation is one way how we can overcome this uncertainty.

If we talk about the model of the system we usually mean *mathematical model* which is the mathematical descrition of the system. This description comes from the balances inside the system and result in the set of algebraic, differential or partial differential equations depending on the type of system.

Very important parts in this procedure of model creation is introducing of the simplifications. Unfortunately, most of the processes has nonlinear behaviour with a lot of variables which can vary in time or space variable. Common simplification is that we expect that some of these variables which are not state variables or vary in small range are expected to be constant during the simulations. The next important step is verification of the simplificated model by the measurements on the real system. We can say that we have found good mathematical model when simulation results are acceptable according to the results from real measurements.

When we have appropriate model, the simulation analyses follow. The static analysis displays behaviour of the system in the system in the steady-state. This study results in the optimal working point. On the other hand, the dynamic analysis provides step, frequency responses etc. which display dynamic behaviour of the system and they are for the choice of the optimal control strategy and designing of the controller. Numerical mathematics is widely used in the solution of these two analyses.

The dynamic behaviour of the observed system often means numerical solution of the ordinary or partial differential equations. The numerical method used in this work is Runge-Kutta's 4th order method. The big advantage of this method is that it is easily programmable or even build-in function in the mathemacal software like Mathematica or Matlab.

This method comes from the Taylor's series

$$y_{i+1} = y_i + \frac{dy}{dx}\Big|_{x_i,y_i} (x_{i+1} - x_i) + \frac{1}{2!} \frac{d^2y}{dx^2}\Big|_{x_i,y_i} (x_{i+1} - x_i)^2 + \ldots$$
$$\ldots + \frac{1}{3!} \frac{d^3y}{dx^3}\Big|_{x_i,y_i} (x_{i+1} - x_i)^3 + \ldots \tag{1}$$

which could be rewritten to the form

$$y_{i+1} = y_i + f(x_i, y_i)(x_{i+1} - x_i) + \frac{1}{2!} f'(x_i, y_i)(x_{i+1} - x_i)^2 + \ldots$$
$$\ldots + \frac{1}{3!} f''(x_i, y_i)(x_{i+1} - x_i)^3 + \ldots \tag{2}$$

The Runge-Kutta's 4th order method uses first five parts of this series which can be transferred to the well-known form:

$$y_{i+1} = y_i + \frac{1}{6} \cdot (k_1 + 2k_2 + 2k_3 + k_4) \tag{3}$$

where variables k_{1-4} are computed from

$$
\begin{aligned}
k_1 &= h \cdot f(x_i, y_i) \\
k_2 &= h \cdot f\left(x_i + \tfrac{1}{2}h, y_i + \tfrac{1}{2}k_1\right) \\
k_3 &= h \cdot f\left(x_i + \tfrac{1}{2}h, y_i + \tfrac{1}{2}k_2\right) \\
k_4 &= h \cdot f(x_i + h, y_i + k_3)
\end{aligned}
\tag{4}
$$

3   Simulated System

The system under the consideration is a Continuous Stirred-Tank Reactor (CSTR) with the so called *Van der Vusse reaction* $A \to B \to C, 2A \to D$ inside and cooling jacket - see the scheme of the CSTR in Fig. 1.

If we introduce common simplifications like the perfect mixture of the reactant, all densities, transfer coefficients, heat capacities and the volume of the

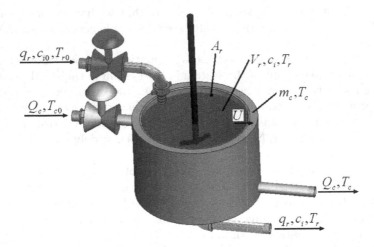

Fig. 1. Continuous Stirred Tank Reactor with Cooling in the Jacket

reactant are constant throughout the reaction, the mathematical model developed with the use of material and heat balances inside has form of the set of Ordinary Differential Equations (ODEs) [9]

$$
\begin{aligned}
\frac{dc_A}{dt} &= \frac{q_r}{V_r}\left(c_{A0} - c_A\right) - k_1 c_A - k_3 c_A^2 \\
\frac{dc_B}{dt} &= -\frac{q_r}{V_r} c_B + k_1 c_A - k_2 c_B \\
\frac{dT_r}{dt} &= \frac{q_r}{V_r}\left(T_{r0} - T_r\right) - \frac{h_r}{\rho_r c_{pr}} + \frac{A_r U}{V_r \rho_r c_{pr}}\left(T_c - T_r\right) \\
\frac{dT_c}{dt} &= \frac{1}{m_c c_{pc}}\left(Q_c + A_r U\left(T_r - T_c\right)\right)
\end{aligned}
\tag{5}
$$

where t in is the time, c are concentrations, T represents temperatures, c_p is used for specific heat capacities, q_r means the volumetric flow rate of the reactant, Q_c is the heat removal of the cooling liquid, V_r is volume of the reactant, ρ stands for densities, A_r is the heat exchange surface and U is the heat transfer coefficient. Indexes $(\cdot)_A$ and $(\cdot)_B$ belong to compounds A and B, respectively, $(\cdot)_r$ denotes the reactant mixture, $(\cdot)_c$ cooling liquid and $(\cdot)_0$ are feed (inlet) values.

The variable h_r and k_{1-3} in (1) denotes the reaction heat and reaction rates which are computed from

$$
\begin{aligned}
h_r &= h_1 \cdot k_1 \cdot c_A + h_2 \cdot k_2 \cdot c_B + h_3 \cdot k_3 \cdot c_A^2 \\
k_j\left(T_r\right) &= k_{0j} \cdot \exp\left(\frac{-E_j}{RT_r}\right), \text{ for } j = 1, 2, 3
\end{aligned}
\tag{6}
$$

where h_i stands for reaction enthalpies. Reaction rates k_{1-3} in the second equation are nonlinear functions of the reactants temperature computed via *Arrhenius law* with k_{0j} as rate constants, E_j are activation energies and R means gas constant.

Equations (5) together with (6) construct the *mathematical model of the plant* used later for simulation studies. Due to simplifications introduced above we can say, that this type of reactor is *a nonlinear lumped-parameters system*. We have four state variables c_A, c_B, T_r and T_c and four input variables the volumetric flow rate of the reactant, q_r, the heat removal of the coolant, Q_c, the input concentration c_{A0} and input temperature of the reactant, T_{r0}. The fixed values of the reactor are shown in Table 1 [9].

Table 1. Fixed parameters of the CSTR

Name of the parameter	Symbol and value of the parameter
Volume of the reactant	$V_r = 0.01 m^{-3}$
Density of the reactant	$\rho_r = 934.2 kg.m^{-3}$
Heat capacity of the reactant	$c_{pr} = 3.01\ kJ.kg^{-1}.K^{-1}$
Weight of the coolant	$m_c = 5\ kg$
Heat capacity of the coolant	$c_{pc} = 2.0\ kJ.kg^{-1}.K^{-1}$
Surface of the cooling jacket	$A_r = 0.215\ m^2$
Heat transfer coefficient	$U = 67.2\ kJ.min^{-1}.m^{-2}.K^{-1}$
Pre-exponential factor for reaction 1	$k_{01} = 2.145 \cdot 10^{10}\ min^{-1}$
Pre-exponential factor for reaction 2	$k_{02} = 2.145 \cdot 10^{10}\ min^{-1}$
Pre-exponential factor for reaction 3	$k_{03} = 1.5072 \cdot 10^8\ min^{-1}.kmol^{-1}$
Activation energy of reaction 1 to R	$E_1/R = 9758.3\ K$
Activation energy of reaction 2 to R	$E_2/R = 9758.3\ K$
Activation energy of reaction 3 to R	$E_3/R = 8560\ K$
Enthalpy of reaction 1	$h_1 = $ -4200 $kJ.kmol^{-1}$
Enthalpy of reaction 2	$h_2 = 11000\ kJ.kmol^{-1}$
Enthalpy of reaction 3	$h_3 = 41850\ kJ.kmol^{-1}$
Input concentration of compound A	$c_{A0} = 5.1\ kmol.m^{-3}$
Input temperature of the reactant	$T_{r0} = 387.05\ K$

The step changes of these inputs are later used in dynamic analysis. The four changes are described by:

$$u_1(t) = \frac{Q_c(t)-Q_c^s}{Q_c^s} \cdot 100 \quad [\%] \quad u_2(t) = \frac{c_{A0}(t)-c_{A0}^s}{c_{A0}^s} \cdot 100 \quad [\%]$$
$$u_3(t) = \frac{T_{r0}(t)-T_{r0}^s}{T_{r0}^s} \cdot 100 \quad [K] \quad u_4(t) = \frac{q_r(t)-q_r^s}{q_r^s} \cdot 100 \quad [\%] \tag{7}$$

The working point is in this case defined by values:

$$c_{A0}^s = 5.1 \quad kmol.m^{-3} \qquad T_{r0}^s = 378.05 K$$
$$q_r^s = 2.365 \cdot 10^{-3} \quad m^3.min^{-1} \qquad Q_c^s = -18.56 kJ.min^{-1} \tag{8}$$

4 Matlab - Simulink

As it is written above, this paper deals is focused mainly on the simulation methods used in mathematical software Matlab and its special graphical language Simulink. This tool offers wide range of the functions and it is widely used for teaching and demonstrating purposes because it is very visual.

If we focus on our case we need to create Simulink scheme which describes the dynamic behaviour of the system described in Chapter 3 generally for various step changes of the input quantities (7). There are several ways how we can implement numerical methods for simulating the system's behaviour. Two methods were discussed here I. use Matlab's M-functions or II. use Simulink's S-functions. Both methods are described in the next chapters. The other way is to create the model of the system with the use of standard Simulink block line gains, transfer functions etc. but this method is not described here because of the length of the contribution.

4.1   Matlab's M-Functions

The first way is to implement the simulation of the mathematical model in the form of Matlab's M-function inside the Simulink scheme. This way is good for cases when we use classical M-functions for simulations and we have finished simulating M-files.

The M-files could be done via block MATLAB function in the *User-Defined Functions* section. We can use our own Matlab code, for example in this case the function which do the numerical solving of the mathematical model which described by the set of ordinary differential equations (ODE) in equations (5).

The Runge-Kutta's 4th order method described in Chapter 2 is used here for numerical solving of this set. This method could be either implemented by the build-in **ode45** function in Matlab or code this method by your own, for example with the use of function **rk45** described in [10].

This function is then executed by:

```
[T,Y] = rk45(@cstr,h,t0,th,y0)
```

where **Y** represents outputs from the system, **T** is corresponding time vector this output, **@cstr** is reference to the M-file which defines the set of ODE (sla), **t0** and **th** are starting and final time of the simulation and y0 are initial values of the outputs at the beginning of the time slot **t0**. Values of this initial vector are in this case the steady-state values of the state variables.

The function **cstr** which describes the mathematical model of this CSTR (5) is

Matlab's function cstr.m

```
function dy=cstr(t,y)
% dy=cstr(t,y)    FUNCTION cstr
%
%              CSTR - function for computing dynamics of the CSTR
%
% inputs:    t - time
%              y - previous outputs
%
% outputs:  dy - computed outputs
```

```
% Author: Jiri Vojtesek, Ph.D.,
%    Tomas Bata University in Zlin 2015, vojtesek@fai.utb.cz

% global variables, u - step changes, wp - working point
global u wp
% fixed parameters
const_cstr;
% working point
ca0s=wp(1);qrs=wp(2);qcs=wp(3);t0s=wp(4);
% step changes
qc = qcs + u(1)*qcs/100;
ca0 = ca0s + u(2)*ca0s/100;
t0v = t0s + u(3);
qs = qss + u(4)*qss/100;
% reaction rates
for j = 1:3
    kr(j) = k0(j) * exp(-g(j) / y(3));
end
% reaction heat
hr = (kr(1) * dh(1) + kr(3) * dh(3) * y(1)) * y(1) + ...
    kr(2) * dh(2) * y(2);
% the set of ODE
dy(1) = qs/v * (ca0 - y(1)) - (kr(1) + kr(3) * y(1)) * y(1);
dy(2) = -qs/v * y(2) + kr(1) * y(1) - kr(2) * y(2);
dy(3) = qs/v * (t0v - y(3)) + ...
        hr/(ro*cp) + a_r*alf/(v*ro*cp) * (y(4) - y(3));
dy(4) = (qc + a_r * alf * (y(3) - y(4)))/(mc*cpc);
dy=dy';
```

The general M-file function which will be executed from the block *MATLAB Fcn* is

Matlab's function cstr_Mfile.m representing block MATLAB Fcn.

```
function    [out]=cstr_Mfile(in)
%[out]=cstr_Mfile(in)     FUNCTION cstr_Mfile
%
% Simulink function for solving dynamics of the CSTR
%
% inputs:   in(1:4) - step changes of Qc, cA0, t0v and qr
%           in(5:8) - values of the ouputs in previous step
%
% ouputs:   out - ouput values of cA, cB, Tr and Tc

% Author: Jiri Vojtesek, Ph.D.,
%    Tomas Bata University in Zlin 2015, vojtesek@fai.utb.cz
```

```
global u wp  % global variables

u = in(1:4);  % action values
y = in(5:8);      % values of output y in the beginning of the step

const_cstr;               % load constants

% WORKING POINT %
ca0s = 5.1;               % input concentration of compound A
qrs = 0.002365;           % input volumetric flow reate of the RC
qcs = -18.56;             % input heat removal of the CL
t0s = 378.05;             % input temperature of the RC
wp = [ca0s qrs qcs t0s];

h=0.1;                    % integration step
[T,Yv] = rk45(@dyn_cstr_allg_ode,h,0,h,y);
y = Yv(length(T),:);

out  =  y';               % output vector
```

Where `const_cstr` represents M-file with the fixed parameters of the reactant from Table.

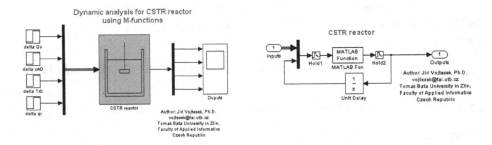

Fig. 2. Simulink scheme for usage of M-function

The Simulink scheme is the viewed in Fig. 2. The scheme on the left side represents whole simulation scheme and the scheme on the right side is unmasked block *CSTR reactor* from the left side figure.

4.2 Simulink's S-Functions

The second option is usage of the special Simulink's S-functions which has `ode45` as a build-in function. The scheme is shown in Fig. 3 The scheme on the left side is similar to the one shown in Fig. 2 but difference is if we look under the mask of the block *CSTR reactor* (right figure) which has only one block S-function. The parameters of this block S-function is the name of the S-function (`cstr_s` in this case) and parameters of the S-function (working point defined by the values of quantities c_{A0}^s, q_r^s, Q_c^s and T_{r0}^s in (8)) see fig on the right side of Fig. 3.

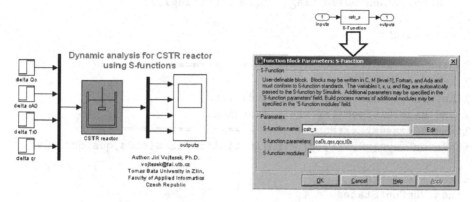

Fig. 3. Simulink scheme for usage of S-function

The S-function `cstr_s` has following form:

Matlab's S-function cstr_s.m

```
function [sys,x0,str,ts]=cstr_s(t,x,u,flag,ca0s,qss,qcs,t0s)
%
% S-function for dynamics of CSTR reactor
%


% Author: Jiri Vojtesek, Ph.D.,
%    Tomas Bata University in Zlin 2015, vojtesek@fai.utb.cz

switch flag
  % Initialization %
  case 0
    [sys,x0,str,ts] = mdlInitializeSizes(ca0s,qss,qcs,t0s);

  % Derivatives %
  case 1
    sys = mdlDerivatives(t,x,u, ca0s,qss,qcs,t0s);

  % Update and Terminate %
  case {2,9}
    sys = []; % do nothing
```

```
% Output %
case 3
   sys = mdlOutputs(t,x,u);

otherwise
   error(['unhandled flag = ',num2str(flag)]);
end
% end limintm


%==========================================================
% mdlInitializeSizes
% Return the sizes, initial conditions,
% and sample times for the S-function.
%==========================================================
function [sys,x0,str,ts] = mdlInitializeSizes(ca0s,qss,qcs,t0s)

sizes = simsizes;
sizes.NumContStates   = 4;
sizes.NumDiscStates   = 0;
sizes.NumOutputs      = 4;  % cA, cB, Tr, Tc
sizes.NumInputs       = 4;  % DQc, DcAO, DTr0, Dqr
sizes.DirFeedthrough  = 0;
sizes.NumSampleTimes  = 1;

sys = simsizes(sizes);
str = [];
% initial conditions => steady state values
x0  = [2.1403 1.0903 387.3397 386.0551];
ts  = [0 0];    % sample time: [period, offset]

% end mdlInitializeSizes


%==========================================================
% mdlDerivatives
% Compute derivatives for continuous states.
%==========================================================
function sys = mdlDerivatives(t,x,u, ca0s,qss,qcs,t0s)

% constants
const_cstr;

% step changes
qc = qcs + u(1)*qcs/100;
ca0 = ca0s + u(2)*ca0s/100;
t0v = t0s + u(3);
```

```
qs = qss + u(4)*qss/100;
% reaction rates
for j = 1:3
    kr(j) = k0(j) * exp(-g(j) / x(3));
end
% reaction heat
hr = (kr(1) * dh(1) + kr(3) * dh(3) * x(1)) * x(1) + ...
    kr(2) * dh(2) * x(2);
% the set of ODE
sys(1) = qs/v * (ca0 - x(1)) - (kr(1) + kr(3) * x(1)) * x(1);
sys(2) = -qs/v * x(2) + kr(1) * x(1) - kr(2) * x(2);
sys(3) = qs/v * (t0v - x(3)) + 1/(ro*cp) * hr + ...
    a_r*alf/(v*ro*cp) * (x(4) - x(3));
sys(4) = 1/(mc*cpc) * (qc + a_r*alf * (x(3) - x(4)));

% end mdlDerivatives


%===========================================================
% mdlOutputs
% Return the output vector for the S-function
%===========================================================
function sys = mdlOutputs(t,x,u)

sys = x;

% end mdlOutputs
```

5 Simulation Experiments

We have created two simulation schemes which could provide theoretically the same results because both schemes have the same inputs step changes $u_{1-4}(t)$ from (7). Practically, results could differ because of numerical solvers the M-file has numerical solver rk45 with fixed integration step and the S-function has Matlab's build-in function ode45. Both represents Runge-Kutta's 4th order function described by the equations (1) - (4).

As it was shown by experiments for example in [10], in some cases the build-in function ode45 could provide different or even wrong results because of variability of the integration step. This could be solved with the use of option odeset where we can fix the integration step if the results are inappropriate.

The following graphs show step responses of the output variables c_A, c_B, T_r and T_c to the different step changes on the input, specifically $u_1(t) = +20\%$ and $u_4(t) = -25\%$. Why only these two intputs? The first reason is the length of this contribution. The second point is that from the practical point of view, only these two inputs which represents step changes of the input volumetric flow rates have significance for the use from the practical point of view.

The inputs are in the real world usually practically represented by twists of valves on the input pipes. On the other hand, changes of the input concentration c_{A0} (input $u_2(t)$) or input temperature of the reactant T_{r0} (input $u_3(t)$) have only theoretical meaning because these quantities are hard to change quickly which could be problem in the control of this system.

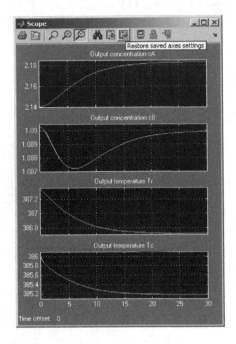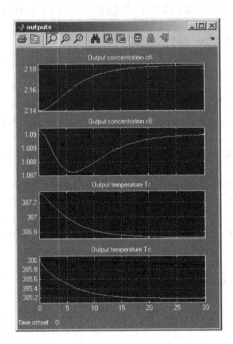

Fig. 4. Results form the scope in the Simulink scheme for simulation with M-file (left graph) and S-function(right graph) for $u_1(t) = +20\ \%$

As it can be seen from the graphs in Figs. 4 and 5 the results are very similar and both schemes can be used for later experiments mainly for the simulation of the control. The modularity where you can easily change, delete or add other blocks, for example controllers, without the additional programming is big advantage of the Simulink.

Other big thing is that once you have designed the controller and get good simulation results by experiments on the mathematical model, this mathematical model could be easily changed by the blocks from the Matlab's *Real-Time Toolbox* which is used for manipulating with the real system or physical model of the real system.

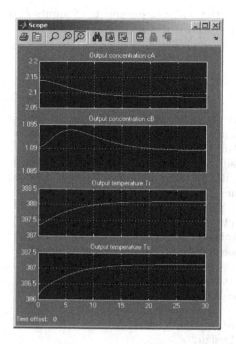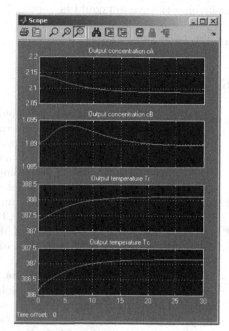

Fig. 5. Results form the scope in the Simulink scheme for simulation with M-file (left graph) and S-function(right graph) for $u_1(t) = $ -25 %

6 Conclusion

The goal of this contribution was to show usability of the simulation software Matlab and its special part Simulink which is used for graphical programming and simulation. The main emphasis is put on the application on the modelling and simulation of the continuous stirred-tank reactor (CSTR) as a typical member of the nonlinear process used frequently in the chemical or biochemical industry for the production of the various materials. The computer modelling and simulation of the behaviour of such processes has big advantages mainly for the security reasons and it could also save a lot of money.

The mathematical model of the examined CSTR here is described by the set of four nonlinear ordinary differential equations which can solved by the numerical methods like Runge-Kutta's methods etc. The paper shows two ways how we can use Simulink for simulating of the CSTR's dynamics. The first way is for the cases where we have finished simulating program in Matlab in the form of classical M-file and the second, more sophisticated, approach uses Simulink's S-functions as a special tool which has Runge-Kutta's numerical methods as a build-in method and we only need to initialize the computation and describe the simulated mathematical model. The proposed Simulink schemes can be used not only for the simulation of the system's dynamics but also for testing of various control techniques which are usually the next step after the dynamic analysis. Moreover, once we have found the appropriate control strategy, the block describing the system could be replaced by the real measurements of inputs and outputs from the real model of the system or even the real model via Matlab's Real-Time toolbox. This feature makes the use of Simulink much more effective.

References

1. Vojtesek, J., Dostal, P., Haber, R.: Simulation and Control of a Continuous Stirred Tank Reactor. In: Proc. of Sixth Portuguese Conference on Automatic Control, CONTROLO 2004, Faro. Portugal, pp. 315–320 (2004)
2. Ingham, J., Dunn, I.J., Heinzle, E., Penosil, J.E.: Chemical Engineering Dynamics. An Introduction to Modelling and Computer Simulation, 2nd Completely Revised Edition. VCH Verlagsgesellshaft, Weinheim (2000)
3. Maria, A.: Introduction to Modeling and Simulation. In: Proceedings of the 1997 Winter Simulation Conference, pp. 7–13 (1997)
4. Mathews, J.H., Fink, K.K.: Numerical Methods Using Matlab. Prentice-Hall (2004)
5. Advanced Numerical Differential Equation Solving in Mathematica. Webpages of Wolfram's Mathematica,
   http://reference.wolfram.com/mathematica/tutorial/NDSolveOverview.html
6. Johnston, R.L.: Numerical Methods. John Wiley & Sons (1982)
7. Matlab's help to function ode45,
   http://www.mathworks.com/help/matlab/ref/ode45.html
8. Simulink - Simulation and Model-Based Design,
   http://www.mathworks.com/products/simulink/
9. Chen, H., Kremling, A., Allgöwer, F.: Nonlinear Predictive Control of a Benchmark CSTR. In: Proceedings of 3rd European Control Conference, Rome, Italy (1995)
10. Vojtesek, J.: Numerical Solution of Ordinary Differential Equations Using Mathematical Software. In: Modern Trends and Techniques in Computing Sciences. Springer Science+Business Media B.V., Dordrecht (2014) ISSN 2194-5357, ISBN 978-3-319-06739-1

Optimization of Access Points in Wireless Sensor Network: An Approach towards Security

Arun Nagaraja[1], Rajesh Kumar Gunupudi[1],
R. Saravana Kumar[2], and N. Mangathayaru[1]

[1] Department of Information Technolgy, VNR VJIET, Hyderabad, India
[2] Software Engineer, Sasken Technologies, Bangalore, India
{arun1611,gunupudirajesh,saravana75}@gmail.com,
mangathayaru_n@vnrvjiet.in

Abstract. Wireless Sensor Networks has lot to do with the technology. With several problems existing, the use of Access Points in a secured way will provide the long lasting connectivity throughout the covered area. The connectivity and the security process are discussed using two different algorithms in this paper. The connectivity to the end user is established with the help of encryption and decryption process of keys. The two algorithms are Optimization algorithm and Craving algorithm. By using these two algorithms, the connectivity is extended outside the range of signals with sensor nodes.

Keywords: Optimization, Access Point, Wireless Sensor Network, Craving Algorithm.

1 Introduction

The latest happenings in the world are because of the Wireless Sensor Networks. Access Points are the one which is used for the Wireless Networks to carry forward the data and the connectivity. The APs gets the connectivity from the base network and transmit the network connections to the desired users in a wireless channel. By broadcasting the frequency using APs in a medium, the connectivity can be achieved to the end users. The APs consisting of sensors require the raw energy to get the power and get connectivity to the network [5].

Wireless Sensor Networks is used by every user in some or the other mode to get connected to the network and update the information. The sensor nodes are used for obtaining the signals, which are light weighted and battery powered. The information is retrieved at the receiver end after obtaining the connectivity to the network. For better performance and stability in network, the sensors require regular maintenance[2].

The survey says that, 802.11 a,b,g,n protocols have higher transmission rates. The transmission rate of 802.11g protocol is 54 Mbps. This protocol works in 2.4 GHz band where OFDM based transmission is used. Compared to 802.11g, 802.11b has 21% reduced throughput with hardware legacy issues [6].

© Springer International Publishing Switzerland 2015
R. Silhavy et al. (eds.), *Intelligent Systems in Cybernetics and Automation Theory,*
Advances in Intelligent Systems and Computing 348, DOI: 10.1007/978-3-319-18503-3_29

Study says that the uses of WLAN are, it has numerous protocols that, to have good performance with the help of real rates of transmission. The transmission rate of the specified protocol is 54 Mbps which covers with a maximum distance of 30mts radius. The whole service will be used in the half duplex channel which shares the access point [9].

The APs connectivity issues occur in the networks is because of the improper positioning. Due to this, some of the cells are over loaded with many request handlers, for which the total area is to be covered [9]. The devices connection with the help of access points happens in a secured way. As the user provides the security key for the AP connection, they can utilize the connectivity in the APs limited range.

For obtaining the desired frequency in the required range, the APs which consists of RF Antennas are placed at the floor. Here we are going to discuss on the connectivity issue and the transmission issue of the APs, during transmission of signals. The study says that, as the frequency increases the transmitting distance gets limited [8]. The paper is discussed on two algorithms.

a) Trying to provide the continuous frequency for larger distance even after increasing the frequency by using Optimization algorithm.

b) To provide security on relocation of users from different APs on different network using Security Algorithm. Level 2 security is provided for the authentication and verification.

In this paper, the Section 2 tells about the earlier work, Section 3 talks on the proposed approach using 2 algorithms, Section 4 on basic results of the algorithms and Conclusion in Section 5.

2   Related Study

Earlier study says that, the Access points those are located on the floor provides the better connectivity to the end users. The Fig.1 shows the WANs model, for which the mobile devices are connected to the APs. Sensors are placed to make the availability of the internet across the region with the help of APs. [8] Says that, the user should provide authentication using private keys of 64 bit of key length to avail the connectivity to the network from the nearby APs those are placed in the floor. In the wireless networks, the false base station attacks and the normal level authentication attacks are common. Based on sensors and the APs, the connectivity can be extended for the desired range.

Network controllers connect to the APs to get the connectivity to the network in the range. The study says that, by increasing the frequency, the APs transmit the higher signals with the help of sensors and users get connected to the same but for only the limited range. The limited range of access is because; sensors that get higher frequency will transmit the signals only to the limited range. The earlier study says that, the user gets the connectivity to the network de-pending on the authentication. The Fig.1 says that, the users are placed on the floor and trying to connect to the network with the help of network. When the signal strengths fall bad in the desired range, the connectivity will be lost. Based on

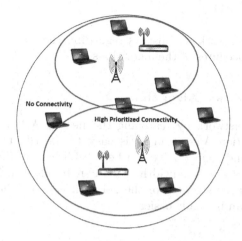

Fig. 1. Access Points Connected to a Network

the sensors which are placed in the floor, the connectivity can be established be-tween the same networks.

The Fig.1 also tells that, even with the good signal strength, if the user fails to connect to the network which is in the desired range then it has to be addressed with physical lookup.

As per the survey made, connectivity is obtained with the help of authentication and the low level encryption. As in Fig.1; the connectivity to the devices can be obtained with the help of sensors that are connected to the nearby APs. When the device is covered in either of the regions, the connectivity is obtained depending on the priority. If the device moves out of range, the connectivity cannot be established with the end user as the user will have less sensor signal strength in the device and connectivity is lost, i.e. by increasing the frequency, the distance coverage is getting reduced. For which the APs should be placed more in the number across the range for lossless connectivity. By this the number of APs usage is increased and said to be more cost effective

The approaches used in this paper are Optimization Algorithm and Craving Algorithm. The optimization algorithm is used for reducing the APs count by increasing the user's range of connectivity. Craving algorithm is the other technique used to perform secure approach. Using these two algorithms, the connectivity to the clients is made secured and the data transfer happens in a secured environment. Hence the connection can be established to the network with low level authentication. This approach with the algorithms helps to analyze that, as the frequency increases the coverage distance is increased with the help of bandwidth.

3   Proposed Work

Here in the proposed work, we discuss regarding the two different algorithms and the working procedures of the algorithms.

3.1   The Optimization Algorithm

Users look for the network by searching the nearby APs in the range. Upon using the Optimization Algorithm, it is easy to identify the AP having network connectivity and connectivity can be established by the devices with the increased frequencies and bandwidth. This can be identified with the help of signal strengths. As the strength of the network is good, the connectivity and data transmission can be made easier.

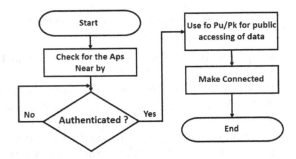

Fig. 2. Flowchart of Optimization Algorithm

The algorithm tells the coverage range and its connectivity which we can look in to flowchart. Main goal of using the algorithm is to get the best solutions with less effort on various computations. This algorithm tells about the area coverage that can be computed in the grid.

This Algorithm Structure as Follows. 1) the computations are performed by knowing the midpoint of the grid and the coverage area of the grid which are done simultaneously across the area.

a) The area of coverage of access points is based on the midpoint of the AP and the depending on the frequency and bandwidth of the radius cycle the connectivity's are verified. The bandwidth range is identified with the help of Bit Error Rate (BER) as by increasing the frequency. By significantly reducing the BER, the bandwidth can be maintained. To achieve connectivity by transferring the packets from node to node and classifying the higher frequency sensors near the nodes, BER is performed. The Fig.3 shows the basic format of the node connectivity and signal range by increasing the frequency and connecting the APs to the nearby sensors. If the mobile device moves out of the range, then

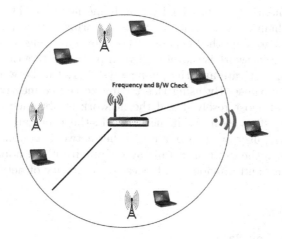

Fig. 3. Optimization of AP using light weight sensor

the sensors which are placed in the outer boundary which has signals of AP and emits signals of 2dB and connectivity is established.

b) The other approach is connectivity issues. Even with higher frequency and bandwidth, when the device does not respond with connectivity, the device is to be physically addressed.

2) If the APs gets disconnected with the network, the APs will ping to the network for every 3 seconds and tries to establish the connection with proper authentication.

3) The diagonal verification for the devices is performed by APs so as to provide connectivity to the devices.

4) Step 3 is continued till the devices are identified within the APs range. The Fig.2 flowchart gives information on usage of private or public keys for network connectivity. As the user knows the authentication procedure, the connectivity to the access points is done authentically by increasing the frequency and bandwidth. By performing this operation, the performance can be boosted up.

3.2 Craving Algorithm

The Craving algorithm mainly works on the frequency calculator. Limited connections are provided for a single AP, to avoid the congestion in the network. Here Sliding window protocol is used for establishing the network connection. If the connection with the AP exceeds the specified limit, the new user needs to either wait for the termination of other Users or to look after other network location.

Proper authentication is to be provided to the mobile devices at the time of connectivity to APs when the user relocates from one place to other. Once the authentication is verified, the devices get connected to the other network with the primary authentication provided in the home network. The home network provides the authentication details to the neighboring network and the connectivity is obtained to the other network. Based on encoding and decoding, the connectivity to the network is established with proper authentication. The algorithm provides the private and the public keys shared in between the network for connectivity purpose. By using this technique the connectivity issues with security can be at most resolved and the network is obtained everywhere. As once the user registers with the home network, the user can use the network anywhere when the user moves around the other network as well. The other network fetches the authentication details by default from the home network and makes more secure and provides the lossless connectivity of network to the end user.

Craving Algorithm

```
Declaration of Key Values
   Initialize a, key
     if(i<k)
      {
           Evaluate ResultKey()
           Increment the Result EKey
      }


Algorithm for Encryption and Decryption
   Result[i] = Result[i] + Private_Key;
  if(EKey==DKey)
      {
           using Key Decrement Result Value
           compute results for Decryption
           compute a[i]
           fetch Ascii Values
           Result[i]-=(Public_Key-D_Key);
           Get Connectivity by getting the values
           Time Stamp for connection is identified
      }
```

The Craving Algorithm, shows the basic encryption and decryption techniques. The network connectivity is obtained with initialization and using the basic encoding and decoding procedures with the help of shared secret keys. First level authentication is provided with the help of public key and then re authentication is provided with the help of private keys. By performing two levels of encryption, the authentication is made more secured. Once the connection is established between the device and the network, the lossless connectivity is maintained, even after the device switches to other network with the help of proper authentication.

A Craving algorithm is a population based algorithm, where users get connected to the nearby APs with shared public keys. Here in this the encoding and decoding techniques are used. The key values are initialized and the computations are made using the variable a. The condition i<k means that, if the count of devices who have linked to the AP is less than the total number supposed to be, then initialize the connectivity to the network with the help of AP. In this algorithm the string is converted in to ASCII to obtain the bit length and the public key is required for the authentication purpose. The computation is performed to establish connection with the AP and even after the user is moved out of the network, it follows the above said procedure to get connected to the other network.

When devices relocates from home network to the visiting network, the method resultkey() will verify keys and make the connection established. The Connectivity time stamp is calculated of the users those are trying to connect to the network. Based on the time stamp the latency and bandwidths of the Access point, connectivity's is observed.

To obtain the count of APs present in the specified range, the ratio is calculated by the formula - the total number of users connected with AP, to the number of APs available in that network.

$$AP = U_t/AP_d$$

Here APs means Access point and Ut is total count of users and AP_d is the number of APs placed in the desired range. Hence the actual number of APs present in this range is calculated.

4   Experimental Studies

Practical experiments are performed with the help of the network simulators to identify the different areas where the network is available. The simulation is performed by emitting the Frequencies and Bandwidths by Access Points are utilized by different users in a circular room. The frequency is calculated diagonally depending on the remittance. As the transmissions of signals are performed using APs it covers one fourth distance of the room area.

Feature	Platform
User Count	100
Connection Time	0.80 sec
Users Relocation	1.20 sec
Selection	1-Method
Group weight	20

Fig. 4. Fixed values used in the practical observation

Bandwidth and frequencies are monitored in a regular interval to identify the performance. With the help of proper authentication, the connectivity to the APs are established which are put up on the floor. Tests are performed with the help of above said algorithm, by changing the frequency from 2.4 GHz to 5.4 GHz and the good performance is obtained. The performance is boosted with the help of Craving algorithm for which the above shown values in Fig.4 are much relevant to the practical experiments made. The time stamp of 1.2 seconds was obtained to make the connectivity to the device. The bandwidth also increased to 2 MHz with the signal strength of 2dB and the performance is monitored time to time. By the experiment made, as the frequency is increased the distance covered is also getting increased in a frequent amount of time after the connectivity.

Hence the count of APs can be reduced to establish the connection to the extended area using Optimization algorithm, where the number of users using the network can be increased with less number of APs using more authentication.

5   Conclusion

The two different algorithms presented in the paper gives the overview on utilization of Access point. Thus the algorithm says that, the APs can be optimized by increasing the frequency and bandwidth. By this the coverage distance is increased and the number of APs usage can be reduced, which is also cost effective. The algorithms discussed in this paper can be made more reliable by performing practical experiments. Thus the optimization of APs can be made and it can be made cost effective.

References

1. Gupta, A.K., Dhyani, P.: Performance indicators in a 802.11 wlan deployment. In: International Conference on Advances in Recent Technologies in Communication and Computing (ARTCom), pp. 490–494. IEEE Press (2009)
2. Bertrand, A., Szurley, J., Ruckebusch, P., Moerman, I., Moonen, M.: Efficient Calculation of Sensor Utility and Sensor Removal in Wireless Sensor Networks for Adaptive Signal Estimation and Beam forming. IEEE Transactions on Signal Processing 60(11), 5857–5869 (2012)
3. Korte, B., Vygen, J.: Combinatorial Optimization: Theory and Algorithms, 4th edn. Springer, Berlin (2010)
4. Chang, B.J., Chen, J.F.: Cross layer based adaptive vertical handoff with predictive RSS in heterogeneous wireless networks. IEEE Transactions Vehicular Technology 57(6), 3679–3692 (2008)
5. Cheng, C.-T., Tse, C.K., Lau, F.C.M.: A Delay Aware Data Collection Network Structure for Wireless Sensor Networks. IEEE Sensors Journal 11(3), 699–710 (2011)
6. wiki link for IEEE 802.11 Document, http://en.wikipedia.org/wiki/IEEE_802.11
7. Kurose, J.F.: Computer Networking - Using A Top down Approach, 5th edn. Pearson Publications (2010)
8. Lima, M., Carrano, E., Takakahashi, R.: Planejamento multicritério deredes wlans uilizando algoritmos genéticos. Technical Report, UFMG (2010)
9. Barbosa, M.A.S., Gouvea, M.M.: Access Point Design with a Genetic Algorithm. In: Sixth International Conference on Genetic and Evolutionary Computing (2012)

The Architecture of Software Interface for BCI System

Roman Žák, Jaromír Švejda, Roman Jašek, and Roman Šenkeřík

Tomas Bata University in Zlín, Faculty of Applied Informatics,
Nam T.G. Masaryka 5555, 760 01 Zlin, Czech Republic
{rzak,svejda,jasek,senkerik}@fai.utb.cz

Abstract. The basic idea of Brain Computer Interface (BCI) is the connection of brain waves with an output device through some interface. Aim of this article is to clarify the potential utilization of complex EEG signal in BCI system. For this purpose, the architecture of the software interface was designed and tested. The main task of the interface is to transfer brain activity signal into commands of intelligent robot.

The paper is organized as follows. Firstly, there is a physiological description of the human brain, which summarizes current knowledge and also points out its complexity. The basic principle of BCI system is also explained.

Secondly, the specification of used technical equipment (hardware component and software tools) is provided.

Thirdly, the transfer operation is explained in the description of proposed software interface. Moreover, results of interface tests are also presented.

Finally, discussion deals with the advantages and disadvantages of BCI system and its usage in real-time applications.

Keywords: Electroencephalography, Bran Computer Interface, Robotics, Neuro-headset.

1 Introduction

The human brain is a complex system, which is an object of our research. It is regarded as the most complex system in the universe. The modern science is currently attempting to understand the complex interconnection among individual parts of the brain. [10] There are many publications, which deal with a description of the brain. [1], [3], [10]

The brain itself is composed of several parts, without which his activity could not be possible. One of its basic structural parts is a neuron. The neuronal cells are characterized by the fact that electrical activity is carried out in them. These cells communicate with each other by electrical signals. According to the last estimate, there are approximately 10^{11} neurons in the brain. Every one of them is connected with thousands of other neurons. The main source of Electroencephalography (EEG) signal is an electric activity of synapse - dendrites membrane located in the surface layer of the cortex. Each active synapse dispatches electromagnetic pulse to the environment during excitation. [6] Due to the high number of these pulses, it is difficult to locate their

© Springer International Publishing Switzerland 2015
R. Silhavy et al. (eds.), *Intelligent Systems in Cybernetics and Automation Theory*,
Advances in Intelligent Systems and Computing 348, DOI: 10.1007/978-3-319-18503-3_30

source by means of multichannel sensor on the skin. This issue could be compared to full amphitheatre, in which there are chanting people and the task is to recognize from the outside, which specific group of fans shouts. A different perspective on this issue may be such that the aim is to identify a uniqueness of the signal for each individual subject. In the example shown above, it is as we would like to recognize the type of the stadium by the mass of chanting people. For example, there is a noticeable difference between hockey and tennis fans. The biometric signatures are different for each creature on the planet Earth.

Many scientific disciplines deal with the human brain; for example numerical neuroscience, neuroinformatics, informatics or medicine. All of them bring theories, which could explain different brain activities. Numerical neuroscience provides mathematical and biophysical models, which are able to model basic processes in neurons and neural networks. The main goal of neuroinformatics is a systematic development of database intended to collect information such as brain morphology, brain parts anatomy and their functional connection, brain electrophysiology, brain states obtained with magnetic resonance and their integration. Further, it seeks to develop tools for modelling, where the aim is the most accurate emulation of brain activity. In Informatics, complex networks are highly suitable to model a complex system among which the brain includes. The contribution of medicine is undisputable especially in brain anatomy research.

This article deals with the Brain - Computer Interface (BCI) technology, which represents the connection of brain waves with the output device through some interface. Figure 1. illustrate the basic principle of a system based on BCI.

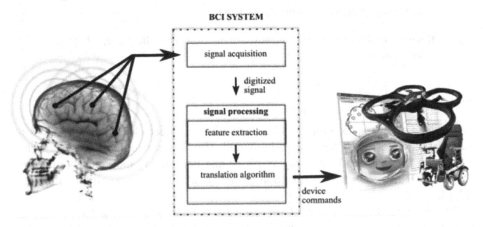

Fig. 1. Basic principle of BCI system

Firstly, the brain activity is obtained from the subject's brain by appropriate device based on some of the technologies, which are currently available for sensing the brain activity; for example fMRI (functional magnetic resonance imaging), EEG (Electroencephalography), ECoG (Electrocorticography) etc. For the purposes of the study described below, EEG technology was chosen to record brain activity.

Further, received signal has to be processed and prepared for translation algorithm. This phase involves a feature extraction during which the most expressive characteristics of obtained signal are discovered. In the case of EEG signal, these characteristics usually relate to some physiological activity such as eye blink, eye movement, raise brow etc.

Finally, the translation algorithm transfers selected features to commands of external device or software.

Currently, there are many known applications of BCI technology, but not enough to each particular field of study. Signal that is sensed from the brain is the key element in the BCI model; therefore the design of an appropriate algorithm for processing of the signal is the most discussed part of BCI model structure. [9]

The aim of this article is to offer an architecture of communication interface between Emotiv EPOC neuro-headset (EEG device) and Mindstorms EV3 (robotic device). Secondary aim was to investigate the reliability of communication between robotic device and neuro-headset.

2   Specification of Technical Equipment

BCI system consists of three main parts: signal sensing, signal processing and external system control. The technical equipment of each part is described separately in the following chapters.

2.1   Signal Sensing

There are several approaches for sensing brain activity. The most widely used is EEG technology, which belongs among the non – invasive methods. Devices based on EEG technology provide signal with very low voltage amplitude because the signal has to pass through the relatively low conductive skull. The amplitude ranges from tens to hundreds microvolts.

Recently, we use Emotiv EPOC neuro-headset to obtain EEG signal from the human brain. Sensing of EEG by Emotiv EPOC neuro-headset has a number of advantages because it already involves solved elementary issues in the processing of the measured signal. Due to this fact, it is not necessary to operate with raw data. It depends on the further usage of the data. Although the spectrum of this data could be used in many applications, it is not simple to understand the entire significance of the whole signal even if the proportion of the noise is minimal. This technology has the greatest expansion and certainly also the priority significance in diagnosis of various diseases in medicine. [1]

Emotiv Corporation developed personal brain - computer interface for human – computer interaction using neuro-technology, which is based on processing of electromagnetic waves occurring in the human brain. The interface has a wide range of possible applications; for example in interactive games, intelligent adaptive environment, audio-visual art and design, medicine, robotics and automotive industry. Moreover, it can be deployed in a large amount of scientific research.

Emotiv EPOC neuroheadset (Figure 2) measures a signal wirelessly transferred to common personal computer. It is a device, which has a set of sensors intended for sensing the activity produced by human brain. Traditional EEG devices requires the use of conductive pasta to improve the conductivity between electrodes and hairs. On the other hand, the neuroheadset do not need any additional tools. It has 14 high resolution sensors, which are placed on optimal positions on the human head (Figure 3).

Fig. 2. Emotiv EPOC neuroheadset [4]

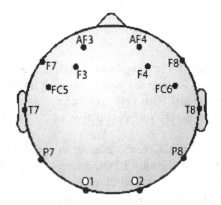

Fig. 3. Placement of electrodes of Emotiv EPOC neuroheadset

Moreover, it also includes gyroscope for determinate the position in the area. Each channel has its own label based on its position on the head: AF3, F7, F3, FC5, T7, P7, O1, O2, P8, T8, FC6, F4, F8, and AF4. Internal sampling frequency of the neuroheadset is 2048 Hz. On the other hand, the neuroheadset provides signal with sampling frequency of 128Hz. More information about neuroheadset can be found in [4].

Emotiv provide basic software set containing many tools, which can be used for recording various signals such as electric potential from all 14 sensors, power spectrum of individual EEG channels in real time and rotational acceleration of the head in horizontal and vertical axis using data from gyroscope. All of these outputs are shown in graphs. Data are also available in raw form, which can be used for further analysis. If it is required special functionality, which is not provided by native software, it is desirable to develop own application using Emotiv SDK (Software Development Kit).

2.2   Signal Processing

Signal processing is important part of BCI system; therefore, this area has to be deeply examined. This part is responsible to either physiological or mental activity recognition. Current recognition methods are able to detect following states of mind:

Instantaneous excitement - introduced as a consciousness or feeling of physiological excitement with positive value. Excitement is characterized by activation of sympathetic nervous system, which is responsible for physiological responses such as dilated pupils, stimulation of the sweat glands, pulse frequency etc.

Long-term excitement - similar to instantaneous excitement. Detection of this state is designed and set to obtain more precise measurement of excitement changes in longer time periods (minutes).

Engagement - known as both alertness and directing attention to suggestions to the tasks. It is characterized as a growth of physiological excitement. It can be observed in beta-waves and alpha waves of EEG record. Contrary of this state is called boredom. The more attention or concentration is performed, the higher value is recorded during detection phase. The writing of text to the paper or writing on computer rises a value of engagement state, while closing eyes almost always rapidly decreases that value.

The most pronounced physiological activity is facial expression; thus, movement of brow, mouth or eyes can be detected. The brain signals of these activities is similar among all people; therefore, universal signatures can be used to detect facial expressions of almost each person.

2.3   External System Control

External system can be either software or hardware model. For the purposes of our research, robotics device Mindstorms EV3 was chosen as an external system (Figure 4.), because it supports most of communication interfaces such as Bluetooth, Wi-Fi, USB connection etc. It consist of many static parts from which it is possible to construct various robotic solutions. Further, robot can be equipped with colour sensor, ultrasonic sensor, gyroscope sensor or touch sensor. Robot's motion is assured by interactive servomotors. Communication and logic of robot's behaviour is controlled by programmable intelligent EV3 Brick. Figure 4. shows an example of one specific robotic device, which is possible to construct from parts described above.

Intelligent EV3 Brick can be programmed in native graphics program language in LabView software. Moreover, there is also an option to develop own software application in some other supported languages such as Java, C# etc.

Fig. 4. Robotic device Mindstorms EV3

3   Results

The whole application interface was realized on the higher abstraction level. Its architecture was designed with emphasis on mutual compatibility among available technical resources.

Parallelization was chosen in order to achieve real time response; thus, proposed application contains special program threads, where individual consuming operations are carried out. Thread B (labelled as Brain part) performs communication with the neuroheadset and listens to events related to physiological activity such as blink of an eye. Thread C (labelled as Computer/device part) was implemented analogously to previous thread, but it is connected to the external system; in our case, it is a robotic device Mindstorms EV3. The main task of thread C is to switch servomotors of a robot according to states, which are transmitted between threads B and C. Time constants for both loops were selected in milliseconds. Finally, Interface part is main thread in which translation algorithm takes care of interconnection between B and C thread. Proposed architecture of communication interface is depicted on Figure 5. Then, whole design and realization with the real devices is a prototype, which has to be subjected to thorough testing on several levels of development. Interface part is main thread of application.

The part of Emotiv research SDK edition software package is also a simulator of brain activity called EmoComposer, which is able to simulate occurrence of brain activities mentioned in previous section of this article. Firstly, the testing of reliability of proposed communication interface were tested using that simulator, which communicates through the same network protocol. At this case, reliability was 100%; thus, each initiation of specific brain activity using the simulator caused a motion of the robotic device. Eye blinking was chosen as test brain activity.

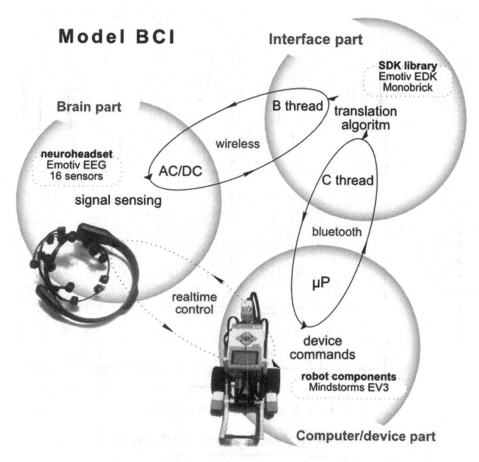

Fig. 5. Architecture of proposed software interface

Experiment with real neuroheadset showed different results. EEG record of eye blinking is depicted in Figure 6, where its activity is bounded by red rectangle. Neuroheadset was set on the head of test subject. Further, the scanned brain activity of the subject was sent to the computer through the wireless connection. Obtained signal was processed using proposed communication interface mentioned above. Finally, an appropriate command was sent to the robotic device. The aim of this experiment was to find out the real efficiency of brain- computer interface when universal signature is used to detect a specific brain activity from the subject.

Eleven individual experiment sets were performed. Table 1. shows the results for each of them. Each set involve certain number of attempts. Whenever a movement of robot appeared after the subject's eyewink, it was counted as successful attempt. Overall, it was measured 330 attempts. The lowest value of reliability was 0.5 (50%), while the highest value was 0.83 (83%). The average reliability was 0.65 (65%).

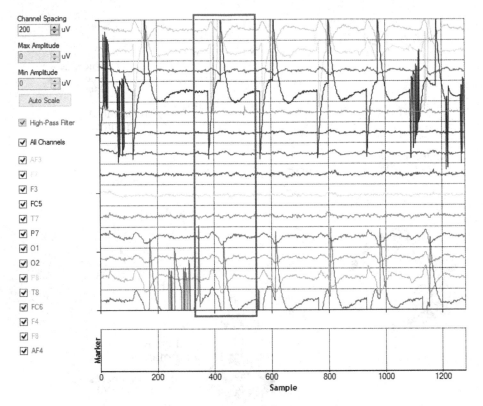

Fig. 6. Example of eye blinking record

Table 1. Reliability of robot response on eye blinking activity

Set ID	Number of attempts [-]	Number of successful attempts [-]	Reliability [%]
1	30	17	56,667
2	30	18	60
3	30	17	56,667
4	40	33	82,5
5	30	25	83,333
6	30	21	70
7	30	18	60
8	30	19	63,333
9	30	15	50
10	20	16	80
11	30	17	56,667

4   Discussions

Human brain is the most complex known system in the universe. Study of its activity is extremely important mainly due to the most precise diagnosis of brain diseases and their treatment. Furthermore, acquired knowledge could be used in modern technologies with BCI systems, where an interaction between brain and computers appears.

Our research deals with BCI system, which was used to control robotic device. We designed the architecture of software communication interface between neuroheadset and robotic equipment. The abilities of proposed software were confirmed on real BCI system. The system consisted of EEG device, computer and a robotic device. Tests proved that proposed architecture of software communication interface meets requirements for real-time control of an external device using brain waves.

Further, our research examined practical issues associated with currently one of the most advanced EEG equipment intended for technical utilization. The time, which one spends with its installation on the subject head, is approximately from 5 to 10 minutes. It depends on whether it is the very first installation of the equipment or whether it is reused in the same day. Even if the device does not need any special gel, which will be applied on subject's head, it still requires application of saline solution on sensor pads. Further, the important part of the installation is also to find the right position for the neuro-headset on the head. This process is usually controlled by software, which provides information about contact quality of each sensor. This issue is not problem in laboratory conditions, but it could bring complications in practical applications; thus, it could be the one of the main reasons of making the whole system unusable because of the time needed to set the system up. On the other hand, current EEG devices provide EEG signal in the highest possible quality depended on the current technical progress.

The aim of our experiment was to demonstrate the real efficiency of communication between EEG and the robotic device using universal signature of selected brain activity. The average reliability of robotic device response on signal of eye blinking was 65.45%. This low reliability could be caused by noise in the electromagnetic signal and different artefacts in biological activities. Moreover, prolonged use of neuroheadset may leads to headache caused by slight pressure of soft pads on the skull; thus, subject's focus on repeating the same activity may be partly affected by this unwanted headache. Finally, it was also the reason for making breaks after each three set of attempts. In addition, inexperienced subject need time to become familiar with how to properly perform an activity, which should be recognized using universal signature. Other investigations shows that even subjects who have no BCI control in the first few sessions can learn the operation by neuro-/biofeedback training. [2], [5], [7], [8]

Our research proved that robot can react on eye blinking activity. The set of universal signatures contains other activities (brow movement, smile etc.). Each of them could be mapped into different robot action. Unfortunately, there are many activities, which are not included in the set of universal signatures e.g. movement imagination, limb movement etc. These activities can currently be recognized only after additional learning of neural network. This kind of activities could be another appropriate subject of further research.

Acknowledgments. This work was supported by Internal Grant Agency of Tomas Bata University under the project No. IGA/FAI/2015/063, further by Grant Agency of the Czech Republic - GACR P103/15/06700S, further by financial support of research project NPU I No. MSMT-7778/2014 by the Ministry of Education of the Czech Republic and also by the European Regional Development Fund under the Project CEBIA-Tech No. CZ.1.05/2.1.00/03.0089.

References

1. Adeli, H.: Wavelet-Chaos-Neural Network Models for EEG-Based Diagnosis of Neurological Disorders. In: Kim, T.-H., Lee, Y.-h., Kang, B.-H., Ślęzak, D. (eds.) FGIT 2010. LNCS, vol. 6485, pp. 1–11. Springer, Heidelberg (2010)
2. Birbaumer, N., Ghanayim, N., Hinterberger, T., Iversen, I., Kotchoubey, B., Kübler, A., Perelmouter, J., Tuab, E., Flor, H.: A spelling device for the paralysed. Nature 398(6725), 297–298 (1858), doi:10.1038/18581 (cit. February 23, 2015)
3. Damasio, H.: Human brain anatomy in computerized images, 303 p. Oxford University Press, New York (1995) ISBN 0195082044
4. Emotiv I EEG System I Electroencephalography (2012), http://www.emotiv.com/index.php
5. Guger, C., Edlinger, G., Harkam, W., Niedermayer, I., Pfurtscheller, G.: How many people are able to operate an eeg-based brain-computer interface (bci)? IEEE Transactions on Neural Systems and Rehabilitation Engineering 11(2), 145–147 (2003), doi:10.1109/tnsre.2003.814481 (cit. February 23, 2015)
6. Kandel, E.R., Schwartz, J.H., Jessell, T.M.: Principles of neural science, 4th edn., vol. xli, 1414 p. McGraw-Hill, Health Professions Division, New York (2000) ISBN 978-0-8385-7701-1
7. Neuper, C., Schlögl, A., Pfurtscheller, G.: Enhancement of Left-Right Sensorimotor EEG Differences During Feedback-Regulated Motor Imagery. Journal of Clinical Neurophysiology 16(4), 251–261 (1999), doi:10.1007/978-4-431-30962-8_23 (cit. February 23, 2015)
8. Pfurtscheller, G., Guger, C., Müller, G., Krausz, G., Neuper, C.: Brain oscillations control hand orthosis in a tetraplegic. Neuroscience Letters 292(3), 211–214 (2000), doi:10.1016/s0304-3940(00)01471-3 (cit. February 23, 2015)
9. Schalk, G., McFarland, D.J., Hinterberger, T., Birbaumer, N., Wolpaw, J.R.: BCI2000: A General-Purpose Brain-Computer Interface (BCI) System. IEEE Transactions on Biomedical Engineering 51(6), 1034–1043 (2004), doi:10.1109/tbme.2004.827072 (cit. February 23, 2015)
10. Sporns, O., Tononi, G., Kötter, R.: The Human Connectome: A Structural Description of the Human Brain. PLoS Computational Biology 1(4) (2005), doi:10.1371/journal.pcbi.0010042 (cit. February 23, 2015)

Author Index